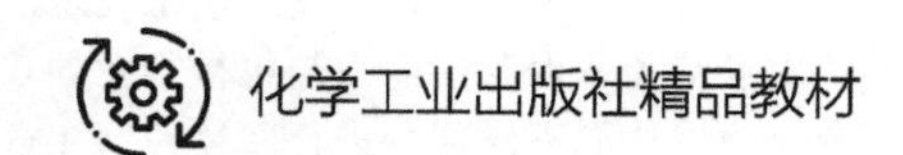

Intelligent Manufacturing and Advanced Electrical Technology

智能制造与先进电工技术

王鹏　谢军　编著

化学工业出版社
·北京·

内容简介

本书针对智能制造与先进电工技术的最新前沿进行介绍，力图使机械类和电气类专业相关技术人员与学生了解和掌握我国电力与机械行业的攻关热点。全书共分为11章，各部分各有侧重，读者可根据自己的专业领域选取不同部分进行阅读。第1章介绍“中国制造2025”，有助于了解国家发展情况，提高民族自信心；第2章介绍智能制造，摘录了周济院士、李培根院士等对国内外智能制造发展现状的评述；第3章介绍纳米制造技术，列举了国内外主流的微加工技术；第4章概述了国内外3D打印的主流技术；第5章是关于机器人的概述；第6章侧重于电动汽车的介绍；第7章系统地阐述了我国电力输送的发展现状和面临的问题；第8章概述了我国电力系统的骄傲——特高压工程；第9~11章属于扩展阅读，分别介绍了新型电力系统外绝缘防护、先进电工装备绝缘老化机理和智能诊断方法。

本书既可作为高校机械类和电气类学生专业英语教材，又可供相关领域工程技术人员学习、参考。

图书在版编目（CIP）数据

智能制造与先进电工技术＝Intelligent Manufacturing and Advanced Electrical Technology：英文/王鹏，谢军编著.—北京：化学工业出版社，2021.11

ISBN 978-7-122-39924-3

Ⅰ.①智… Ⅱ.①王…②谢… Ⅲ.①智能制造系统-英文②电工技术-英文 Ⅳ.①TH166②TM

中国版本图书馆CIP数据核字（2021）第188451号

责任编辑：金林茹　　装帧设计：王晓宇

责任校对：刘　颖

出版发行：化学工业出版社（北京市东城区青年湖南街13号　邮政编码100011）

印　　装：北京虎彩文化传播有限公司

787mm×1092mm　1/16　印张15¾　字数405千字　2022年1月北京第1版第1次印刷

购书咨询：010-64518888　　售后服务：010-64518899

网　　址：http://www.cip.com.cn

凡购买本书，如有缺损质量问题，本社销售中心负责调换。

定　　价：98.00元

Preface

With the emergence of new technologies, new processes and new products, the content about machinery industry should be updated and improved in time. Especially, there are new requirements and targets for mechanical engineering teaching with the proposal of "Made in China 2025". Based on "Made in China 2025", this book not only meets the professional English learning needs of mechanical engineering students in universities, but also further highlights the characteristics of the power industry. On the occasion of the beginning of the 14th Five-Year Plan, this book focuses on the latest research frontier of intelligent manufacturing and advanced electrical technology to make the students of mechanical and electrical engineering understand and master the hot issues of intelligent manufacturing.

This book is divided into 11 chapters, and each part has its own emphasis. Readers can choose different parts to read according to their own areas of expertise. The first chapter is the English version of "Made in China 2025", which can be used to carry out ideological and political courses to improve national confidence. The second chapter is about intelligent manufacturing, which summarizes the comments of academician Zhou Ji and Li Peigen on the development of intelligent manufacturing. The third chapter is about nanomanufacturing. Professor J. Alexander Liddle's overview of the mainstream micro/nano fabrication technology at home and abroad is summarized. The fourth chapter briefly summarizes the mainstream technology of 3D printing. The fifth chapter is a brief overview of robots. The sixth chapter focuses on electric vehicles. The seventh chapter systematically expounds the development status and problems of our country's electric power transmission. The eighth chapter summarizes the pride of China's power system—ultrahigh voltage project. The chapters from nine to eleven are extended reading, which introduces the advanced electrical equipment's insulation protection strategy, insulation ageing mechanism, and intelligent diagnosis method, respectively. Through the study of this book, the readers not only can understand the relevant technical knowledge background, but also can be familiar with and master the relevant technical words, phrases and specific usage.

This book is edited by Wang Peng and Xie Jun. Graduate students Chen Tao, Zhang Xuesong, Zhao Hui, Yang Mengyu, Han Lei, Sun Bin, Huai Jiru and Chen Qigong have done a lot of work in literature search, grammar check and format proofreading. I would like to extend my sincere thanks here!

Thanks to the favor of relevant experts and leaders in the Graduate School of North China Electric Power University, this book was honored to be supported by the "Double First-Class" Postgraduate Talent Training and Construction Project. Here, I would like to express my heartfelt thanks to North China Electric Power University for its love and strong support!

Due to the limited level and the short time to write the book, it is inevitable that there are omissions and mistakes, I sincerely ask the readers to criticize and correct.

Wang Peng

Contents

Chapter 1 Made in China 2025 ······ **001**

1.1 Brief Overview ······ 001

1.2 Background and Stated Goals ······ 002

1.3 Key Industries ······ 002

1.4 Key Industries MIC 2025 Changes the Terms of the Game ······ 003

Notes and References ······ 004

Chapter 2 Intelligent Manufacturing ······ **006**

2.1 Introduction of Intelligent Manufacturing ······ 006

2.2 Three Basic Paradigms of Intelligent Manufacturing ······ 006

2.2.1 Digital Manufacturing ······ 007

2.2.2 Digital-Networked Manufacturing ······ 008

2.2.3 New-Generation Intelligent Manufacturing ······ 009

2.3 New-Generation Intelligent Manufacturing Leads and Promotes the New Industrial Revolution ······ 010

2.3.1 Development Background ······ 010

2.3.2 New-Generation Intelligent Manufacturing as a Core Technology of the New Industrial Revolution ······ 010

2.3.3 Vision ······ 011

2.4 The Technological Mechanism of New-Generation Intelligent Manufacturing: the Human-Cyber-Physical System ······ 011

2.4.1 Traditional Manufacturing and the Human-Physical System ······ 012

2.4.2 Digital Manufacturing, Digital-Networked Manufacturing and the Human-Cyber-Physical System (HCPS) ······ 012

2.4.3 New-Generation Intelligent Manufacturing and the New-Generation HCPS ······ 013

Notes and References ······ 014

Chapter 3 Nanomanufacturing ······ **016**

3.1 Top-Down Versus Bottom-Up Processes ······ 018

3.2 Top-Down Fabrication ······ 019

3.3 Photolithography ······ 019

3.4 Nanoimprint Lithography ······ 020

3.5 Other Top-Down Techniques ······ 022

3.6 Bottom-Up Fabrication ······ 023

3.7 Colloidal Self-Assembly ······ 024

3.8 DNA-Based Self-Assembly ······ 024

3.9 Directed Self-Assembly: Top-Down Combined with Bottom Up ······ 025

3.9.1 Directed Self-Assembly of Block Copolymers ······ 025

3.9.2 Fluidic Assembly ······ 026

3. 9. 3 Damped-Driven Systems …… 027
3. 9. 4 Design for Nanomanufacturing …… 027
Notes and References …… 028

Chapter 4 3D Printing …… 030
4. 1 Overview …… 030
4. 2 Terminology …… 030
4. 3 History …… 031
4. 3. 1 1950 …… 031
4. 3. 2 1970s …… 031
4. 3. 3 1980s …… 032
4. 3. 4 1990s …… 033
4. 3. 5 2000s …… 033
4. 3. 6 2010s …… 034
4. 4 General Principles …… 034
4. 4. 1 Modeling …… 034
4. 4. 2 Printing …… 035
4. 4. 3 Finishing …… 036
4. 4. 4 Materials …… 036
4. 4. 5 Multi-Materials 3D Printing …… 036
4. 5 Processes and Printers …… 037
4. 6 Applications …… 039
Notes and References …… 043

Chapter 5 Robot Manufacturing …… 045
5. 1 Etymology …… 046
5. 2 Robotic Aspects …… 046
5. 3 Applications …… 047
5. 4 Components …… 048
5. 4. 1 Power Source …… 048
5. 4. 2 Actuation …… 048
5. 4. 3 Sensing …… 049
5. 4. 4 Manipulation …… 050
5. 4. 5 Locomotion …… 051
5. 4. 6 Environmental Interaction and Navigation …… 054
5. 4. 7 Human-Robot Interaction …… 055
5. 5 Control …… 056
5. 6 Research …… 057
5. 7 Education and Training …… 059
5. 7. 1 Career Training …… 059
5. 7. 2 Certification …… 059
5. 7. 3 Summer Robotics Camp …… 059

5. 7. 4 Robotics Competitions …… 059
5. 7. 5 Robotics Afterschool Programs …… 060
5. 7. 6 Decolonial Educational Robotics …… 060
5. 8 Employment …… 060
5. 9 Occupational Safety and Health Implications …… 061
Notes and References …… 061

Chapter 6 Electric Car …… 063
6. 1 Terminology …… 063
6. 2 History …… 064
6. 3 Economics …… 066
6. 3. 1 Total Cost of Ownership …… 066
6. 3. 2 Purchase Cost …… 066
6. 3. 3 Operating Cost …… 067
6. 3. 4 Manufacturing Cost …… 067
6. 4 Environmental Aspects …… 067
6. 5 Performance …… 067
6. 6 Energy Efficiency …… 068
6. 7 Safety …… 069
6. 7. 1 Risk of Fire …… 069
6. 7. 2 Vehicle Safety …… 069
6. 8 Controls …… 070
6. 9 Batteries …… 070
6. 9. 1 Range …… 070
6. 9. 2 Charging …… 071
6. 9. 3 Lifespan …… 071
6. 9. 4 Future …… 071
6. 10 Electric Vehicle Charging Patents …… 072
6. 11 Infrastructure …… 072
6. 11. 1 Charging Station …… 072
6. 11. 2 Vehicle-to-Grid: Uploading and Grid Buffering …… 073
6. 12 Currently Available Electric Cars …… 073
6. 12. 1 Highway Capable …… 073
6. 12. 2 Retrofitted Electric Vehicles …… 074
6. 12. 3 Electric Cars by Country …… 074
6. 13 Government Policies and Incentives …… 074
6. 14 EV Plans From Major Manufacturers …… 075
6. 15 Psychological Barriers to Adoption …… 075
6. 15. 1 Range Anxiety …… 075
6. 15. 2 Identity Concerns …… 075
Notes and References …… 076

Chapter 7 Electric Power Transmission ······ **078**
7. 1 System ······ 079
7. 2 Overhead Transmission ······ 079
7. 3 Underground Transmission ······ 080
7. 4 History ······ 080
7. 5 Bulk Power Transmission ······ 082
7. 5. 1 Grid Input ······ 084
7. 5. 2 Losses ······ 084
7. 5. 3 Transposition ······ 085
7. 5. 4 Subtransmission ······ 086
7. 5. 5 Transmission Grid Exit ······ 086
7. 6 Advantage of High-Voltage Power Transmission ······ 086
7. 7 High-Voltage Direct Current ······ 086
7. 8 Capacity ······ 087
7. 9 Control ······ 088
7. 9. 1 Load Balancing ······ 088
7. 9. 2 Failure Protection ······ 089
7. 10 Communications ······ 089
7. 11 Electricity Market Reform ······ 090
7. 12 Cost of Electric Power Transmission ······ 090
7. 13 Merchant Transmission ······ 090
7. 14 Health Concerns ······ 091
7. 15 Policy by Country ······ 092
7. 16 Special Transmission ······ 093
7. 16. 1 Grids for Railways ······ 093
7. 16. 2 Superconducting Cables ······ 093
Notes and References ······ 093

Chapter 8 China Ultrahigh Voltage Project ······ **095**
8. 1 Research History and Background of UHV Transmission ······ 096
8. 1. 1 Russia (The Former USSR) ······ 096
8. 1. 2 Japan ······ 097
8. 1. 3 The USA ······ 097
8. 1. 4 Italy ······ 099
8. 1. 5 Canada ······ 099
8. 1. 6 Brazil ······ 100
8. 1. 7 China ······ 100
8. 2 Target Design and Research Background of China's UHV System ······ 101
8. 2. 1 The Demand and Goals Analysis of UHV Transmission in China ······ 101
8. 2. 2 Important Innovations and Progress ······ 104
8. 2. 3 UHV AC and DC Key Technologies Researches and Achievements ······ 105
8. 3 Equipment Manufacturing ······ 111

8. 3. 1 UHV AC Equipment Manufacture …… 111
8. 3. 2 UHV DC Equipment Manufacture …… 113
Notes and References …… 114

Chapter 9 Advanced External Insulation Protection System …… 116
9. 1 Transparent and Superhydrophobic Coating for the Solar Panels …… 116
9. 1. 1 Materials …… 117
9. 1. 2 Preparation of Silica Nanoparticle Suspension …… 117
9. 1. 3 Preparation of Superhydrophobic Coating …… 117
9. 1. 4 Characterization …… 118
9. 1. 5 Results and Discussion …… 118
9. 2 The Superhydrophobic Graphene Coating for Anti-Corrosion Application …… 123
9. 2. 1 Materials …… 124
9. 2. 2 Preparation of EEG …… 124
9. 2. 3 Preparation of Superhydrophobic Composite Coating …… 124
9. 2. 4 Characterization …… 125
9. 2. 5 Results and Discussion …… 125
9. 3 The Superhydrophobic Steel for Anti-Corrosion Application …… 129
9. 3. 1 Materials …… 130
9. 3. 2 The Preparation of Micro Nano Roughness …… 130
9. 3. 3 Ultrasonic Treatment …… 130
9. 3. 4 Surface Modification …… 130
9. 3. 5 Characterization …… 130
9. 3. 6 Results and Discussion …… 131
9. 4 The Graphene Semiconductor Superhydrophobic Coating for Anti-Icing Application …… 135
9. 4. 1 Materials …… 136
9. 4. 2 Preparation of Hydrophobic Powders …… 137
9. 4. 3 Dissolution and Resolidifcation Process to Construct Superhydrophobic Sample …… 137
9. 4. 4 Characterization …… 137
9. 4. 5 Icing/Deicing Test …… 137
9. 4. 6 Results and Discussion …… 138
9. 5 The Self-Healable Graphene Coating for Anti-Icing Application …… 143
9. 5. 1 Materials …… 145
9. 5. 2 The Synthesis of Prepolymer A …… 145
9. 5. 3 The Synthesis of Prepolymer B …… 145
9. 5. 4 Preparation of Hydrophobic CNT Powders …… 145
9. 5. 5 Preparation of Hydrophobic CNT Solution …… 145
9. 5. 6 Preparation of Self-Healing Superhydrophobic Coating …… 146
9. 5. 7 Characterization …… 146
9. 5. 8 Icing/Deicing Test …… 146
9. 5. 9 Results and Disscussion …… 146

Notes and References ······ 152

Chapter 10 Thermal Ageing Mechanism of Advanced Electrical Insulation Materials —Taking Oiled Paper Insulation as an Example ······ 154
10. 1 Introduction of Oiled Paper Insulation ······ 154
10. 2 Experimental Plan Design ······ 155
10. 2. 1 Typical Defect Model and Experimental Wiring of Oil-Paper Insulation ······ 155
10. 2. 2 Selection and Treatment of Oil-Paper Insulation Samples ······ 157
10. 2. 3 Accelerated Ageing of Insulating Paperboard ······ 158
10. 3 Surface Discharge Deterioration Law of Transformer Oil-Paper Insulation ······ 160
10. 3. 1 Initial Discharge Voltage and Discharge Endurance Time of Transformer Oil-Paper Insulation ······ 160
10. 3. 2 Description of Surface Discharge Deterioration Phenomenon of Oil-Paper Insulation of Transformer and Division of Discharge Development Degree ······ 161
10. 3. 3 Variation Law of Characteristic Quantity of Surface Discharge of Transformer Oil-Paper Insulation ······ 164
10. 4 Development Law of Tip Discharge Deterioration of Transformer Oil-Paper Insulation ······ 174
10. 4. 1 Initial Discharge Voltage and Discharge Endurance Time of the Transformer Oil-Paper Insulation Tip Discharge ······ 174
10. 4. 2 Phenomenon Description of Discharge Development of Transformer Oil-Paper Insulation Tip and Division of Discharge Development Process ······ 174
10. 4. 3 Variation Law of Characteristic Quantities of Discharge at the Tip of Transformer Oil-Paper Insulation ······ 177
10. 5 Reasons for the Influence of Ageing of Insulating Paperboard on the Development of Partial Discharge of Oil-Paper Insulation ······ 184
10. 5. 1 Basic Structure of Insulation Paperboard ······ 184
10. 5. 2 Analysis of the Microscopic Properties of Insulating Paperboard with Different Ageing Degrees ······ 186
10. 5. 3 Explanation of the Influence of Ageing of Insulating Paperboard on the Development of Discharge ······ 189
10. 6 Summary of This Chapter ······ 191
Notes and References ······ 191

Chapter 11 Intelligent Diagnosis Method of Advanced Electrical Equipment ······ 193
11. 1 Technology and Application of Intelligent Sensing and State Sensing for Transformation Equipment ······ 193
11. 1. 1 Introduction ······ 193
11. 1. 2 Data Situation ······ 195
11. 1. 3 Key Technology ······ 197
11. 1. 4 Application Scenarios ······ 202
11. 1. 5 Facing Challenges and Future Trends ······ 208

11. 2 Typical Application and Prospect of Digital Twin Technology in Power Grid Operation ········ 209
11. 2. 1 Introduction of Digital Twin Technology ········ 209
11. 2. 2 Digital Twin ········ 211
11. 2. 3 Digital Twin Power Grid System ········ 213
11. 2. 4 Typical Applications and Prospects of Digital Twin Power Grids ········ 218
11. 2. 5 Conclusion ········ 227
11. 3 Technologies and Solutions of Blockchain Application in Power Equipment Ubiquitous Internet of Things ········ 228
11. 3. 1 Introduction ········ 228
11. 3. 2 Ubiquitous IoT Architecture Design for Power Equipment based on Blockchain ········ 230
11. 3. 3 Key Technology ········ 235
11. 3. 4 Technology Outlook ········ 238
11. 3. 5 Conclusion ········ 239
Notes and References ········ 240

Chapter 1
Made in China 2025

1.1 Brief Overview

"Made in China 2025" (MIC 2025) is a national strategic plan to further develop the manufacturing sector of the People's Republic of China, issued by Premier Li Keqiang and his cabinet in May 2015. As part of the Thirteenth and Fourteenth Five-Year Plan, China aims to move away from being the "world's factory" —a producer of cheap low-tech goods facilitated by lower labour costs and supply chain advantages. The plan aims to upgrade the manufacturing capabilities of Chinese industries, growing from labor-intensive workshops into a more technology-intensive powerhouse[1].

The stated goals of Made in China 2025 include increasing the Chinese-domestic content of core materials to 40 percent by 2020 and 70 percent by 2025 (Figure 1-1). To help achieve indepen-

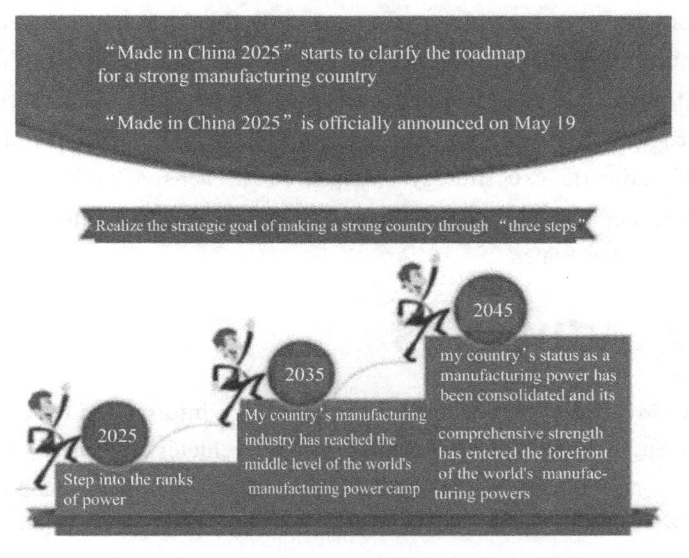

Figure 1-1 The roadmap of Made in China 2025

dence from foreign suppliers, the initiative encourages increased production in high-tech products and services, with its semiconductor industry central to the plan, partly because advances in chip technology may "lead to breakthroughs in other areas of technology, handing the advantage to whoever has the best chips—an advantage that currently is out of Beijing's reach."

Since 2018, following a backlash from America, Europe, and elsewhere, the phrase "MIC 2025" has been de-emphasized in government and other official communications, while the program remains in place. The Chinese government continues to invest heavily in identified technologies. In 2018, the Chinese government committed to investing roughly $300 billion into achieving the plan. In the wake of the COVID-19 pandemic, at least additional $1.4 trillion was also invested into MIC 2025 initiatives. Given China's current middle income country status, the practicality of its disproportionate expenditure on pioneering new technologies has been called into question.

1.2 Background and Stated Goals

Since the 2010s, China has become an emerging superpower as the second largest economy and the largest one on a purchasing power parity basis. It faces manufacturing competition from countries with lower wages, like Vietnam, as well as from highly industrialized countries. In order to maintain economic growth, standard of living, and meet the demand of its increasingly educated workforce, China undertook stimulating the potential of its economic and technological competitiveness with MIC 2025, to become a "world-leading manufacturing power". Alan Wheatley of British think tank Chatham House indicated, in 2018, that a broad and growing Chinese middle class is necessary for the country's economic and political stability.

China believes in its industrial policy programs, which it sees as key to its economic success. Chinese leaders hope that government investment in crucial technology sectors will lead to a strong position in the Fourth Industrial Revolution. The key objective of the Made in China 2025 program is, in a world which it views as increasingly dominated by US-China competition, to identify key technologies, such as AI, 5G, aerospace, semiconductors, electric vehicles and biotech, indigenize those technologies with the help of national champions, secure market share domestically within China, and ultimately capture foreign markets globally[2].

The Center for Strategic and International Studies in Washington, D. C. described MIC 2025 as an "initiative to comprehensively upgrade Chinese industry", which is directly inspired by Germany's proposed Industrie 4.0 strategy. It is a comprehensive undertaking to move China's manufacturing base higher up the value chain and become a major manufacturing power in direct competition with the United States[3].

1.3 Key Industries

Industries integral to MIC 2025 include aerospace, biotech, information technology, smart manufacturing, maritime engineering, advanced rail, electric vehicles, electrical equipment, new materials, biomedicine, agricultural machinery and equipment, pharmaceuticals, and robotics manufacturing, many of which have been dominated by foreign companies, as shown in Figure 1-2[4].

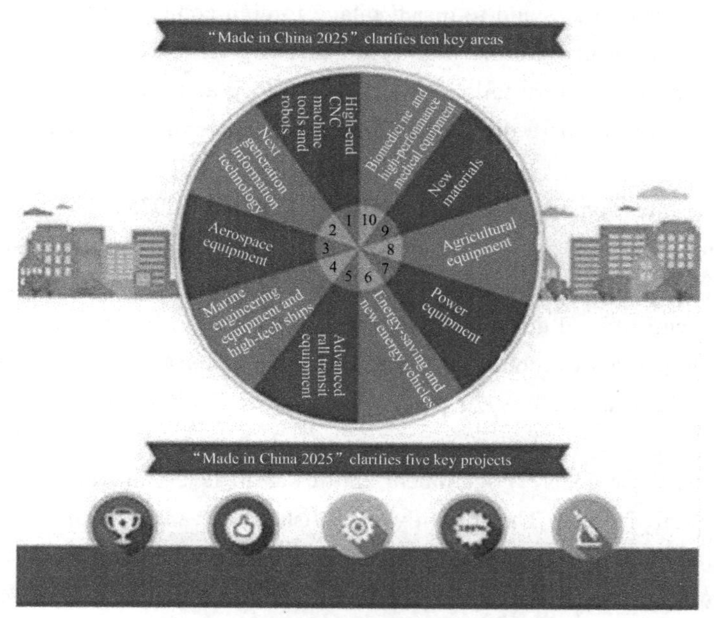

Figure 1-2 The key areas of MIC 2025

Advanced standards in industries are absolutely essential to foster innovation and eliminate bottlenecks in industrial development. China has a growing middle class who are demanding higher quality goods and services. Compared with overseas competition, the quality and innovation of Chinese goods have not caught up. The quality revolution revolves around entrepreneurship and craftsmanship. It will involve embracing a culture of continuous innovations and refinement in quality of goods produced[5].

1.4 Key Industries MIC 2025 Changes the Terms of the Game

The political push for industrial modernisation in China creates an enormous demand for smart manufacturing products like industrial robots, smart sensors, wireless sensor networks and radio frequency identification chips. For many foreign enterprises, this initially provides highly attractive business opportunities: the transformation of China's manufacturing base requires advanced technologies that Chinese suppliers are not able to provide at their current technological level. China's industrial upgrading, in the short-term, can mean tremendous proits for international companies. For China's economic partners in Europe and the United States, it could even open opportunities for a mutually beneficial deepening of economic, technological as well as political cooperation[6]. In principle, the global economy has good reasons to welcome China's quest for increased innovation capacity, provided that China abides by the principles and rules of open markets and fair competition.

However, Made in China 2025 in its current form represents exactly the opposite: China's leadership systematically intervenes in domestic markets so as to benefit and facilitates the economic dominance of Chinese enterprises and to disadvantage foreign competitors. This is visible in smart manufacturing as well as in many other high-tech industries targeted by the strategy. In essence, Made in China 2025 aims for substitution: China seeks to gradually replace foreign with Chinese technology at home and to prepare the ground for Chinese technology companies entering international markets[7].

Indications of this intention are omnipresent in Made in China 2025. The strategy stresses terms like "indigenous innovations" and "self-suiciency". It intends to increase the domestic market share of Chinese suppliers for "basic core components and important basic materials" to 70 percent by the year 2025. Semi-official documents related to the strategy set very concrete benchmarks for certain segments: 40 percent of mobile phone chips on the Chinese market are supposed to be produced in China by 2025, as well as 70 percent of industrial robots and 80 percent of renewable energy equipment[8].

In order to achieve these goals, government entities at all levels funnel large amounts of money into China's industrial future. The recently established Advanced Manufacturing Fund alone amounts to 20 billion CNY (2.7 billion EUR). The National Integrated Circuit Fund even received 139 billion CNY (19 billion EUR). These national level funds are complemented by a plethora of provincial level inancing vehicles. The inancial resources are enormous compared to, for instance, the 200 million EUR of federal funding that the German government has provided for research on Industrie 4.0 technologies so far[9].

While Chinese high-tech companies enjoy massive state backing, their foreign competitors in China face a whole set of barriers to market access and obstacles to their business activities: the closing of the market for information technology, the exclusion from local subsidy schemes, the low level of data security and the intensive collection of digital data by the Chinese state[10]. As China's own smart manufacturing capabilities mature, it is likely that the Chinese state will further step up its discriminatory practices and restrictions of market access in the field of smart manufacturing[11].

At the moment, however, these barriers are not yet as established in smart manufacturing as in other areas such as the service sector and the aviation industry. Made in China 2025 is in its early days and there are still opportunities to adjust its direction and targets, at least in some sectors. If the administration in the United States implements the protectionist agenda announced during the election campaign, Europe's negotiation position will be potentially improved. Keeping global trade and investment lows open will become an overarching shared interest between Europe and China. Europe's economic importance for China will increase and vice-versa. Despite all current frictions, this mid-term shift in the global economy will potentially open new avenues for negotiating the conditions of Sino-European economic relations, including in smart-manufacturing.

Notes and References

[1] Jost Wübbeke, Björn Conrad. Industrie 4.0: will German technology help China catch up with the West. MERICS China Monitor, 2016 (23).

[2] Stepan, Matthias, Lea Shih. These are the super-rich people shaping China. Fortune. 2016.

[3] 周济，朱高峰. 制造强国战略研究：智能制造卷. 北京：电子工业出版社，2015.

[4] Kiel D, et al. Sustainable industrial value creation in SMEs: a comparison between Industrie 4.0 and Made in China 2025. International Journal of Precision Engineering and Manufacturing-Green Technology, 2018, 5: 659-670.

[5] Li Ling. China's manufacturing locus in 2025: with a comparison of "Made in China 2025" and "Industrie 4.0". Technological Forecasting and Social Change, 2018, 135: 66-74.

[6] 广东将加大机器人采购补贴：重点扶持15家龙头企业. 人民网，2016，http://news.youth.cn/gn/201601/t20160126_7571490.htm.

[7] 苏力. 广东产机器人迎来产品创新升级潮，产量年增4000台. 2016，http://it.people.com.cn/n1/2016/0629/c1009-28508566.html.

[8] Wang Jian, Wu Huiqin, Chen Yan. Made in China 2025 and manufacturing strategy decisions with reverse QFD. International Journal of Production Economics, 2020, 224: 107539.

[9] Wilfried Kubinger, Roland Sommer. Industrie 4.0-auswirkungen von digitalisierung und Internet auf den Industriestandort. Elektrotechnik und Informationstechnik, 2016, 330-333.

[10] Liu Xihui. Innovation design: Made in China 2025. Design Management Review, 2016, 27 (1): 52-58.

[11] Tian Shubin, Pan Zhi. "Made in China 2025" and "Industrie 4.0" in motion together. The Internet of Things, 2017, 87-113.

Chapter 2
Intelligent Manufacturing

2.1 Introduction of Intelligent Manufacturing

Countries around the world are actively engaging in the new industrial revolution[1]. The United States has launched the Advanced Manufacturing Partnership, Germany has developed the strategic initiative Industrie 4.0, and the United Kingdom has put forward the UK Industry 2050 strategy[2]. In addition, France has unveiled the New Industrial France program[3], Japan has a Society 5.0 strategy[4], and Korea has started the Manufacturing Innovation 3.0 program[5]. The development of intelligent manufacturing is regarded as a key measure to establish competitive advantages for the manufacturing industry of major countries around the world. The Made in China 2025 plan, formerly known as China Manufacturing 2025, has specifically set the promotion of intelligent manufacturing as its main direction, with a focus on the in-depth integration of new-generation information technology within the manufacturing industry.

Since the beginning of the 21st century, new-generation information technology has shown explosive growth. The broad application of digital networked and intelligent technologies in the manufacturing industry, and the continuous development of integrated manufacturing innovations have been the main driving forces of the new industrial revolution. In particular, new generation intelligent manufacturing, which serves as the core technology of the current industrial revolution, incorporates major and profound changes in the development philosophy, manufacturing modes, and other aspects of the manufacturing industry. Intelligent manufacturing is now reshaping the development paths, technical systems, and industrial forms of the manufacturing industry, and is thereby pushing the global manufacturing industry into a new stage of development[6-10].

2.2 Three Basic Paradigms of Intelligent Manufacturing

Intelligent manufacturing is a general concept that covered wide range of specific topics[11,12].

New-generation intelligent manufacturing represents an in-depth integration of new-generation artificial intelligence (AI) technology and advanced manufacturing technology. It runs through every link in the full life-cycle of design, production, product and service. The concept also relates to the optimization and integration of corresponding systems. It aims to continuously raise enterprises' product quality, performance, and service levels while reducing resources consumption, thus promoting the innovative, green, coordinated, open, and shared development of the manufacturing industry.

For decades, intelligentization for manufacturing has involved many different paradigms as it continues to develop in practice. These paradigms include lean production, flexible manufacturing, concurrent engineering, agile manufacturing, digital manufacturing, computer-integrated manufacturing, networked manufacturing, cloud manufacturing, intelligent manufacturing, and more. All of these paradigms have played an active role in guiding technology upgrading in the manufacturing industry. However, there are too many paradigms to form a unified intelligent manufacturing technology roadmap. This lack of unity causes enterprises to experience many perplexities in their practice of pushing forward intelligent upgrading. Considering the continuously emerging new technologies, new ideas and new modes of intelligent manufacturing, we consider it necessary to summarize the basic paradigms of intelligent manufacturing.

Intelligent manufacturing has developed in parallel with the progress of informatization. There are three stages in the development of informatization worldwide[13]:

(1) From the middle of the 20th century to the mid-1990s, informatization was in a digital stage with computing, communications and control applications as the main features.

(2) Starting in the mid-1990s, the Internet came into large-scale popularization and application, and informatization entered a networked stage with the interconnection of all things as its main characteristic.

(3) At present, on the basis of cluster breakthroughs in and integrated applications of big data, cloud computing, the mobile Internet, and the Industrial Internet, strategic breakthroughs have been made in AI. As a result, informatization has entered an intelligent stage, with new-generation AI technology as its main feature.

Taking the various intelligent manufacturing-related paradigms into account and considering the characteristics of the integration of information technology and the manufacturing industry through different stages, it is possible to generalize three basic paradigms of intelligent manufacturing: digital manufacturing, digital networked manufacturing, and new-generation intelligent manufacturing (Figure 2-1).

2.2.1 Digital Manufacturing

Digital manufacturing is the first basic paradigm of intelligent manufacturing. It may also be referred to as first-generation intelligent manufacturing. The concept of intelligent manufacturing first appeared in the 1980s. Because the first-generation AI technology that was in application at that time could hardly solve specific engineering problems, first-generation intelligent manufacturing was essentially digital manufacturing. Starting in the second half of the 20th century, as demand for technological progress in the manufacturing sector became increasingly urgent, digital information technologies were widely applied in the manufacturing industry, driving forward revolu-

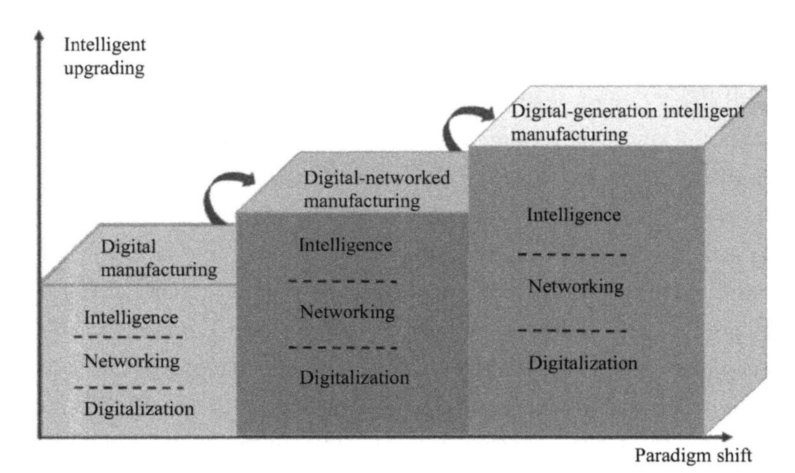

Figure 2-1 The evolution of three basic paradigms of intelligent manufacturing

tionary changes in the industry. Against a background of the integration of digital technology with manufacturing technology, digital manufacturing undertook the digital description, analysis, decision-making and control of product information, process information, and resources information. In this way, digital manufacturing remarkably shortened the time required for designing and manufacturing products to meet specific customer requirements.

The key features of digital manufacturing are as follows:

① Digital technology is widely used in products, forming a "digital generation" of innovative products;

② Digital design, modeling and simulations, and digital equipment information management are widely applied;

③ Production process integration and optimization are achieved.

The point that needs to be clarified here is that digital manufacturing is the foundation of intelligent manufacturing. Digital manufacturing continues to evolve, and runs throughout the three basic paradigms and all the development processes of intelligent manufacturing. The digital manufacturing being defined here is the digital manufacturing of the first basic paradigm, which positions digital manufacturing in a relatively narrow sense. On an international level, several types of positioning and theories on digital manufacturing have also been developed in a broad sense[14].

2.2.2 Digital-Networked Manufacturing

Digital-networked manufacturing is the second basic paradigm of intelligent manufacturing. It may also be referred to as "Internet+Manufacturing" or as second-generation intelligent manufacturing.

In the end of the 20th century, Internet technology started to gain popularity. "Internet+" has continuously pushed forward the integrated development of the Internet and the manufacturing industry. The network connects humans, processes, data and things. Through intra- and inter-enterprise collaborations and the sharing and integration of all kinds of social resources, "Internet+" reshapes the value chain of the manufacturing industry and drives the transformation from digital manufacturing to digital-networked manufacturing.

The main characteristics of digital-networked manufacturing are as follows:

(1) At the product level, digital technology and network technology are widely applied. Products are connected through the network, while collaborative and shared design and R&D are achieved.

(2) At the manufacturing level, horizontal integration, vertical integration, and end-to-end integration are completed, thereby connecting the data flows and information flows of the entire manufacturing system.

(3) At the service level, enterprises and users connect and interact through the network platforms, while enterprises begin to transform from product-centered production to user-centered production.

Both Germany's Industrie 4.0 report and General Electric's Industrial Internet report present very informative and well-structured descriptions of the digital-networked manufacturing paradigm, and put forward technology roadmaps for digital networked manufacturing.

2.2.3 New-Generation Intelligent Manufacturing

Digital-networked-intelligent manufacturing is the third basic paradigm of intelligent manufacturing. It may also be referred to as new-generation intelligent manufacturing.

Jointly driven by a strong demand for economic and social development, the penetration of the Internet, the emergence of cloud computing and big data, the development of the Internet of Things (IoT), and rapid changes in the information environment, there has been an accelerating development of strategic breakthroughs in new-generation AI technologies. These include big data intelligence, human-machine hybrid-augmented intelligence, crowd intelligence, and cross-media intelligence. The in-depth integration of new-generation AI technology and advanced manufacturing technology leads to the formation of new-generation intelligent manufacturing. New-generation intelligent manufacturing will reshape all the processes of the full product cycle, including design, manufacture and services, as well as the integration of these processes. It will promote the emergence of new technologies, new products, new business forms, and new models, and it will profoundly influence and change the production structure, production modes, lifestyles, and thinking models of humankind. It will ultimately result in a great improvement of social productive forces. New-generation intelligent manufacturing will bring revolutionary changes to the manufacturing industry and will become the main driving force for the future development of the industry.

The three basic paradigms of intelligent manufacturing reflect the inherent development pattern of intelligent manufacturing. On the one hand, the three basic paradigms developed in sequence, each with their own characteristics and key problems to solve. In this way, they embody the characteristics of different developmental stages of advanced information technology and advanced manufacturing technology. On the other hand, the three basic paradigms cannot be technologically separated from each other, rather, they are interconnected and iteratively upgraded, thus showing the integrated development characteristics of intelligent manufacturing. China and other emerging industrial countries must leverage their late-mover advantages and adopt a technology roadmap for the "parallel promotion and integrated development" of the three basic paradigms.

2.3 New-Generation Intelligent Manufacturing Leads and Promotes the New Industrial Revolution

2.3.1 Development Background

At present, manufacturing enterprises in all countries are faced with an urgent need to improve quality, boost efficiency, lower cost, and have quick responses. There is also a need for enterprises to continuously adapt to the growing personalized consumption demand of users and to address the greater challenges of resources, energy and environmental constraints. However, existing manufacturing systems and levels can scarcely meet the value appreciation and upgrade requirements of high-end, personalized, and intelligent products and services. The further development of the manufacturing industry faces huge bottlenecks and difficulties. To solve these problems and meet these challenges, there is an urgent need for the manufacturing industry to promote technological innovation and complete intelligent upgrading.

The new industrial revolution is still emerging, and its fundamental driving force lies in a new revolution of science and technology. Since the start of the 21st century, the mobile Internet, supercomputing, big data, cloud computing, IoT and other new-generation information technologies have developed rapidly, and they have achieved swift penetration and applications, resulting in mass breakthroughs. These historical technological advancements are concentrated in strategic breakthroughs in new-generation AI technology, which has taken fundamental strides forward. New-generation AI possesses new features such as deep learning, crossover collaboration, human-machine hybrid-augmented intelligence, and crowd intelligence. With these, new-generation AI provides humankind with new ways of thinking that can help us to understand complex systems and new technologies with the capability to reconstruct both nature and society. Of course, new-generation AI technology is still under development and will continue to develop from "narrow AI" to "general AI". In doing so, it will expand the "brainpower" of humankind and achieve ubiquitous applications. New-generation AI has become the core technology of a new science and technology revolution. It provides historical opportunities for the revolutionary upgrading of the manufacturing industry into a powerful engine that will boost economic and social development. Most of the major countries in the world have made developing new-generation AI into a top priority.

The in-depth integration of new-generation AI technology and advanced manufacturing technology is leading to the formation of new-generation intelligent manufacturing technology, and is becoming the major driving force of the new industrial revolution.

2.3.2 New-Generation Intelligent Manufacturing as a Core Technology of the New Industrial Revolution

Science and technology are the first productive force, and they are also the fundamental driving force for economic and social development. The first and second industrial revolutions were respectively marked by the invention and application of the steam engine, and by electric power, and both innovations greatly improved productive force and helped to usher human society into the modern industrial age. Highlighted by the innovation and application of computing, communications, control and other information technologies, the third industrial revolution has continuously pushed industrial development to a new height[15].

Since the start of the 21st century, digitalization and networking have made information acquisition, use, control and sharing extremely rapid and widespread. Furthermore, breakthroughs in and applications of new-generation AI have further raised the levels of digitalization, networking and intelligence in the manufacturing industry. The most fundamental features of new-generation AI are its cognitive and learning capabilities, which can generate and better use knowledge. In this way, new-generation AI can fundamentally improve the efficiency of industrial knowledge generation and utilization, greatly liberate the physical power and brainpower of humans, enormously speed up the pace of innovation, and make applications more ubiquitous. Thus, it can push the manufacturing industry forward into a new stage of development: new-generation intelligent manufacturing. If digital-networked manufacturing is considered to be the start of the new industrial revolution, the breakthroughs in and wide application of new-generation intelligent manufacturing will advance the new industrial revolution to its peak, reshape the technological system, production models and industrial forms of the manufacturing industry; and usher in Industrie 4.0 in its real sense.

2.3.3 Vision

New-generation intelligent manufacturing systems will acquire increasingly powerful intelligence and, in particular, increasingly powerful cognitive and learning capabilities. The mutually heuristic growth of human intelligence and machine intelligence will shift knowledge-based work in the manufacturing industry toward the direction of autonomous intelligence and then solve the bottlenecks and difficulties that hinder the current development of the manufacturing industry. In new-generation intelligent manufacturing, products are highly intelligent and human-friendly. At the same time, production processes feature high quality, flexibility, high efficiency and environmental friendliness. The industrial model will undergo revolutionary changes. The service-oriented manufacturing industry and the production-based service industry will achieve greater development and will then optimize and integrate new manufacturing systems together, thus fully rebuilding the value chain of the manufacturing industry and greatly improving the innovativeness and competitiveness of the manufacturing industry. New-generation intelligent manufacturing will bring revolutionary changes to human society. On the one hand, the boundary between humans and machines will shift dramatically, with intelligent machines taking over a huge amount of manual labor and a considerable amount of brainwork from humans. This shift will leave humans to be more engaged in creative work. On the other hand, our working and living environments and modes will become more people-centered. Meanwhile, new-generation intelligent manufacturing will effectively reduce the consumption and waste of resources and energy while continuously promoting the green and harmonious development of the manufacturing industry.

2.4 The Technological Mechanism of New-Generation Intelligent Manufacturing: the Human-Cyber-Physical System

Intelligent manufacturing involves intelligent products, intelligent production, intelligent services, and many other aspects. The optimization and integration of these aspects are also included. Although they differ in terms of technological mechanisms, these aspects are consistent in their essence. Here, we take the production process as an example for analysis.

2.4.1 Traditional Manufacturing and the Human-Physical System

The traditional manufacturing system includes two major parts: humans and physical systems. Machine operation controls are completely manual in order to complete all kinds of work tasks [Figure 2-2(a)]. The power revolution greatly improved the production efficiency and quality of physical systems (i. e. machines). From then on, physical systems began to replace humans by taking over the majority of work. The traditional manufacturing system requires humans to complete tasks such as information sensing, analysis, decision-making, operation, control, cognition, and learning. It not only has high requirements for humans but also carries high labor intensity. Moreover, the work efficiency, quality and capability of the system to perform complex work tasks are still rather limited. The traditional manufacturing system can be abstractly described as a human-physical system (HPS) [Figure 2-2(b)].

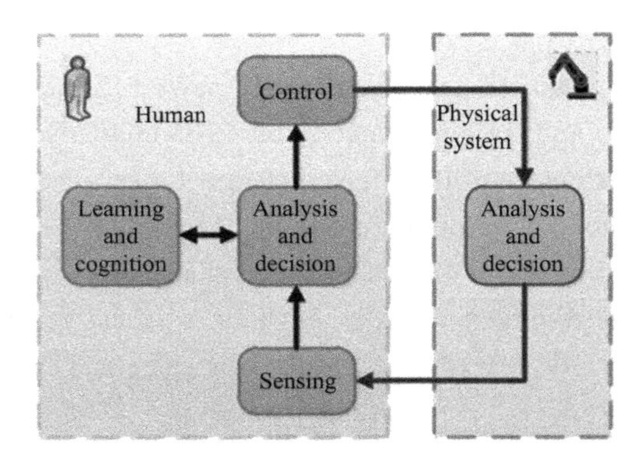

(a) The technological mechanism of the traditional manufacturing system

(b) Schematic of a human-physical system(HPS)

Figure 2-2 Traditional manufacturing and the human-physical system

2.4.2 Digital Manufacturing, Digital-Networked Manufacturing and the Human-Cyber-Physical System (HCPS)

First-and second-generation intelligent manufacturing systems differ from traditional manufacturing systems in their addition of cyber systems between humans and physical systems. A cyber system can replace humans in order to complete some of the brain-work. A considerable portion of humans' sensing, analysis and decision-making functions are reproduced and migrated to the cyber system. The physical systems are controlled through the cyber system in order to replace humans and complete more manual labor (Figure 2-3).

By integrating the advantages of humans, cyber systems, and physical systems, first-and second-generation intelligent manufacturing systems acquire great capability enhancement, especially in computing analysis, precision control and sensing capabilities. On the one hand, the work

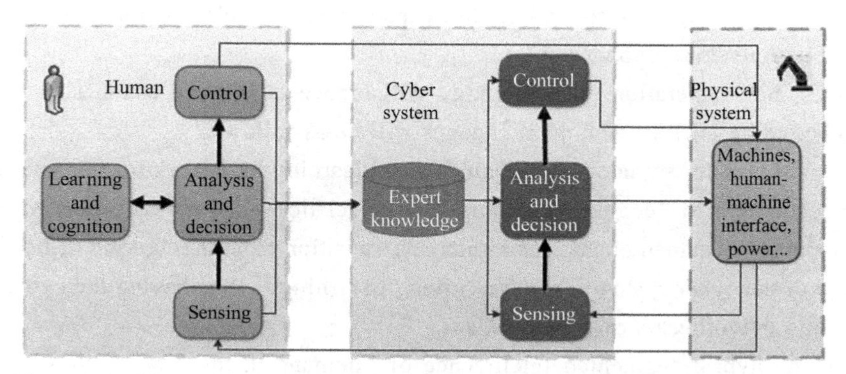

Figure 2-3 First-and second-generation intelligent manufacturing systems incorporate cyber systems between human and physical systems

efficiency, quality and stability of the systems are markedly improved. On the other hand, by transferring humans' relevant manufacturing experience and knowledge to the cyber system, the efficiency of human knowledge management, transfer and application is effectively improved.

The evolution of manufacturing systems from traditional HPSs to human-cyber-physical systems (HCPSs) is abstractly described in Figure 2-4.

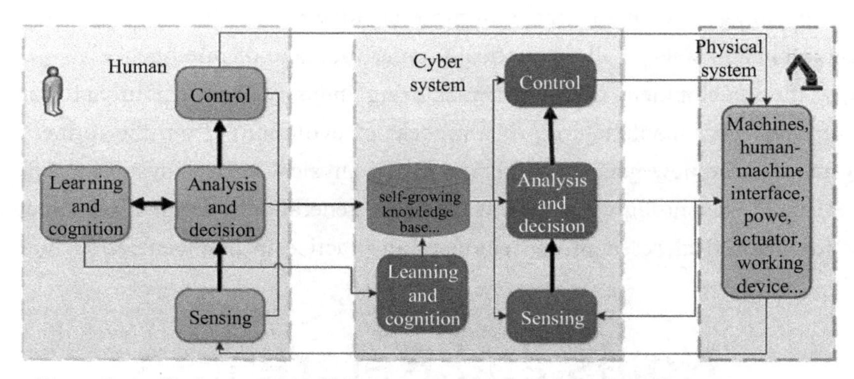

Figure 2-4 Basic mechanism of new-generation intelligent manufacturing systems

The introduction of the cyber system concurrently adds human-cyber systems (HCSs) and cyber-physical systems (CPSs) to manufacturing systems. In particular, the CPS is a very important part of an intelligent manufacturing system. The United States put forward CPS theory at the beginning of the 21st century, and Germany has taken the CPS as a core technology of Industrie 4.0. The application of the CPS in engineering aims to achieve the perfect mapping and in-depth integration of cyber systems with physical systems, with the concept of the digital twin as the most fundamental and essential technology. As a result, the performance and efficiency of manufacturing systems can be improved significantly

2.4.3 New-Generation Intelligent Manufacturing and the New-Generation HCPS

The most fundamental feature of new-generation intelligent manufacturing systems is that cognitive and learning functions are added into the cyber systems. Cyber systems not only possess powerful sensing, computing analysis and control capabilities, but also acquire the capability to improve learning and generate knowledge.

At this stage, new-generation AI technology will induce qualitative changes to the HCPS and form a new-generation HCPS. The main changes will be as follows.

① Humans will transfer some of their cognitive and learning brainwork to the cyber system, enabling the cyber system to "cognize and learn". The relationship between humans and the cyber system will undergo fundamental changes as humans transition from the "giving of fish" (i. e. passing knowledge to the cyber system), to the "giving of fishing" (i. e. having the cyber system cognize, learn, and obtaining its own knowledge).

② Through the hybrid-augmented intelligence of "humans in the loop", in-depth human-machine integration will fundamentally improve the capability of manufacturing systems to handle complex and uncertain problems, and will greatly optimize the performance of manufacturing systems.

In the new-generation HCPS, the HCS, HPS and CPS will all take great strides forward. New-generation intelligent manufacturing further highlights the central position of humans. It is a grand integrated system that coordinates humans, cyber systems and physical systems. It will bring quality and efficiency in the manufacturing industry to a higher level, and strengthen the foundation of human civilization. It will free humankind from intensive and tiring manual labor and low-level thinking, thus enabling humans to engage in more creative work. With new-generation intelligent manufacturing, human society will authentically enter the "age of intelligence".

To sum up, the development of the manufacturing industry from traditional manufacturing to new-generation intelligent manufacturing is a process of evolution: from the former human-physical binary systems to the new-generation human-cyber-physical tertiary systems. The new-generation HCPS reveals the technological mechanism of new-generation intelligent manufacturing and can effectively guide theoretical research and engineering practice in new-generation intelligent manufacturing.

Notes and References

[1] Executive Office of the President, National Science and Technology Council. A national strategic plan for advanced manufacturing. Washington, DC: Office of Science and Technology Policy, 2012.

[2] Kagermann H, Wahlster W, Helbig J. Recommendations for implementing the strategic initiative Industrie 4.0: Final report of the Industrie 4.0 Working Group. Munich: National Academy of Science and Engineering (acatech), 2013.

[3] Daniel Bittighofer, Maike Dust, Alexander Irslinger, et al. State of Industrie 4.0 across German companies. 2018 IEEE International Conference on Engineering, Technology and Innovation (ICE/ITMC), 2018.

[4] Taki H. Towards technological innovation of Society 5.0. J Inst Electr Eng Jpn, 2017, 137(5): 275.

[5] Khalid Hasan Tantawi, Alexandr Sokolov, Omar Tantawi. Advances in industrial robotics: from Industry 3.0 automation to Industrie 4.0 collaboration. 2019 4th Technology Innovation Management and Engineering Science International Conference (TIMES-iCON), 2019.

[6] Evans P C, Annunziata M. Industrial Internet: pushing the boundaries of minds and machines. Boston: General Electric, 2012.

[7] National Manufacturing Strategy Advisory Committee, Center of Strategic Studies of the Chinese Academy of Engineering. Intelligent manufacturing. Beijing: Publishing House of Electronics Industry, 2016.

[8] Hu H, Zhao M, Ning Z, et al. Three-body intelligence revolution. Beijing: China Machine Press, 2016.
[9] Lee J, Ni J, Wang A Z. From big data to intelligent manufacturing. Shanghai: Shanghai Jiao Tong University Press, 2016.
[10] Lee J, Qiu B, Liu Z, et al. Cyber-physical system: The new-generation of industrial intelligence. Shanghai: Shanghai Jiao Tong University Press, 2017.
[11] National Manufacturing Strategy Advisory Committee, Center of Strategic Studies of the Chinese Academy of Engineering. Intelligent manufacturing. Beijing: Publishing House of Electronics Industry, 2016.
[12] Zhou J. Intelligent manufacturing—Main direction of "Made in China 2025". China Mech Eng, 2015, 26 (17): 2273-2284.
[13] Pan Y. Heading toward artificial intelligence 2.0. Engineering, 2016; 2(4): 409-413.
[14] Wu D, Rosen D W, Wang L, et al. Cloud-based design and manufacturing: A new paradigm in digital manufacturing and design innovation. ComputAided Des, 2015, 59: 1-14.
[15] Brynjolfsson E, Mcafee A. The second machine age: work, progress, and prosperity in a time of brilliant technologies. New York: W. W. Norton and Company, Inc., 2014.

Chapter 3 Nanomanufacturing

Nanotechnology has the potential to make a significant impact in a multitude of diverse areas[1]. It offers precise control over the composition and nanostructure of materials enabling the production of multifunctional devices with unique properties, ranging from multiferroics to nanocomposites. For example, coatings containing nanoparticles can act as thermal barriers[2] and flame retardants[3], confer resistance to ultraviolet light-induced degradation, be self-cleaning, antibacterial, scratch-resistant, improve the look of your skin, keep food fresh, and, by combining superhydrophobic and superoleophobic properties, keep the screen of your smartphone clean. Nanostructured materials can resolve the mismatch between the generally large absorption length of light in organic photovoltaic materials and the generally small charge-carrier diffusion distance 18 to enable more efficient energy generation. Similarly, the high surface-to-volume ratios of nanostructured materials can enable rapid charge/discharge cycles in batteries and prevent strain-induced electrode degradation. Quantum confinement effects permit the light emission characteristics of nanoparticles made of a single material to be tuned across a wide range of wavelengths for applications in lighting and displays. However, to realize the potential benefits of all of these diverse applications, we must develop efficient, cost-effective and robust nanomanufacturing methods.

Nanomanufacturing is a term whose usage varies with the approach, and rationale for the choice of approach, for fabricating, or 3D nanostructures. It can mean making small features on larger objects (e. g., integrated circuit (IC) fabrication)[4], making nanoscale objects with special properties (e. g., quantum dot synthesis)[4], assembling nano-scale objects into more complex structures (e. g., DNA origami-directed assembly), 30 incorporating nanoscale objects into larger objects to enable special functionality (e. g., graphene into electronic devices or into liquor distillation apparatuses)[5,6], and using nanotechnology to manufacture nanoscale structures (e. g., dip-pen nanolithography). Given these varied interpretations, we must define our use of the terms "nanomanufacturing" and "nanofabrication" before proceeding further. They are often used interchangeably, but in the interest of adding a level of precision to the conversation, we will draw a distinction between them. Firstly, we note that here we employ a broad definition of nanofabrication that includes both conventional and top-down methods, such as those used in the production of semiconductors, as well as bottom-up methods, such as chemical synthesis and self-assembly. For the purposes of this discussion, we distinguish between nanofabrication and nanomanufacturing using the criterion of economic viability, suggested by the connotations of industrial scale and profitability associated with the word "manufacturing". Nano-manufacturing, as we define it here, has the salient characteristic of being a source of money, while nano-fabrication[7,8] is often a sink. In all cases, for a process or technology to be considered manufacturable, the cost of manufacturing and the volumes that can be produced must be consistent with the selling price and total addressable sales market. In other words, if it is possible only to produce something in small volumes and at high cost, then it must command a high price. Conversely, if the product fetches a low price, then not only must the cost of production be correspondingly low, but the volumes required by the market[9] must be large enough in order to make the enterprise economically self-sustaining[10,11] (Figure 3-1).

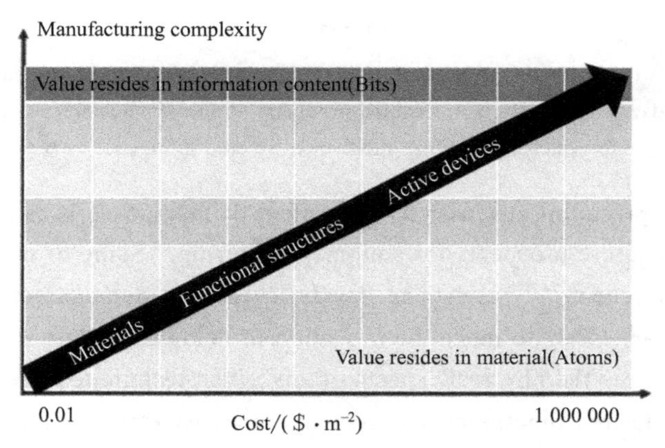

Figure 3-1 The relationship between the cost and the manufacturing complexity

Mathematically, this relationship follows from the fact that the yearly revenue generated by a given tool or process is the selling price of the product multiplied by the amount of product generated per year. The rate of product generation is most conveniently represented in terms of the throughput. A majority of the products generated by or processes used in nano-manufacturing concern what are essentially thin-film or quasi two-dimensional structures. We therefore choose to represent the amount of product generated in units of area, specifically meters squared. For example, the thickness of integrated circuits, hard drives, photovoltaics, sensors, and coatings is extremely small compared to their area and so the amount of area correlates directly to the amount of product. While this is not always the case, such as for catalysts, nanoparticles, and nanotubes, which can be made volumetrically, these are often used to cover or coat a given area and so the area metric for production capacity in this case can be taken to refer to the "as used" or "as applied" area. It is clear that either a high throughput or a high selling price is required to achieve a given yearly revenue. For example, a yearly revenue of $1 million can be obtained with a throughput of 10^{-12} $m^2 \cdot s^{-1}$ only if the selling price is on the order of 10 billion $ $\cdot$ m^{-2}, but the same revenue follows from a few cents per meter squared if the throughput is on the order of 1 $m^2 \cdot s^{-1}$. Of course, multiple tools and/or processes can be used to increase revenue, but this is economically viable only if money can be made on each tool or process individually.

To supply some background and indicate the scale of the nanomanufacturing challenge, Figure 3-2 shows the selling price ($\$ \cdot m^{-2}$) versus the annual production (m^2) for a variety of nanoenabled or potentially nanoenabled products. The overall global market sizes are also indicated.

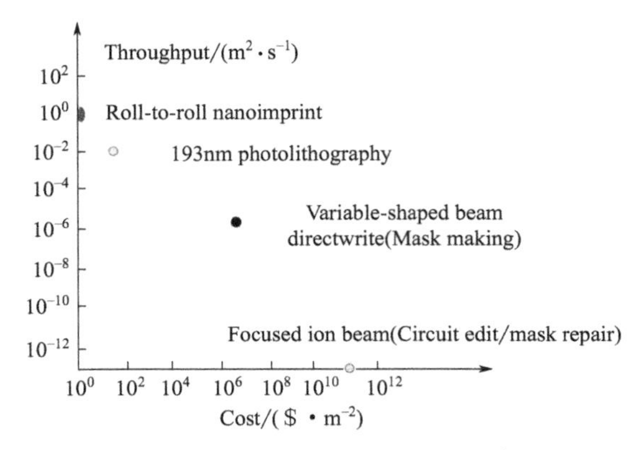

Figure 3-2 Log-log plot of the approximate product selling price ($\$ \cdot m^{-2}$) versus global annual production (m^2) for a variety of nanoenabled, or potentially nanoenabled products. Approximatemarket sizes (2014) are shown next to each point (the supporting information contains the information we used to estimate each datapoint)

In the context of the preceding discussion, the intent of this article is to provide an overview of the current state and of future prospects for nanomanufacturing. Some of the many nanofabrication techniques under development[12-16] may one day be used in nanomanufacturing, but in order to understand which ones are likely to make the transition to being a revenue source and not a sink, it is necessary to identify how the physical aspects of any given technique affect its ability to generate products that meet the desired functional requirements in a cost-effective manner. Every product in the market place has a set of "functional requirements", i. e. , things it must do to be useful to the consumer. The sophistication of these functional requirements drives the product specifications in terms of structural complexity, dimensional and compositional accuracy and precision, tolerable defect levels, and the degree to which the product can be classified as being active as opposed to passive. These product specifications can be used to determine which fabrication approaches potentially have the necessary capabilities. The final choice of nanomanufacturing technology must be driven by the cost of ownership, which depends critically on the characteristics of the manufacturing process, including yield and throughput.

3.1 Top-Down Versus Bottom-Up Processes

Before we begin our discussion of these two approaches, we must clarify our use of the words "deterministic" and "stochastic". Deterministic manufacturing processes are designed to produce outcomes that exhibit very narrow distributions of the product performance mean and variation. However, this does not mean that there cannot be significant randomness at the atomic or nanometer scale. Stochastic processes, while statistical in nature, can, when averaged over the large numbers of nanostructures typically involved, also exhibit narrow distributions of the product performance mean and variation. Another way is to consider the length scales over which variations occur. Short-range precision and accuracy may be excellent for some stochastic processes, such as protein synthesis and folding, but tend to degrade rapidly at length scales larger than an individual unit. In contrast, deterministic methods may be disordered at the atomic or nanometer scale, but can have exceptional long-range accuracy and precision. As noted above, the product's functional requirements are used to determine if a given approach might be suitable.

3.2 Top-Down Fabrication

Top-down processes are fundamentally deterministic, in that order is imposed on the system by the action of external forces. As Gordon Moore observed in his seminal 1965 paper, this means that yield is not dictated by equilibrium thermodynamic considerations, in contrast to, for example, chemical reactions. Yield in top-down processes therefore can be engineered essentially to any desired degree consistent with physical laws: current IC manufacturing processes produce functioning devices with error rates below roughly 1 in 10^{12} and have generally far fewer than 0.1 defects/cm^2. It follows that, as long as there is no fundamental physical limit and the relevant economic drivers apply, the capabilities of top-down fabrication processes will tend to evolve according to typical learning curves.

3.3 Photolithography

Integrated circuit manufacturing represents the apogee of top-down control over matter, yielding devices with unprecedented and ever-increasing levels of functionality in ever smaller spaces. The six decades following the invention of the IC have seen the evolution of photolithography to the point where a modern photolithography tool, operating with an immersion lens at a wavelength of 193nm, is capable of printing 10^{12} features/s, at a resolution of 38nm. It does this while maintaining control over the feature size to within 10%, and the ability to overlay 30 or more layers with respect to one another to within an uncertainty of less than 2.5nm. This degree of control has of course also been enabled by concomitant progress in etch and deposition technologies and photoresist chemistry. The ability to print features which are so much smaller than the wavelength of the exposing radiation requires careful engineering of the mask pattern which in the end has no simple relationship to the features being printed on the wafer. This type of mask engineering must be combined with simultaneous optimization and precise control of the illumination incident on the mask to create the required intensity distribution at the wafer. In fact, the complexity of the coupled illumination-mask diffraction problem is so great that each future generation of chips relies on the computing power made available by the current generation of devices to solve it and, for all but the highest-volume devices, the mask cost is the dominant factor in the cost of ownership. In addition, multiple masks may be required to print the features for a single level when double- or multiple-patterning approaches are used to achieve the desired feature density. As noted above, it is the economic advantage gained by increasing integration that drives the technological progress in IC production, and it is this economic advantage that will determine whether the current incarnation of photolithography gives way to extreme ultraviolet lithography (EUV), which operates at a wavelength of 13. 5nm. The use of this much shorter wavelength brings many additional complexities, and therefore higher costs, as far as the lithography tool is concerned, but can, in principle, reduce the mask complexity thereby reducing the overall cost of ownership. The final result will depend on whether a suitable combination of resist sensitivity and illumination power can be reached to deliver economically viable throughputs. Alternatively, the drive for greater circuit densities and functionalities may be satisfied by the use of 3D approaches.

As impressive as the current state of the art is, it is important to point out that photolithography is only as good as it needs to be. In particular, for integrated circuits to function, it is only the relative placement of individual circuit levels across the chip that must be controlled. Variations in the overall size and shape of the chip up to a few percent are acceptable as long as the level-to-level overlay is maintained. Thus, for ICs, absolute accuracy is not critical. On the other hand, with the increasing emphasis on combining nanophotonic and nanomechanical devices with conventional ICs, absolute accuracy will be extremely important since both nanophotonic and nanomechanical devices typically are required to operate at a fixed external wavelength or frequency. For example, an on-chip nanophotonic device structure such as an add-drop filter comprising ring resonators, waveguides, and/or gratings operating in the 1550nm telecommunications band with a 30GHz channel separation (~0. 3nm wavelength difference) nominally requires dimensions and/or periodicities accurate to within a fraction of the wavelength difference, i. e., a few picometers. These requirements are beyond the current capabilities of photolithography and necessitate an increase in the overall device complexity to include systems which can tune them to match a given external wavelength or frequency. Even a perfectly fabricated device will require tuning to compensate for thermal effects which shift the operating wavelength.

3. 4 Nanoimprint Lithography

Embossing processes have been extended to the nanoscale, opening up a range of new applications. Nanoimprint lithography not only can, in principle, produce features at the size needed in integrated circuit production, but also because the imprint template is replicated precisely, can do so without those complexities inherent in subwavelength optical lithography mentioned above: the features on the template look exactly like the features on the substrate. However, this means that the patterned area on the imprint mask is the same size as the patterned area on the wafer (it is therefore described as a 1× mask) and fabricating a 1× mask to the required degree of precision is not trivial. Pattern placement capabilities close to those of photolithography tools have been achieved by introducing schemes that use controlled deformation of the imprint template. Such schemes, coupled with improvements in template fabrication processes, are currently suitable for IC manufacturing at feature sizes of~20nm (i. e., the 2× nm nodes). Another potential benefit comes from the ability of nanoimprint to print multiple pattern levels simultaneously, which can result in a significant reduction in the number of process steps needed to complete a device. Aside from meeting overlay requirements, the principal obstacle facing the technology is for it to achieve the same throughput and low level of defects as photolithography.

Nanoimprint's ability to fabricate almost arbitrarily small features precisely and accurately means that it is almost uniquely suited to the production of bit-patterned magnetic storage media, which have tolerances on feature size and size variation that are significantly more stringent than those for ICs. Interestingly, although the local pattern placement specification for bit-patterned media (BPM) must be better than 1.25nm, because the flying read head cannot track glitches in position, the long-range pattern placement requirement is only 10μm or approximately 0.3%, because the head can track long-range, slowly varying placement errors.

For nanoimprint to be an appropriate choice of manufacturing technology for bit-patterned media, it is not enough that it can produce small features competitively with photolithography because the price per unit area must be 2 orders of magnitude lower. Part of the necessary fabrication cost reduction may be achieved because only a single level needs to be patterned, which reduces the tool complexity and cost as well as the number of fabrication steps, and part may be achieved by dramatically lowering the amortized mask costs, by using a single master to generate up to 10000 copies, which can then each generate 10000 disk-drive platters. Additional cost reductions come from the much higher defect density that can be tolerated for BPM as opposed to ICs (1 in 10^4 versus 1 in 10^{12}) because of the availability of read-channel error correction schemes. A less demanding specification for defect density translates into a reduced need for costly defect inspection during manufacturing. As the technology evolves and defect levels decrease, it is even being considered for the production of flash memory, which has relatively relaxed overlay requirements, though defect levels will need to be reduced to ~0.1cm^{-2}.

An attractive feature of imprint or embossing processes in general is that they can be adapted for use in continuous, roll-to-roll (R2R) manufacturing, which reduces the cost of fabrication and increases the throughput. The cost per meter squared may range between 0.1$ • m^{-2} and 10$ • m^{-2}, depending on whether direct embossing or a thin-film ultraviolet light curing process is used. It is difficult to achieve precise overlay and long-range placement accuracy in R2R because of the tendency of the flexible substrate (called the "web") to deform during processing. Single-level structures, used in applications such as large-area reflective coatings, and holographic wrapping paper, do not suffer from these problems. Extending single-level R2R technology to the nanoscale would enable applications such as organic photovoltaics, light-management films and wire-grid polarizers for high-contrast displays antireflective coatings and superhydrophobic (philic) surfaces, as well as those requiring plasmonic activity. Self-aligned imprint lithography uses a multilevel template that produces prealigned structures, avoiding level-to-level alignment problems. It can be used to generate large-area, low-cost electronics such as display backplanes with minimum feature sizes of 1μm at web speeds of up to 5m/min. We also note that, while stitching defects are important in displays, other forms of defects are much less significant. Nanostructured surfaces produced by R2R imprint may also be useful for energy generation and storage if high aspect ratio features can be generated.

Nanoimprint is already used to produce nanoparticles for diagnostic and therapeutic applications. At first sight, it might appear that high-volume particle-production techniques such as milling would be by far more economical. However, medical and biological applications generally require very precise control over particle size and shape, and the small amount of product needed to achieve the desired effect (e.g., drug delivery, imaging contrast), and the high value associated with medical treatments make this an economically viable approach. In addition, nanoimprint can be a relatively gentle process, creating patterns without the need for high temperatures or aggressive chemicals, and thus allowing the safe handling of fragile biomolecules.

Perhaps the most exciting aspect of nanoimprint is that it is heir to all the numerous macroscale embodiments of the printing process. At the most fundamental level, these all transfer material from a patterned surface to a substrate. Functional inks are already enabling a revolution in printed electronics for flexible displays, wearable electronics and the "Internet of Things", and there is every reason to suppose that one as profound will occur as inks are developed for nanoscale applications. As an example, microcontact printing, the small-scale analogue of flexographic printing, has recently been demonstrated at the nanoscale. Similarly, nanotransfer printing, analogous to transfer printing on ceramics, enables the fabrication and heterogeneous integration of complex nanostructures made from a wide variety of materials. This gives the family of nanoimprint methods the potential to be a true nanomanufacturing platform technology, limited only by the availability of suitable templates, inks, and surface energy control.

3.5 Other Top-Down Techniques

Optical lithography and nanoimprint are the dominant top-down nanomanufacturing methods, despite there being a large number of other nanofabrication approaches available. At this point, it is worth asking why these other techniques have not made the transition into nanomanufacturing. One of the principle obstacles that must be overcome is reaching an economically viable throughput. Electron-beam lithography, for example, can generate sub-10nm features, over large areas, with good placement and overlay, but because of its relatively low throughput, it is limited commercially to the production of masks for use in photo print and nanoimprint lithography and device development, and noncommercially to the production of nanostructures for research and defense purposes. The prospects for increasing the throughput of electron-beam systems are severely limited because of the fundamental physics of space-charge effects: the repulsion between neighboring electrons in a single electron column leads to a loss of resolution or blurring of the beam as the beam current increases. This limits the maximum beam current, and hence the throughput, that can be attained at a given resolution. This inability to scale led to the demise of early programs focused on creating electron-beam systems for IC production. However, the economic pressures associated with leading-edge optical lithography still make this an active area of research, with newer efforts relying on groups of columns, or targeted toward low-volume applications, under development.

A common response to this type of scaling problem is to propose the use of massively parallel arrays of columns or tips to boost the throughput. Achieving the required feature size control means that the beam size and dose delivered must be well calibrated or dynamically controlled across the array. However, the electron source for each column is typically a field emitter for which the beam current is exponentially dependent on the emission area and geometry. Even if each emitter can be made identical, as soon as they are put into operation, the evolution of the nanoscale emission area resulting from phenomena such as surface diffusion, ion bombardment, absorption of contaminants, etc. will cause the emission characteristics for the individual tips to diverge from one another. Feedback control to remedy this problem is possible in principle, but has so far proven to be extremely difficult to implement in practice. Similar considerations apply to many forms of scanning probe lithography, where attempts at parallelization are frustrated by, for example, tip wear. A notable exception to this is dip-pen nanolithography and its variants, which, through its ability to directly pattern different types of chemistries, including biological ones, targets and satisfies a set of constraints different from those relevant to IC device fabrication. This characteristic qualifies it as a disruptive technology, and the same may be true of other fabrication processes that also offer benefits orthogonal to those provided by IC fabrication methods.

3.6 Bottom-Up Fabrication

In contrast to the deterministic nature of top-down processes, bottom-up processes are driven by a combination of thermodynamics and kinetics which then determines the yield of the desired structure. The most attractive features of bottom-up nanomanufacturing processes are that there is typically no need for expensive tooling to create nanoscale structures, and scaling to large volumes is potentially straightforward. With the application of the tools of chemical synthesis, quantum dots, plasmonically active particles, carbon nano-tubes, metallic nanowires and multifunctional particles for medical applications have been successfully produced in manufacturing quantities. Efforts to develop purely bottom-up self-assembly methods to create more complex devices typically rely on engineering the interactions between the various components, placing them in a simple environment and then letting the system evolve to a final state.

For example, consider the case where there is one relative arrangement of the elemental units that meets the product's functional requirements. If the energy of this particular arrangement is E and, out of all the possible arrangements, there are N other arrangements all with energy E_N close to E which are not the desired structure, and all other arrangements have much higher energy, then the odds of getting the desired structure, assuming the system is in thermal equilibrium, are on the order of

$$\frac{\exp\left(-\frac{E}{k_B T}\right)}{N\exp\left(-\frac{E_N}{k_B T}\right)} \approx \frac{1}{N}$$

Where $k_B T$ is the thermal energy at the end of the production cycle. In general, as the number of units needed to build the target structure increases, there are likely to be more arrangements that have energy close to E, i. e., N increases and so the odds of the getting the "right" result decrease. Hence, unlike top-down manufacturing, the product yield is statistically determined. This means that, when high yields (i. e., high purity) are needed, costly separation and purification steps are required.

The expression above represents the best case for a single-step or one-pot process, and is based on the assumption that thermodynamic equilibrium can be reached within a relevant time scale. However, as systems become more complex, the phenomenon of kinetic trapping can prevent them from reaching the desired equilibrium state. This effect is most easily understood in terms of the potential energy landscape of the elemental units. If the element-to-element potential energy depends on the relative position and orientation of each unit with respect to each other and there are n elemental units, then, up to an overall rotation and translation, the potential energy landscape is a function of all the $3n+3n=6n$ position and rotation coordinates of each elemental unit. In general, the $6n$ dimensional potential energy landscape will have numerous metastable local minima, and only one global minimum which defines the desired assembled structure. The challenge in the bottom-up assembly of complex structures is then to engineer the element-to-element interactions, formation of intermediate structures, and process conditions (e. g., annealing schedule) so that there is a clear path for the components to follow through the potential energy landscape to the global minimum. Alternatively, applications must be targeted that require, for example, only short-range order and/or allow for a high defect level.

3.7 Colloidal Self-Assembly

Colloidal self-assembly has been investigated intensively for many years because of the potential for colloidal structures to produce photonic bandgap materials and high-density magnetic recording media. Early work was directed toward trying to use photonic bandgap materials generated in this way for nanophotonic applications. However, the difficulty of avoiding kinetic trapping to achieve the requisite structural perfection, engineering in features such as waveguides and programmed point defects, and finding cost-effective ways to integrate the structures produced with other photonic devices proved too great to overcome. Nanophotonic structures are now typically fabricated using top-down methods. The same is true of bit-patterned media (BPM). In contrast, self-assembled colloidal structures are ideal candidates for the generation of large-area, low-cost, structural-color materials because the degree of perfection required to meet the functional specification is so much less and the ability to scale production to large areas through roll-to-roll processing is so much greater. Colloidal self-assembly specifically at the nanoscale has exciting possibilities in terms of generating novel materials by combining nanoparticles with different properties into well-defined crystalline structures. Here again, it is important to identify applications for which such materials can be integrated into a manufacturing process flow.

3.8 DNA-Based Self-Assembly

DNA is the archetypal self-assembling system, with tremendous flexibility in the types of structures that can be produced, based on single-stranded (ssDNA), double-stranded or duplex (dsDNA), and more complex supra-molecular assemblies. One-, two-and three-dimensional structures can be made, and the ability of other nanoscale objects to be functionalized with DNA, combined with the specificity conferred by complementary sequence recognition, means that DNA can connect and organize disparate nanostructures to make relatively complex constructs, including well-controlled nanoparticle crystal lattices, and even active systems. DNA origami is a prime example of the power of DNA to control the arrangement of nanoscale objects, providing a molecularly precise "breadboard" to which nanostructures can be attached. In addition, DNA structures can be responsive to variations in temperature, ionic species/concentration and pH. It is also possible to vary the number and strength of DNA-mediated interactions between nanoparticles which can lead to interesting stimulus-dependent responses, allowing the creation of new, environmentally responsive nanostructures.

The development of DNA-based self-assembly is still at a relatively early stage, though progressing rapidly, and there are few studies on the yield, speed, and ultimate levels of complexity that can be achieved in single units and assemblages on substrates. In addition, although the underlying arrangement of DNA may be precise, the presence of linker molecules that must be used between the nanostructure and the DNA, and the fact that DNA structures are not perfectly rigid inevitably lead to a reduction in placement precision. The diffusional nature of the assembly process and the typical rate constants for the reactions involved mean that it takes a long time to create complex structures and that the yields for those structures are going to be consistent with chemical synthesis, not top-down fabrication, even though substantial progress is being made in developing optimized annealing schedules and buffer compositions. Additional improvements may be achieved by controlling the energetics of the structure to guide the assembly process and formation of secondary structure. Applications, such as the fabrication of vaccines and other biomedically active structures, for which 100% yield and purity, precise placement, and control of multiple levels of structural hierarchy are not prerequisites, therefore, need to be identified for this technology to be used in nanomanufacturing. Finally, it is important to note that, while DNA itself is not particularly robust, recent work has shown how structures produced using DNA can subsequently be encapsulated in silica-based materials to dramatically improve their environmental stability.

3.9 Directed Self-Assembly: Top-Down Combined with Bottom Up

So far we have discussed the strengths and weaknesses of top-down and bottom-up approaches to nanomanufacturing, but combining the two together can yield the best of both worlds. Guided or templated self-assembly typically makes use of boundaries created by top-down methods that interact with a system that has an intrinsic structural length scale. This latter can arise from the balance between long-range magnetic, electrostatic, or strain energy, or, as in the case of block copolymers, can come from local interactions built into the molecular structure of the material.

3.9.1 Directed Self-Assembly of Block Copolymers

Block copolymers phase separate on the nanoscale, with an intrinsic length scale determined by the molecular weights of the components and a structure determined by their relative volume fractions. Nanostructures formed in this way can be functional themselves, can be used to template the formation or arrangement of other nanostructures, produce materials responsive to their environment, or can be used to pattern an underlying material. While using a self-assembled structure as an intermediate step in a patterning process, as opposed to the final functional structure, may seem to introduce unnecessary process complexity, there are other challenges involved in using functional materials directly. In the case of diblocks, these are creating a material that phase separates at the requisite length scale: has the necessary surface energies to assemble in the desired orientation with respect to the substrate; has the correct kinetic behavior to minimize defects; can be coated in thin-film form, all while maintaining the sought-for functionality; which gives some idea of the difficulty involved. Given these factors, it becomes clear that the investment required to develop a suitable material system is only worthwhile for high-volume/high-value applications.

Long-range order can be introduced by using a sparse templating pattern generated by top-down methods, and is a very attractive route to making well-controlled nanoscale features: it greatly relaxes the requirements for the top-down process in terms of feature spacing and throughput and deals with the limitations of the bottom-up assembly process. In addition, this approach can even be used to generate relatively complex three-dimensional structures, and is therefore being considered for the manufacture of ICs, BPM and other structures.

Photolithography is limited, not in terms of the smallest feature size that can be produced, but in its ability to place them close together. With the use of photolithography to create a guiding pattern and the assembly of a diblock to fill in the details, it is possible to make dense, nanoscale patterns with excellent control. Similarly, cylindrical or spherical diblocks can be templated by sparse patterns of posts made by electron-beam lithography to form well-ordered arrays of features for BPM. In this case, the benefit lies in both dramatically reducing the time needed for the electron-beam lithography step and in the ability of the diblock to effectively repair patterning defects, and generate much more uniform and dense patterns than could otherwise be produced. An interesting feature of this approach is the interplay between the templating pattern and the diblock which, by altering the energetics of the system, can lead to either a reduction or an increase in the defect level.

There are limits to how sparse a templating pattern can be employed. In the BPM case, if the guiding features are too far apart, then there is a degeneracy because more than one orientation of the diblock can match the templating pattern, leading to the formation of domain boundaries. In lamellar diblocks, as the distance from a directing boundary increases, undulations in the interfaces increase to the point where they lead to an unacceptable level of line-edge roughness for IC fabrication. These limitations represent design constraints, but are unlikely to impede the adoption of this technology in manufacturing. The time taken for diblock systems to order can be relatively long and increases with increasing molecular weight, but recent work using high-temperature, solvent annealing or a combination indicates that this is unlikely to be a serious issue. Reducing chain entanglement with brush block copolymers is also an effective strategy. However, these types of directed self-assembly processes are restricted in terms of the amount of information that can be added to the system and are capable only of producing single harmonics of the templating structure. Additional patterning steps will therefore always be needed to create the kind of structural complexity necessary for logic devices. Finally, although the number of equilibrium defects is expected to be negligible, eliminating defects related to the templating structures is still challenging, and is becoming more so as feature sizes decrease, requiring smaller molecular weight diblocks with smaller domain sizes operating closer to the order-disorder transition. This latter effect is leading to the search for materials with higher Flory-Huggins interaction parameters (χ), or for small-molecule additives that can be used to drive phase separation.

3.9.2 Fluidic Assembly

One difficulty in controlling the assembly of nanoscale objects is finding interactions that are strong enough to manipulate them, and that scale well to small dimensions. Capillary interactions can satisfy these requirements and have been used to create a variety of interesting structures. In particular, the capillary interactions that occur at a fluid interface on a patterned substrate can be used to assemble nanoparticles precisely and with high yield onto lithographically patterned features. The convective flows that are set up at a meniscus can make the process quite efficient, by concentrating nanoparticles at the fluid-substrate contact line, the so-called coffee-stain effect. Beyond the need to control the contact angle between substrate and fluid within a fairly forgiving range, this process is agnostic with regard to the nature of the substrate and the nanoparticles and can so be used to assemble a wide variety of materials without the need for any kind of harsh processing involving solvents, acids/bases, or energetic plasmas. Additionally, once assembled onto a templating substrate, the nanoparticle structures can readily be transferred to a different material. Unfortunately, the maximum linear contact line speeds achieved so far are only $\sim 1m \cdot s^{-1}$ to $\sim 1mm \cdot s^{-1}$, which, even if used in a roll-to-roll process with a meter wide web, translate into an areal throughput of $10^{-6}m^2 \cdot s^{-1}$ to $10^{-3}m^2 \cdot s^{-1}$. The limiting factors are the overall concentration of nanoparticles, which cannot be increased indefinitely without causing deposition in unpatterned areas, and the evaporation rate of the carrier fluid, which also cannot be increased dramatically. Interestingly, the combination of a roll-to-roll patterned substrate with inkjet printing has been used to create color filters with a precision far better than can be achieved with inkjet alone. Although currently only being used for micrometer-sized features, this is potentially a highly extensible approach.

3.9.3 Damped-Driven Systems

Unlike the self-assembling systems described above, which simply "fall down" a free-energy landscape to a stable equilibrium, damped driven systems require energy input in order to form and maintain a self-organized structure. The Belousov-Zhabotinsky reaction is a well-known example, but the archetype is a living being. All living things require energy input (the driver) or they die and decay. The input energy is dissipated as work and heat (the damping). This constant flow of energy through the system maintains it in its self-organized form, for example, in all aspects of intracellular transport. Although we have learned to harness living systems to manufacture everything from alcohol to spidersilk, in contrast to systems at equilibrium, very little is known about the general principles governing damped-driven, self-organizing systems, preventing us from creating our own. In this context, one particular biological process is worth discussing: protein folding. Key parts of the folding process require energy input which is dissipated as heat and so this is a damped driven process. Folding occurs much more rapidly than would be expected if it was purely a stochastic approach to equilibrium as when colloidal particles are annealed into crystalline structures. Protein folding is not a random walk but rather a quasi-deterministic trajectory across the potential energy landscape directed both by the internal structure of the molecule as well as by the action of external, ATP-driven, "chaperone" molecules, known as chaperonins. The concept of using both internal structure combined with external control may be the most effective way forward for creating more complex, functional nanostructures cost-effectively.

3.9.4 Design for Nanomanufacturing

So far, we have discussed how various nanomanufacturing approaches may or may not be suited to the economically sustainable production of functional structures and devices. It is also important to remember that there are often a number of different structures that will yield similar functionality. Choosing the right form can make the difference between something remaining a laboratory curiosity or becoming a product. As a case in point, consider optical metamaterials: the first demonstrations of negative index behavior in the microwave were achieved using macroscale features, such as split-ring resonators, fabricated using simple, scalable, printed-circuit board methods. Subsequent attempts to make materials active at visible wavelengths used the same geometries, replicated at scales 10^{-4} to 10^{-5} smaller using electron-beam lithography. Although these successfully demonstrated the desired functionality, fabrication, integration, and scaling remain challenging, and the cost of such structures is prohibitive ($10^6 \$ \cdot m^{-2}$ to $10^9 \$ \cdot m^{-2}$, depending on whether variable shaped-beam or Gaussian beam patterning is used). A more complete understanding of nanoscale light-matter interactions has led to a new generation of optical metamaterials that comprise alternating layers of metals and dielectrics. These are eminently manufacturable over large areas and at low cost using conventional thin-film deposition processes. As a point of comparison, the types of films made this way, such as those used for touch screens, are approximately $10 \$ \cdot m^{-2}$. A similar development process has occurred in the quest for gecko-type dry adhesives. Early attempts focused on faithful replication of the biological structure via complex lithography to achieve the desired function. More recent work has used the same design principles as the natural system, but in the form of a composite textile that can be made simply, using standard methods.

This type of device structure/fabrication process evolution is familiar in the semiconductor industry. Circuit size reduction often takes place first via a "dumb shrink" or through the use of representative test structures, rapidly followed by an optimized device structure and process redesign that incorporates any new physics and fabrication constraints whose importance has been elucidated during the initial learning phase.

Notes and References

[1] Martin-Palma R J, Lakhtakia A. Nanotechnology: a crash course. SPIE: Bellingham, WA, 2010.

[2] Kriven W M, Lin H T, Zhu D, et al. Defect clustering and nano-phase structure characterization of multi-component rare earth oxide doped Zirconia-Yttria Thermal barrier coatings. The 27th annual Cocoa Beach conference on advanced ceramics and composites: A: Ceramic Engineering and Science Proceedings, Wiley: New York, 2008, 24.

[3] Wang Z Y, et al. Fire-resistant effect of nanoclay on intumescent nanocomposite coatings. J. Appl. Polym. Sci., 2007, 103, 1681-1689.

[4] Talapin D V, Lee J S, Kovalenko M V, et al. Prospects of colloidal nanocrystals for electronic and optoelectronic applications. Chem. Rev. 2010, 110, 389-458.

[5] Novoselov K S, Fal'ko V I, Colombo L, et al. A roadmap for graphene. Nature, 2012, 490, 192-200.

[6] Nair R R, Wu H A, Jayaram P N, et al. Unimpeded permeation of water through helium-leak-tight graphene-based membranes. Science, 2012, 335, 442-444.

[7] Henzie J, Barton E B, Stender C L, et al. Large area nanoscale patterning: chemistry meets fabrication. Acc. Chem. Res., 2006, 39, 249-257.

[8] Wiley B J, Qin D, Xia Y. Nanofabrication at high throughput and low cost. ACS Nano, 2010, 4, 3554-3559.

[9] Swenson B C. Macro media's total addressable market model: methodology and data sources. Macromedia Market Research，2004.

[10] Liddle J A，Gallatin G M. Lithography metrology and nanomanufacturing. Nanoscale，2011，3，2679-2688.

[11] Velev O，Gupta S. Materials fabricated by micro- and nanoparticle assembly-the challenging path from science to engineering. Adv. Mater，2009，21，1897-1905.

[12] Saavedra H M，Mullen T J，Zhang P，et al. Hybrid strategies in nanolithography. Rep. Prog. Phys.，2010，73，036501.

[13] Xia Y，Rogers J A，Paul K E，et al. Unconventional methods for fabricating and patterning nanostructures. Chem. Rev.，1999，99，1823-1848.

[14] Gates B D，Xu Q，Stewart M，et al. New approaches to nanofabrication：molding，printing，and other techniques. Chem. Rev.，2005，105，1171-1196.

[15] Kim P，Epstein A K，Khan M，et al. Structural transformation by electrodeposition on patterned substrates (STEPS)：a new versatile nanofabrication method. Nano Lett.，2012，12，527-533.

[16] Biswas A，Bayer I S，Biris A S，et al. Advances in top-down and bottom-up nanofabrication：techniques，applications & future prospects. Adv. Colloid Interface Sci.，2012，170，2-27.

Chapter 4
3D Printing

4.1 Overview

3D printing, or additive manufacturing, is the construction of a three-dimensional object from a CAD model or a digital 3D model. The term "3D printing" can refer to a variety of processes in which material is deposited, joined or solidified under computer control to create a three-dimensional object[1], with material being added together (such as liquid molecules or powder grains being fused together), typically layer by layer.

In the 1980s, 3D printing techniques were considered suitable only for the production of functional or aesthetic prototypes, and a more appropriate term for it at the time was rapid prototyping. As of 2019, the precision, repeatability and material range of 3D printing has increased to the point that some 3D printing processes are considered viable as an industrial-production technology, whereby the term additive manufacturing can be used synonymously with 3D printing[2]. One of the key advantages of 3D printing is the ability to produce very complex shapes or geometries that would be otherwise impossible to construct by hand, including hollow parts or parts with internal truss structures to reduce weight. Fused deposition modeling, or FDM, is the most common 3D printing process in use as of 2020.

4.2 Terminology

The umbrella term additive manufacturing (AM) gained popularity in the 2000s, inspired by the theme of material being added together (in any of various ways). In contrast, the term subtractive manufacturing appeared as a retronym for the large family of machining processes with material removal as their common process. The term 3D printing still referred only to the polymer technologies in most minds, and the term AM was more likely to be used in metalworking and end use part production contexts than among polymer, inkjet or stereo lithography enthusiasts. Inkjet was the least familiar technology even though it was invented in 1950 and poorly understood because of its complex nature. The earliest inkjets were used as recorders and not printers. As late as the 1970s the term recorder was associated with inkjet. Continuous inkjet later evolved to on-demand or drop-on-demand inkjet. Inkjets were single nozzle at the start and now have thousands of nozzles for printing in each pass over a surface.

By early 2010s, the terms 3D printing and additive manufacturing evolved senses in which they were alternate umbrella terms for additive technologies, one being used in popular language by consumer-maker communities and the media, and the other used more formally by industrial end-use part producers, machine manufacturers, and global technical standards organizations. Until recently, the term 3D printing has been associated with machines low in price or in capability. 3D printing and additive manufacturing reflect that the technologies share the theme of material addition or joining throughout a 3D work envelope under automated control. Peter Zelinski, the editor-in-chief of Additive Manufacturing magazine, pointed out in 2017 that the terms are still often synonymous in casual usage but some manufacturing industry experts are trying to make a distinction whereby additive manufacturing comprises 3D printing plus other technologies or other aspects of a manufacturing process.

Other terms that have been used as synonyms or hypernyms have included desktop manufacturing, rapid manufacturing (as the logical production-level successor to rapid prototyping), and on-demand manufacturing (which echoes on-demand printing in the 2D sense of printing). Such application of the adjectives rapid and on-demand to the noun manufacturing was novel in the 2000s reveals the prevailing mental model of the long industrial era in which almost all production manufacturing involved long lead times for laborious tooling development. Today, the term subtractive has not replaced the term machining, instead complementing it when a term that covers any removal method is needed. Agile tooling is the use of modular means to design tooling that is produced by additive manufacturing or 3D printing methods to enable quick prototyping and responses to tooling and fixture needs. Agile tooling uses a cost-effective and high-quality method to quickly respond to customer and market needs, and it can be used in hydro-forming, stamping, injection molding and other manufacturing processes.

4.3 History

4.3.1 1950

3D printing was first described by Raymond F. Jones in his story, "Tools of the Trade", published in the November 1950 issue of Astounding Science Fiction magazine. He referred to it as a "molecular spray" in that story.

4.3.2 1970s

In 1971, Johannes F. Gottwald patented the liquid metal recorder, US3596285A, a continuous inkjet metal material device to form a removable metal fabrication on a reusable surface for immediate use or salvaged for printing again by remelting. This appears to be the first patent describing

3D printing with rapid prototyping and controlled on-demand manufacturing of patterns.

The patent states "As used herein the term printing" is not intended in a limited sense but includes writing or other symbol, character or pattern formation with an ink. The term ink as used in is intended to include not only dye or pigment-containing materials, but also any flowable substance or composition suited for application to surface for forming symbols, characters, or patterns of intelligence by marking. The preferred ink is of a hot melt type. The range of commercially available ink compositions which could meet the requirements of the invention are not known at the present time. However, satisfactory printing according to the invention has been achieved with the conductive metal alloy as ink.

In 1974, David E. H. Jones laid out the concept of 3D printing in his regular column Ariadne in the journal New Scientist[3].

4.3.3 1980s

Early additive manufacturing equipment and materials were developed in the 1980s. In April 1980, Hideo Kodama of Nagoya Municipal Industrial Research Institute invented two additive methods for fabricating three-dimensional plastic models with photo-hardening thermoset polymer, where the UV exposure area is controlled by a mask pattern or a scanning fiber transmitter. He filed a patent for this XYZ plotter, which was published on November 10, 1981 (JP S56-144478). His research results as journal papers were published in April and November in 1981. However, there was no reaction to the series of his publications. His device was not highly evaluated in the laboratory and his boss did not show any interest. His research budget was just 60,000 yen or $545 a year. Acquiring the patent rights for the XYZ plotter was abandoned, and the project was terminated.

On July 2, 1984, American entrepreneur Bill Masters filed a patent for his Computer Automated Manufacturing Process and System (US 4665492). This filing is on record at the USPTO as the first 3D printing patent in history, and it was the first of three patents belonging to Masters that laid the foundation for the 3D printing systems used today[4].

On 16 July 1984, Alain Le Méhauté, Olivier de Witte and Jean Claude Andréfiled patented for the stereolithography process[5]. The application of the French inventors was abandoned by the French General Electric Company (now Alcatel-Alsthom) and CILAS (the Laser Consortium)[6]. The claimed reason was "for lack of business perspective".

In early 1984, Robert Howard started R. H. Research, later named Howtek, Inc. to develop a color 2D printer using thermoplastic (hot-melt) plastic ink. A team was put together, some from Exxon Office Systems, Danbury Systems Division, an inkjet printer startup and some members of Howtek, Inc. group became popular figures in 3D printing industry. Richard Helinski patent US5136515A, method and means for constructing three-dimensional articles by particle deposition, formed a New Hampshire Company C. A. D-Cast, Inc. , name later changed to Visual Impact Corporation (VIC) on August 22, 1991. A prototype of the VIC 3D printer for this company is available with video presentation showing a 3D model printed with single nozzle inkjet. Another employee Herbert Menhennett formed a New Hampshire Company HM Research in 1991 and introduced the Howtek, Inc. , inkjet technology and thermoplastic materials to Royden Sanders of SDI and Bill Masters of Ballistic Particle Manufacturing (BPM) where he worked for a number of years. Both BPM 3D printers and SPI 3D printers use Howtek, Inc. style inkjets and Howtek,

Inc. style materials. Royden Sanders licensed the Helinksi patent prior to manufacturing the Modelmaker 6 Pro at Sanders Prototype, Inc. (SPI) in 1993. James K. McMahon was hired by Howtek, Inc. to help develop the inkjet, he later worked at Sanders Prototype and now operates Layer Grown Model Technology, a 3D service provider specializing in Howtek single nozzle inkjet and printer support. James K. McMahon worked with Steven Zoltan at Exxon and has a patent in 1978 that expanded the understanding of the single nozzle design inkjets and help perfect the Howtek, Inc. inkjets.

Three weeks later in 1984, Chuck Hull of 3D Systems Corporation filed his own patent for a stereolithography fabrication system, in which layers are added by curing photopolymers with ultraviolet light lasers. Hull defined the process as a "system for generating three-dimensional objects by creating a cross-sectional pattern of the object to be formed"[7]. Hull's contribution was the STL (Stereolithography) file format and the digital slicing and infill strategies common to many processes today.

In 1986, Charles "Chuck" Hull was granted a patent for his system, and his company, 3D Systems Corporation released the first commercial 3D printer, the SLA-1.

The technology used by most 3D printers to date—especially hobbyist and consumer-oriented models—is fused deposition modeling, a special application of plastic extrusion, developed in 1988 by S. Scott Crump and commercialized by his company Stratasys, which marketed its first FDM machine in 1992.

4.3.4 1990s

AM processes for metal sintering or melting (such as selective laser sintering, direct metal laser sintering, and selective laser melting) usually went by their own individual names in the 1980s and 1990s. At the time, all metalworking was done by processes that are now called non-additive (casting, fabrication, stamping and machining). Although plenty of automation was applied to those technologies (such as by robot welding and CNC), the idea of a tool or head moving through a 3D work envelope transforming a mass of raw material into a desired shape with a toolpath was associated in metalworking only with processes that removed metal (rather than adding it), such as CNC milling, CNC EDM and many others. But the automated techniques that added metal, which would later be called additive manufacturing, were beginning to challenge that assumption. By the mid-1990s, new techniques for material deposition were developed at Stanford and Carnegie Mellon University, including microcasting and sprayed materials[8]. Sacrificial and support materials had also become more common, enabling new object geometries[9].

The term 3D printing originally referred to a powder bed process employing standard and custom inkjet print heads, developed at MIT by Emanuel Sachs in 1993 and commercialized by Soligen Technologies, Extrude Hone Corporation, and Z Corporation.

The year 1993 also saw the start of an inkjet 3D printer company initially named Sanders Prototype, Inc. and later named Solidscape, introducing a high-precision polymer jet fabrication system with soluble support structures (categorized as a "dot-on-dot" technique).

In 1995, the Fraunhofer Society developed the selective laser melting process.

4.3.5 2000s

Fused deposition modeling (FDM) printing process patents expired in 2009.

4.3.6 2010s

As the various additive processes matured, it became clear that soon metal removal would no longer be the only metalworking process done through a tool or head moving through a 3D work envelope, transforming a mass of raw material into a desired shape layer by layer. The 2010s were the first decade in which metal end use parts such as engine brackets and large nuts would be grown (either before or instead of machining) in job production rather than obligately being machined from bar stock or plate. It is still the case that casting, fabrication, stamping and machining are more prevalent than additive manufacturing in metalworking, but AM is now beginning to make significant inroads, and with the advantages of design for additive manufacturing, it is clear to engineers that much more is to come.

As technology matured, several authors had begun to speculate that 3D printing could aid in sustainable development in the developing world[10].

In 2012, Filabot developed a system for closing the loop with plastic and allowed for any FDM or FFF 3D printer to be able to print with a wider range of plastics.

In 2014, Benjamin S. Cook and Manos M. Tentzeris demonstrated the first multi-material, vertically integrated printed electronics additive manufacturing platform (VIPRE) which enabled 3D printing of functional electronics operating up to 40GHz.

The term "3D printing" originally referred to a process that deposits a binder material onto a powder bed with inkjet printer heads layer by layer. More recently, the popular vernacular has started using the term to encompass a wider variety of additive-manufacturing techniques such as electron-beam additive manufacturing and selective laser melting. The United States and global technical standards use the official term additive manufacturing for this broader sense.

The most-commonly used 3D printing process (46% as of 2018) is a material extrusion technique called fused deposition modeling, or FDM. While FDM technology was invented after the other two most popular technologies, stereolithography (SLA) and selective laser sintering (SLS), FDM is typically the most inexpensive of the three by a large margin, which lends to the popularity of the process.

But in terms of material requirements for such large and continuous displays, if consumed at theretofore known rates, but increased in proportion to increase in size, the high cost would severely limit any widespread enjoyment of a process or apparatus satisfying the foregoing objects.

It is therefore an additional object of the invention to minimize use to materials in a process of the indicated class.

It is a further object of the invention that materials employed in such process could be salvaged for reuse.

According to another aspect of the invention, a combination for writing and the like comprises a carrier for displaying an intelligence pattern and an arrangement for removing the pattern from the carrier.

4.4 General Principles

4.4.1 Modeling

3D printable models may be created with a computer-aided design (CAD) package, via a 3D scanner, or by a plain digital camera and photogrammetry software. 3D printed models created with CAD result in relatively fewer errors than other methods. Errors in 3D printable models can be identified and corrected before printing[11]. The manual modeling process of preparing geometric data for 3D computer graphics is similar to plastic arts such as sculpting. 3D scanning is a process of collecting digital data on the shape and appearance of a real object, creating a digital model based on it (Figure 4-1).

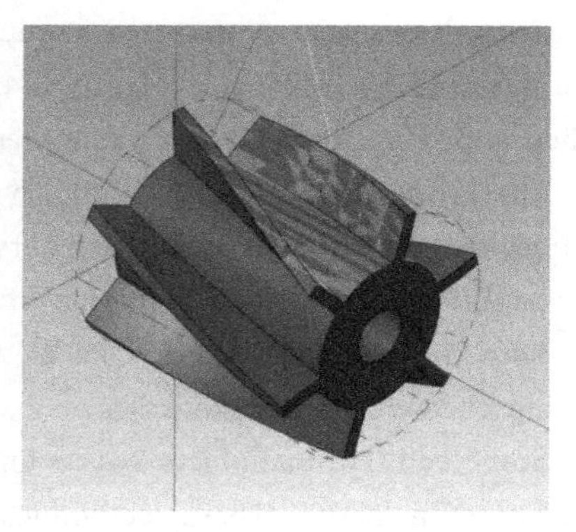

Figure 4-1 One typical model for 3D printing

CAD models can be saved in the stereolithography file format (STL), a de facto CAD file format for additive manufacturing that stores data based on triangulations of the surface of CAD models. STL is not tailored for additive manufacturing because it generates large file sizes of topology optimized parts and lattice structures due to the large number of surfaces involved. A newer CAD file format, the Additive Manufacturing File format (AMF) was introduced in 2011 to solve this problem. It stores information using curved triangulations.

4.4.2 Printing

Before printing a 3D model from an STL file, it must first be examined for errors. Most CAD applications produce errors in output STL files, of the following types: holes, faces normals, self-intersections, noise shells, manifold errors.

A step in the STL generation known as "repair" fixes such problems in the original model. Generally, STLs that have been produced from a model obtained through 3D scanning often have more of these errors as 3D scanning is often achieved by point to point acquisition/mapping. 3D reconstruction often includes errors[12].

Once completed, the STL file needs to be processed by a piece of software called a "slicer", which converts the model into a series of thin layers and produces a G-code file containing instructions tailored to a specific type of 3D printer (FDM printers). This G-code file can then be printed with 3D printing client software (which loads the G-code, and uses it to instruct the 3D printer during the 3D printing process).

Printer resolution describes layer thickness and *X-Y* resolution in dots per inch (DPI) or micrometers (μm). Typical layer thickness is around 100μm (250 DPI), although some machines can print layers as thin as 16μm (1600 DPI). *X-Y* resolution is comparable to that of laser printers. The particles (3D dots) are around 50 to 100μm (510 to 250 DPI) in diameter. For that printer resolution, specifying a mesh resolution of 0.01-0.03 mm and a chord length$\leqslant$0.016mm generate an optimal STL output file for a given model input file. Specifying higher resolution results in larger files without increase in print quality.

Construction of a model with contemporary methods can take anywhere from several hours to several days, depending on the method used and the size and complexity of the model. Additive systems can typically reduce this time to a few hours, although it varies widely depending on the type of machine used and the size and number of models being produced simultaneously.

4.4.3 Finishing

Though the printer-produced resolution is sufficient for many applications, greater accuracy can be achieved by printing a slightly oversized version of the desired object in standard resolution and then removing material using a higher-resolution subtractive process.

The layered structure of all additive manufacturing processes leads inevitably to a stair-stepping effect on part surfaces which are curved or tilted in respect to the building platform. The effects strongly depend on the orientation of a part surface inside the building process[13].

Some printable polymers such as ABS, allow the surface finish to be smoothed and improved using chemical vapor processes[14] based on acetone or similar solvents.

Some additive manufacturing techniques are capable of using multiple materials in the course of constructing parts. These techniques are able to print in multiple colors and color combinations simultaneously, and would not necessarily require painting.

Some printing techniques require internal supports to be built for overhanging features during construction. These supports must be mechanically removed or dissolved upon completion of the print.

All of the commercialized metal 3D printers involve cutting the metal component off the metal substrate after deposition. A new process for the GMAW 3D printing allows for substrate surface modifications to removealuminum[15] or steel[16].

4.4.4 Materials

Traditionally, 3D printing focused on polymers for printing, due to the ease of manufacturing and handling polymeric materials. However, the method has rapidly evolved to not only print various polymers but also metals and ceramics[17], making 3D printing a versatile option for manufacturing. Layer-by-layer fabrication of three-dimensional physical models is a modern concept that "stems from the ever-growing CAD industry, more specifically the solid modeling side of CAD". Before solid modeling was introduced in the late 1980s, three-dimensional models were created with wire frames and surfaces "but in all cases the layers of materials are controlled by the printer and the material properties. The three-dimensional material layer is controlled by deposition rate as set by the printer operator and stored in a computer file. The earliest printed patented material was a hot melt type ink for printing patterns using a heated metal alloy. See 1970s history above.

4.4.5 Multi-Materials 3D Printing

A drawback of many existing 3D printing technologies is that they only allow one material to be printed at a time, limiting many potential applications which require the integration of different materials in the same object. Multi-material 3D printing solves this problem by allowing objects of complex and heterogeneous arrangements of materials to be manufactured using a single printer. Here, a material must be specified for each voxel (or 3D printing pixel element) inside the final object volume.

The process can be fraught with complications, however, due to the isolated and monolithic algorithms. Some commercial devices have sought to solve these issues, such as building a Spec 2

Fab translator, but the progress is still very limited. Nonetheless, in the medical industry, a concept of 3D printed pills and vaccines has been presented. With this new concept, multiple medications can be combined, which will decrease many risks. With more and more applications of multi-material 3D printing, the costs of daily life and high technology development will become inevitably lower.

Metallographic materials of 3D printing is also being researched. By classifying each material, CIMP-3D can systematically perform 3D printing with multiple materials.

4.5 Processes and Printers

There are many different branded additive manufacturing processes, which can be grouped into seven categories:

(1) Vat photopolymerization.

(2) Material jetting.

(3) Binder jetting.

(4) Powder bed fusion.

(5) Material extrusion.

(6) Directed energy deposition.

(7) Sheet lamination.

The main differences between processes are in the way layers are deposited to create parts and in the materials that are used. Each method has its own advantages and drawbacks, which is why some companies offer a choice of powder and polymer for the material used to build the object. Others sometimes use standard, off-the-shelf business paper as the build material to produce a durable prototype. The main considerations in choosing a machine are generally speed, costs of the 3D printer and the printed prototype, choice and cost of the materials, and color capabilities. Printers that work directly with metals are generally expensive. However less expensive printers can be used to make a mold, which is then used to make metal parts.

ISO/ASTM52900-15 defines seven categories of additive manufacturing (AM) processes within its meaning: binder jetting, directed energy deposition, material extrusion, material jetting, powder bed fusion, sheet lamination and vat photopolymerization.

Some methods melt or soften the material to produce the layers. In fused filament fabrication, also known as fused deposition modeling (FDM), the model or part is produced by extruding small beads or streams of material which harden immediately to form layers. A filament of thermoplastic, metal wire or other material is fed into an extrusion nozzle head (3D printer extruder), which heats the material and turns the flow on and off. FDM is somewhat restricted in the variation of shapes that may be fabricated. Another technique fuses parts of the layer and then moves upward in the working area, adding another layer of granules and repeating the process until the piece has built up. This process uses the unfused media to support overhangs and thin walls in the part being produced, which reduces the need for temporary auxiliary supports for the piece. Recently, FFF/FDM has expanded to 3D print directly from pellets to avoid the conversion to filament. This process is called fused particle fabrication (FPF) or fused granular fabrication (FGF) and has the potential to use more recycled materials[18].

Powder bed fusion techniques, or PBF, include several processes such as DMLS, SLS, SLM, MJF and EBM. Powder bed fusion processes can be used with an array of materials and their flexibility allows for geometrically complex structures, making it a go to choice for many 3D printing projects. These techniques include selective laser sintering, with both metals and polymers, and direct metal laser sintering. Selective laser melting does not use sintering for the fusion of powder granules but will completely melt the powder using a high-energy laser to create fully dense materials in a layer-wise method that has mechanical properties similar to those of conventional manufactured metals. Electron beam melting is a similar type of additive manufacturing technology for metal parts (e. g. titanium alloys). EBM manufactures parts by melting metal powder layer by layer with an electron beam in a high vacuum. Another method consists of an inkjet 3D printing system, which creates the model one layer at a time by spreading a layer of powder (plaster or resins) and printing a binder in the cross-section of the part using an inkjet-like process. With laminated object manufacturing, thin layers are cut to shape and joined together. In addition to the previously mentioned methods, HP has developed the Multi Jet Fusion (MJF) which is a powder base technique, though no lasers are involved. An inkjet array applies fusing and detailing agents which are then combined by heating to create a solid layer.

Other methods cure liquid materials using different sophisticated technologies, such as stereolithography. Photopolymerization is primarily used in stereolithography to produce a solid part from a liquid. Inkjet printer systems like the Objet Poly Jet system spray photopolymer materials onto a build tray in ultra-thin layers (between 16 and 30μm) until the part is completed. Each photopolymer layer is cured with UV light after it is jetted, producing fully cured models that can be handled and used immediately, without post-curing. Ultra-small features can be made with the 3D micro-fabrication technique used in multiphoton photopolymerisation. Due to the nonlinear nature of photo excitation, the gel is cured to a solid only in the places where the laser was focused while the remaining gel is then washed away. Feature sizes of under 100nm are easily produced, as well as complex structures with moving and interlocked parts. Yet another approach uses a synthetic resin that is solidified using LEDs.

In mask-image-projection-based stereolithography, a 3D digital model is sliced by a set of horizontal planes. Each slice is converted into a two-dimensional mask image. The mask image is then projected onto a photocurable liquid resin surface and light is projected onto the resin to cure it in the shape of the layer. Continuous liquid interface production begins with a pool of liquid photopolymer resin. Part of the pool bottom is transparent to ultraviolet light (the "window"), which causes the resin to solidify. The object rises slowly enough to allow resin to flow under and maintain contact with the bottom of the object. In powder-fed directed-energy deposition, a high-power laser is used to melt metal powder supplied to the focus of the laser beam. The powder fed directed energy process is similar to selective laser sintering, but the metal powder is applied only where material is being added to the part at that moment.

As of December 2017, additive manufacturing systems were on the market that ranged from $99 to $500000 in price and were employed in industries including aerospace, architecture, automotive, defense, and medical replacements, among many others. For example, General Electric uses high-end 3D printers to build parts for turbines. Many of these systems are used for rapid prototyping, before mass production methods are employed. Higher education has proven to be a major buyer of desktop and professional 3D printers which industry experts generally view as a positive indicator. Libraries around the world have also become locations to house smaller 3D printers for educational and community access. Several projects and companies are making efforts to develop affordable 3D printers for home desktop use. Much of this work has been driven by and targeted at DIY/maker/enthusiast/early adopter communities, with additional ties to the academic and hacker communities.

Computed axial lithography is a method for 3D printing based on computerised tomography scans to create prints in photo-curable resin. It was developed by collaboration between the University of California, Berkeley and Lawrence Livermore National Laboratory. Unlike other methods of 3D printing, it does not build models through depositing layers of material like fused deposition modelling and stereolithography, instead it creates objects using a series of 2D images projected onto a cylinder of resin. It is notable for its ability to build an object much more quickly than other methods using resins and the ability to embed objects within the prints.

Liquid additive manufacturing (LAM) is a 3D printing technique which deposits a liquid or high viscose material (e. g. liquid silicone rubber) onto a build surface to create an object which then is vulcanised using heat to harden the object. The process was originally created by Adrian Bowyer and was then built upon by German RepRap (Figure 4-2).

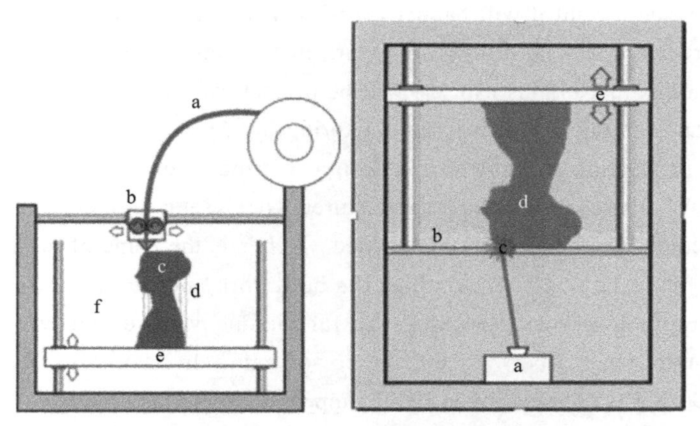

Figure 4-2 The schematic of liquid additive manufacturing

4.6 Applications

In the current scenario, 3D printing or additive manufacturing has been used in manufacturing, medical, industry and sociocultural sectors which facilitate 3D printing or additive manufacturing to become successful commercial technology. More recently, 3D printing has also been used in the humanitarian and development sector to produce a range of medical items, prosthetics, spares and repairs. The earliest application of additive manufacturing was on the toolroom end of the manufacturing spectrum. For example, rapid prototyping was one of the earliest additive variants, and its mission was to reduce the lead time and cost of developing prototypes of new parts and devices, which was earlier only done with subtractive toolroom methods such as CNC milling, turning, and precision grinding. In the 2010s, additive manufacturing entered production to a much greater extent.

Additive manufacturing of food is being developed by squeezing out food, layer by layer, into three-dimensional objects. A large variety of foods are appropriate candidates, such as chocolate and candy, and flat foods such as crackers, pasta and pizza. NASA is looking into the technology in order to create 3D printed food to limit food waste and to make food that is designed to fit an astronaut's dietary needs. In 2018, Italian bioengineer Giuseppe Scionti developed a technology to generate fibrous plant-based meat analogues using a custom 3D bioprinter, mimicking meat texture and nutritional values.

3D printing has entered the world of clothing, with fashion designers experimenting with 3D printed bikinis, shoes and dresses. In commercial production, Nike is using 3D printing to prototype the 2012 Vapor Laser Talon football shoe for players of American football, and New Balance is 3D manufacturing custom-fit shoes for athletes. 3D printing has come to the point where companies are printing consumer grade eyewear with on-demand custom fit and styling (although they cannot print the lenses). On-demand customization of glasses is possible with rapid prototyping.

Vanessa Friedman, fashion director and chief fashion critic at the New York Times, says 3D printing will have a significant value for fashion companies down the road, especially if it transforms into a print-it-yourself tool for shoppers. "There's real sense that this is not going to happen anytime soon," she says, "but it will happen, and it will create dramatic change in how we think both about intellectual property and how things are in the supply chain." She adds: "Certainly some of the fabrications that brands can use will be dramatically changed by technology."

In cars, trucks and aircraft, additive manufacturing is beginning to transform both unibody + fuselage design + production and powertrain design + production. For example:

(1) In early 2014, Swedish supercar manufacturer Koenigsegg announced the One: a supercar that utilizes many components that were 3D printed. Urbee is the name of the first car in the world car mounted using the technology 3D printing (its bodywork and car windows were "printed").

(2) In 2014, Local Motors debuted Strati: a functioning vehicle that was entirely 3D printed using ABS plastic and carbon fiber, except the power train. In May 2015 Airbus announced that its new Airbus A350 XWB included over 1000 components manufactured by 3D printing.

(3) In 2015, a Royal Air Force Euro fighter Typhoon fighter jet flew with printed parts. The United States Air Force has begun to work with 3D printers, and the Israeli Air Force has also purchased a 3D printer to print spare parts.

(4) In 2017, GE Aviation revealed that it had used design for additive manufacturing to create a helicopter engine with 16 parts instead of 900, with great potential impact on reducing the complexity of supply chains.

AM's impact on firearms involves two dimensions: new manufacturing methods for established companies, and new possibilities for the making of do-it-yourself firearms. In 2012, the US-based group Defense Distributed disclosed plans to design a working plastic 3D printed firearm "that could be downloaded and reproduced by anybody with a 3D printer." After Defense Distributed released their plans, questions were raised regarding the effects that 3D printing and widespread consumer-level CNC machining may have on gun control effectiveness.

Surgical uses of 3D printing-centric therapies have a history beginning in the mid-1990s with anatomical modeling for bony reconstructive surgery planning. Patient-matched implants were a natural extension of this work, leading to truly personalized implants that fit one unique individual. Virtual planning of surgery and guidance using 3D printing, personalized instruments have been applied to many areas of surgery including total joint replacement and craniomaxillofacial reconstruction with great success. One example of this is the bioresorbable trachial splint to treat newborns with tracheobronchomalacia[19] developed at the University of Michigan. The use of additive manufacturing for serialized production of orthopedic implants (metals) is also increasing due to the ability to efficiently create porous surface structures that facilitate osseointegration. The hearing aid and dental industries are expected to be the biggest area of future development using the custom 3D printing technology.

In March 2014, surgeons in Swansea used 3D printed parts to rebuild the face of a motorcyclist who had been seriously injured in a road accident. In May 2018, 3D printing has been used for the kidney transplant to save a three-year-old boy. As of 2012, 3D bio-printing technology has been studied by biotechnology firms and academia for possible use in tissue engineering applications in which organs and body parts are built using inkjet printing techniques. In this process, layers of living cells are deposited onto a gel medium or sugar matrix and slowly built up to form three-dimensional structures including vascular systems. Recently, a heart-on-chip has been created which matches properties of cells.

In 3D printing, computer-simulated microstructures are commonly used to fabricate objects with spatially varying properties. This is achieved by dividing the volume of the desired object into smaller subcells using computer aided simulation tools and then filling these cells with appropriate microstructures during fabrication. Several different candidate structures with similar behaviours are checked against each other and the object is fabricated when an optimal set of structures are found. Advanced topology optimization methods are used to ensure the compatibility of structures in adjacent cells. This flexible approach to 3D fabrication is widely used across various disciplines from biomedical sciences where they are used to create complex bone structures and human tissue to robotics where they are used in the creation of soft robots with movable parts[20,21].

3D printing has also been employed by researchers in the pharmaceutical field. During the last few years, there's been a surge in academic interest regarding drug delivery with the aid of AM techniques. This technology offers a unique way for materials to be utilized in novel formulations. AM manufacturing allows for the usage of materials and compounds in the development of formulations, in ways that are not possible with conventional/traditional techniques in the pharmaceutical field, e. g. tableting, cast-molding, etc. Moreover, one of the major advantages of 3D printing, especially in the case of fused deposition modelling (FDM), is the personalization of the dosage form that can be achieved, thus, targeting the patient's specific needs. In the not-so-distant future, 3D printers are expected to reach hospitals and pharmacies in order to provide on demand production of personalized formulations according to the patients' needs.

In 2018, 3D printing technology was used for the first time to create a matrix for cell immobilization in fermentation. Propionic acid production by propionibacterium acidipropionici immobilized on 3D printed nylon beads was chosen as a model study. It was shown that those 3D printed beads were capable of promoting high density cell attachment and propionic acid production, which could be adapted to other fermentation bioprocesses.

In 2005, academic journals had begun to report on the possible artistic applications of 3D printing technology. As of 2017, domestic 3D printing was reaching a consumer audience beyond hobbyists and enthusiasts. Off the shelf machines were increasingly capable of producing practical household applications, for example, ornamental objects. Some practical examples include a working clock and gears printed for home woodworking machines among other purposes. Web sites associated with home 3D printing tended to include backscratchers, coat hooks, door knobs, etc.

3D printing, and open source 3D printers in particular, are the latest technology making inroads into the classroom. Some authors have claimed that 3D printers offer an unprecedented "revolution" in STEM education. The evidence for such claims comes from both the low-cost ability for rapid prototyping in the classroom by students, and the fabrication of low-cost high-quality scientific equipment from open hardware designs forming open-source labs. Future applications for 3D printing might include creating open-source scientific equipment.

In the last several years, 3D printing has been intensively used in the cultural heritage field for preservation, restoration and dissemination purposes. Many Europeans and North American museums have purchased 3D printers and actively recreate missing pieces of their relics and archaeological monuments such as Tiwanaku in Bolivia. The Metropolitan Museum of Art and the British Museum have started using their 3D printers to create museum souvenirs that are available in the museum shops. Other museums, like the National Museum of Military History and Varna Historical Museum, have gone further and sell through the online platform 3D digital models of their artifacts, created by using Artec 3D scanners, in 3D printing friendly file format, which everyone can 3D print at home.

3D printed soft actuators are a growing application of 3D printing technology which has found its place in the 3D printing applications. These soft actuators are being developed to deal with soft structures and organs especially in biomedical sectors and where the interaction between human and robot is inevitable. The majority of the existing soft actuators are fabricated by conventional methods that require manual fabrication of devices, post processing/assembly, and lengthy iterations until maturity of the fabrication is achieved. Instead of the tedious and time-consuming aspects of the current fabrication processes, researchers are exploring an appropriate manufacturing approach for effective fabrication of soft actuators.

Thus, 3D printed soft actuators are introduced to revolutionise the design and fabrication of soft actuators with custom geometrical, functional and control properties in a faster and inexpensive approach. They also enable incorporation of all actuator components into a single structure eliminating the need to use external joints, adhesives, and fasteners. Circuit board manufacturing involves multiple steps which include imaging, drilling, plating, soldermask coating, nomenclature printing and surface finishes. These steps include many chemicals such as harsh solvents and acids. 3D printing circuit boards remove the need for many of these steps while still producing complex designs. Polymer ink is used to create the layers of the build while silver polymer is used for creating the traces and holes used to allow electricity to flow. Current circuit board manufacturing can be a tedious process depending on the design. Specified materials are gathered and sent into inner layer processing where images are printed, developed and etched. The etched cores are typically punched to add lamination tooling. The cores are then prepared for lamination. The stack-up, the buildup of a circuit board, is built and sent into lamination where the layers are bonded. The boards are then measured and drilled.

Many steps may differ from this stage however for simple designs. The material goes through a plating process to plate the holes and surface. The outer image is then printed, developed and etched. After the image is defined, the material must get coated with soldermask for later soldering. Nomenclature is then added so components can be identified later. Then the surface finish is added. The boards are routed out of panel form into their singular or array form and then electrically tested. Aside from the paperwork which must be completed which proves the boards meet specifications, the boards are then packed and shipped. The benefits of 3D printing would be that the final outline is defined from the beginning, no imaging, punching or lamination is required and electrical connections are made with the silver polymer which eliminates drilling and plating. The final paperwork would also be greatly reduced due to the lack of materials required to build the circuit board. Complex designs which may takes weeks to complete through normal processing can be 3D printed, greatly reducing manufacturing time.

Notes and References

[1] Excell J. The rise of additive manufacturing. The Engineer, 2013.

[2] Lam Hugo K S, Ding Li, Cheng T C E, et al. The impact of 3D printing implementation on stock returns: a contingent dynamic capabilities perspective. International journal of operations & production management, 2019, 39 (6/7/8): 935-961.

[3] Evans B. Practical 3D Printers: The Science and Art of 3D Printing [M]. Apress, 2012.

[4] Wang B. Concurrent Design of Products, Manufacturing Processes and Systems. CRC Press. 1999.

[5] Ligon S C, Liska R, Stampfl J, et al. Polymers for 3D printing and customized additive manufacturing. Chem Rev, 2017, 117: 10212-90.

[6] Mohamed O A, Masood S H, Bhowmik J L. Optimization of fused deposition modeling process parameters: a review of current research and future prospects. Adv. Manuf., 2015, 3: 42-53.

[7] Freedman D H. Layer by layer. Technology Review, 2012, 115(1): 50-53.

[8] Jasveer S, Xue J. Comparison of different types of 3D printing technologies. International Journal of Scientific and Research Publications, 2018, 8 (4).

[9] Prinz F B, Merz R, et al. Building parts you could not build before. Proceedings of the 8th international conference on production engineering. London, UK: Chapman & Hall. 1997, 40-44.

[10] Ishengoma F, et al. 3D printing: developing countries perspectives. International Journal of Computer Applications, 2014, 104(11): 30.

[11] Sochol R D, Sweet E, Glick C C, et al. 3D printed microfluidics and microelectronics. Microelectronic Engineering, 2017, 189: 52-58.

[12] Bernardini F, et al. The 3D model acquisition pipeline gas. Computer Graphics. Forum, 2002, 21 (2): 149-172.

[13] Delfs P, Tows M, Schmid H J. Optimized build orientation of additive manufactured parts for improved surface quality and build time. Additive Manufacturing, 2016, 12: 314-320.

[14] Wallin T J, Pikul J, Shepherd R F. 3D printing of soft robotic systems. Nature Publishing Group, 2018.

[15] Haselhuhn A, et al. Substrate release mechanisms for gas metal arc weld 3D aluminum metal printing. 3D Printing and Additive Manufacturing, 2014, 1(4): 2014.

[16] Haselhuhn A S, Wijnen B, Anzalone G C, et al. In situ formation of substrate release mechanisms for gas metal arc weld metal 3D printing. Journal of Materials Processing Technology, 2015, 226: 50.

[17] Galante R, et al. Additive manufacturing of ceramics for dental applications. Dental Materials, 2019, 35 (6): 825-846.

[18] Aubrey W, et al. Fused particle fabrication 3D printing: recycled materials' optimization and mechanical properties. Materials, 2018, 11 (8): 1413.

[19] Zopf D A, et al. Bioresorbable airway splint created with a three-dimensional printer. New England Journal of Medicine, 2013, 368 (21): 2043-5.

[20] Cho K J, et al. Review of manufacturing processes for soft biomimetic robots. International Journal of Precision Engineering and Manufacturing, 2009, 10(3): 171-181.

[21] Rus D, et al. Design, fabrication and control of soft robots. Nature, 2015, 521(7553): 467-475.

Chapter 5
Robot Manufacturing

Robotics is an interdisciplinary field that integrates computer science and engineering. Robotics involves design, construction, operation and use of robots. The goal of robotics is to design machines that can help and assist humans. Robotics integrates fields of mechanical engineering, electrical engineering, information engineering, mechatronics, electronics, bioengineering, computer engineering, control engineering, software engineering, among others.

Robotics develops machines that can substitute for humans and replicate human actions. Robots can be used in many situations and for many purposes, but today many are used in dangerous environments (including inspection of radioactive materials, bomb detection and deactivation), manufacturing processes, or where humans cannot survive (e. g. in space, underwater, in high heat, and clean-up and containment of hazardous materials and radiation). Robots can take on any form but some are made to resemble humans in appearance. This is said to help in the acceptance of a robot in certain replicative behaviors usually performed by people. Such robots attempt to replicate walking, lifting, speech, cognition, or any other human activities. Many of today's robots are inspired by nature, contributing to the field of bio-inspired robotics.

Certain robots require user input to operate while other robots function autonomously. The concept of creating robots that can operate autonomously dates back to classical times, but research into the functionality and potential uses of robots did not grow substantially until the 20th century. Throughout history, it has been frequently assumed by various scholars, inventors, engineers, and technicians that robots will one day be able to mimic human behavior and manage tasks in a human-like fashion. Today, robotics is a rapidly growing field, as technological advances continue. Researching, designing and building new robots serve various practical purposes, whether domestically, commercially, or militarily. Many robots are built to do jobs that are hazardous to people, such as defusing bombs, finding survivors in unstable ruins, and exploring mines and shipwrecks. Robotics is also used in STEM (science, technology, engineering and mathematics) as a teaching aid.

5.1 Etymology

The word "robotics" was derived from the word "robot", which was introduced to the public by Czech writer Karel Čapek in his play R. U. R. (Rossum's Universal Robots), which was published in 1920. The word robot comes from the Slavic word robota, which means slave/servant. The play begins in a factory that makes artificial people called robots, creatures that can be mistaken for humans-very similar to the modern ideas of androids. Karel Čapek himself did not coin the word. He wrote a short letter in reference to an etymology in the Oxford English Dictionary in which he named his brother Josef Čapek as its actual originator.

According to the Oxford English Dictionary, the word robotics was first used in print by Isaac Asimov, in his science fiction short story "Liar", published in May 1941 in Astounding Science Fiction. Asimov was unaware that he was coining the term. Since the science and technology of electrical devices is electronics, he assumed robotics already referred to the science and technology of robots. In some of Asimov's other works, he stated that the first use of the word robotics was in his short story Runaround (Astounding Science Fiction, March 1942), where he introduced his concept of "The Three Laws of Robotics". However, the original publication of "Liar" predates that of "Runaround" by ten months, so the former is generally cited as the word's origin.

5.2 Robotic Aspects

There are many types of robots, and they are used in many different environments and for many different uses. Although being very diverse in application and form, they all share three basic similarities when it comes to their construction.

(1) Robots all have some kind of mechanical construction, a frame, form or shape designed to achieve a particular task. For example, a robot designed to travel across heavy dirt or mud, might use caterpillar tracks. The mechanical aspect is mostly the creator's solution to completing the assigned task and dealing with the physics of the environment around it. Form follows function.

(2) Robots have electrical components that power and control the machinery. For example, the robot with caterpillar tracks would need some kind of power to move the tracker treads. That power comes in the form of electricity, which will have to travel through a wire and originate from a battery, a basic electrical circuit. Even petrol powered machines that get their power mainly from petrol still require an electric current to start the combustion process which is why most petrol powered machines like cars, have batteries. The electrical aspect of robots is used for movement (through motors), sensing (where electrical signals are used to measure things like heat, sound, position, and energy status) and operation (robots need some level of electrical energy supplied to their motors and sensors in order to activate and perform basic operations).

(3) All robots contain some level of computer programming code. A program is how a robot decides when or how to do something. In the caterpillar track example, a robot that needs to move across a muddy road may have the correct mechanical construction and receive the correct amount of power from its battery, but would not go anywhere without a program telling it to move. Programs are the core essence of a robot, it could have excellent mechanical and electrical construc-

tion, but if its program is poorly constructed its performance will be very poor (or it may not perform at all). There are three different types of robotic programs: remote control, artificial intelligence and hybrid. A robot with remote control programming has a preexisting set of commands that it will only perform if and when it receives a signal from a control source, typically a human being with a remote control. It is perhaps more appropriate to view devices controlled primarily by human commands as falling in the discipline of automation rather than robotics. Robots that use artificial intelligence interact with their environment on their own without a control source, and can determine reactions to objects and problems they encounter using their preexisting programming. Hybrid is a form of programming that incorporates both AI and RC functions in them.

5.3 Applications

As more and more robots are designed for specific tasks this method of classification becomes more relevant. For example, many robots are designed for assembly work, which may not be readily adaptable for other applications. They are termed as "assembly robots". For seam welding, some suppliers provide complete welding systems with the robot i. e. the welding equipment along with other material handling facilities like turntables, etc. as an integrated unit. Such an integrated robotic system is called a "welding robot" even though its discrete manipulator unit could be adapted to a variety of tasks. Some robots are specifically designed for heavy load manipulation, and are labeled as "heavy-duty robots"[1].

Current and potential applications include:

(1) Military robots.

(2) Industrial robots. Robots are increasingly used in manufacturing (since the 1960s). According to the Robotic Industries Association US data, in 2016 automotive industry was the main customer of industrial robots with 52% of totalsales[2]. In the auto industry, they can amount for more than half of the "labor". There are even "lights off" factories such as an IBM keyboard manufacturing factory in Texas that was fully automated as early as 2003.

(3) Cobots (collaborative robots).

(4) Construction robots. Construction robots can be separated into three types—traditional robots, robotic arm and robotic exoskeleton.

(5) Agricultural robots (Ag Robots)[3]. The use of robots in agriculture is closely linked to the concept of AI-assisted precision agriculture and drone usage. The research in 1996-1998 also proved that robots can perform a herding task.

(6) Medical robots of various types (such as da Vinci Surgical System and Hospi).

(7) Kitchen automation. Commercial examples of kitchen automation are Flippy (burgers), Zume Pizza (pizza), Cafe X (coffee), Makr Shakr (cocktails), Frobot (frozen yogurts) and Sally (salads). Home examples are Rotimatic (flatbreads baking)[4] and Boris (dishwasher loading).

(8) Robot combat for sport-hobby or sport event where two or more robots fight in an arena to disable each other. This has developed from a hobby in the 1990s to several TV series worldwide.

(9) Clean-up of contaminated areas, such as toxic waste or nuclear facilities.

(10) Domestic robots.

(11) Nanorobots.

(12) Swarm robotics.

(13) Autonomous drones.

(14) Sports field line marking.

5.4 Components

5.4.1 Power Source

At present, mostly (lead-acid) batteries are used as a power source. Many different types of batteries can be used as a power source for robots. They range from lead-acid batteries, which are safe and have relatively long shelf lives but are rather heavy compared to silver-cadmium batteries that are much smaller in volume and are currently much more expensive. Designing a battery-powered robot needs to take into account factors such as safety, cycle lifetime and weight. Generators, often some type of internal combustion engine, can also be used. However, such designs are often mechanically complex and need a fuel, require heat dissipation and are relatively heavy. A tether connecting the robot to a power supply would remove the power supply from the robot entirely. This has the advantage of saving weight and space by moving all power generation and storage components elsewhere. However, this design does come with the drawback of constantly having a cable connected to the robot, which can be difficult to manage. Potential power sources could be:

(1) Pneumatic (compressed gases).

(2) Solar power (using the sun's energy and converting it into electrical power).

(3) Hydraulics (liquids).

(4) Flywheel energy storage.

(5) Organic garbage (through anaerobic digestion).

(6) Nuclear.

5.4.2 Actuation

Actuators are the "muscles" of a robot, the parts which convert stored energy into movement[5]. By far the most popular actuators are electric motors that rotate a wheel or gear, and linear actuators that control industrial robots in factories. There are some recent advances in alternative types of actuators, powered by electricity, chemicals or compressed air.

(1) Electric motors. The vast majority of robots use electric motors, often brushed and brushless DC motors in portable robots or AC motors in industrial robots and CNC machines. These motors are often preferred in systems with lighter loads, and where the predominant form of motion is rotational.

(2) Linear actuators. Various types of linear actuators move in and out instead of by spinning, and often have quicker direction changes, particularly when very large forces are needed such as with industrial robotics. They are typically powered by compressed and oxidized air (pneumatic actuator) or an oil (hydraulic actuator). Linear actuators can also be powered by electricity which

usually consists of a motor and a leadscrew. Another common type is a mechanical linear actuator that is turned by hand, such as a rack and pinion on a car.

(3) Series elastic actuators. Series elastic actuation (SEA) relies on the idea of introducing intentional elasticity between the motor actuator and the load for robust force control. Due to the resultant lower reflected inertia, series elastic actuation improves safety when a robot interacts with the environment (e. g. humans or workpiece) or during collisions. Furthermore, it also provides energy efficiency and shock absorption (mechanical filtering) while reducing excessive wear on the transmission and other mechanical components. This approach has successfully been employed in various robots, particularly advanced manufacturing robots and walking humanoid robots[6-8]. The controller design of a series elastic actuator is most often performed within the passivity framework as it ensures the safety of interaction with unstructured environments. Despite its remarkable stability robustness, this framework suffers from the stringent limitations imposed on the controller which may trade-off performance. The reader is referred to the following survey which summarizes the common controller architectures for SEA along with the corresponding sufficient passivity conditions[9]. One recent study has derived the necessary and sufficient passivity conditions for one of the most common impedance control architectures, namely velocity-sourced SEA[10]. This work is of particular importance as it drives the non-conservative passivity bounds in an SEA scheme for the first time which allows a larger selection of control gains.

(4) Air muscles. Pneumatic artificial muscles, also known as air muscles, are special tubes that expand (typically up to 40%) when air is forced inside them. They are used in some robot applications[11].

(5) Muscle wire. Muscle wire, also known as shape memory alloy, Nitinol or Flexinol wire, is a material which contracts (under 5%) when electricity is applied. They have been used for some small robot applications.

(6) Electroactive polymers. EAPs or EPAMs are a plastic material that can contract substantially (up to 380% activation strain) from electricity, and have been used in facial muscles and arms of humanoid robots, and to enable new robots to float, fly, swim or walk.

(7) Piezo motors. Recent alternatives to DC motors are piezo motors or ultrasonic motors. These work on a fundamentally different principle, whereby tiny piezoceramic elements, vibrating many thousands of times per second, cause linear or rotary motion. There are different mechanisms of operation; one type uses the vibration of the piezo elements to step the motor in a circle or a straight line. Another type uses the piezo elements to cause a nut to vibrate or to drive a screw. The advantages of these motors are nanometer resolution, speed and available force for their size. These motors are already available commercially, and being used on somerobots[12].

(8) Elastic nanotubes. Elastic nanotubes are a promising artificial muscle technology in early-stage experimental development. The absence of defects in carbon nanotubes enables these filaments to deform elastically by several percent, with energy storage levels of perhaps 10 J/cm^3 for metal nanotubes. Human biceps could be replaced with an 8mm diameter wire of this material. Such compact "muscle" might allow future robots to outrun and outjump humans.

5.4.3 Sensing

Sensors allow robots to receive information about a certain measurement of the environment, or internal components. This is essential for robots to perform their tasks, and act upon any changes in the environment to calculate the appropriate response. They are used for various forms of measurements, to give the robots warnings about safety or malfunctions, and to provide real-time information of the task it is performing.

(1) Touch.

Current robotic and prosthetic hands receive far less tactile information than the human hand. Recent research has developed a tactile sensor array that mimics the mechanical properties and touch receptors of humanfingertips[13]. The sensor array is constructed as a rigid core surrounded by conductive fluid contained by an elastomeric skin. Electrodes are mounted on the surface of the rigid core and are connected to an impedance-measuring device within the core. When the artificial skin touches an object the fluid path around the electrodes is deformed, producing impedance changes that map the forces received from the object. The researchers expect that an important function of such artificial fingertips will be adjusting robotic grip on held objects.

Scientists from several European countries and Israel developed a prosthetic hand in 2009, called SmartHand, which functions like a real one—allowing patients to write with it, type on a keyboard, play piano and perform other fine movements. The prosthesis has sensors which enable the patient to sense real feeling in its fingertips.

(2) Vision.

Computer vision is the science and technology of machines that see. As a scientific discipline, computer vision is concerned with the theory behind artificial systems that extract information from images. The image data can take many forms, such as video sequences and views from cameras.

In most practical computer vision applications, the computers are pre-programmed to solve a particular task, but methods based on learning are now becoming increasingly common.

Computer vision systems rely on image sensors which detect electromagnetic radiation which is typically in the form of either visible light or infra-red light. The sensors are designed using solid-state physics. The process by which light propagates and reflects off surfaces is explained using optics. Sophisticated image sensors even require quantum mechanics to provide a complete understanding of the image formation process. Robots can also be equipped with multiple vision sensors to be better able to compute the sense of depth in the environment. Like human eyes, robots' "eyes" must also be able to focus on a particular area of interest, and also adjust to variations in light intensities.

There is a subfield within computer vision where artificial systems are designed to mimic the processing and behavior of biological system, at different levels of complexity. Also, some of the learning-based methods developed within computer vision have their background in biology.

(3) Others.

Other common forms of sensing in robotics use lidar, radar and sonar. Lidar measures distance to a target by illuminating the target with laser light and measuring the reflected light with a sensor. Radar uses radio waves to determine the range, angle or velocity of objects. Sonar uses sound propagation to navigate, communicate with or detect objects on or under the surface of the water.

5.4.4 Manipulation

A definition of robotic manipulation has been provided by Matt Mason as "manipulation refers to an agent's control of its environment through selective contact".

Robots need to manipulate objects. Pick-up, modifying, destroying or otherwise have an effect. Thus the functional end of a robot arm intended to make the effect (whether a hand or tool) are often referred to as end effectors, while the "arm" is referred to as amanipulator[14]. Most robot arms have replaceable end-effectors, each allowing them to perform some small range of tasks. Some have a fixed manipulator that cannot be replaced, while a few have one very general purpose manipulator, for example, a humanoid hand.

(1) Mechanical grippers. One of the most common types of end-effectorsare "grippers". In its simplest manifestation, it consists of just two fingers that can open and close to pick up and let go of a range of small objects. Fingers can, for example, be made of a chain with a metal wire run through it. Hands that resemble and work more like a human hand include the Shadow hand and the Robonaut hand. Hands that are of a mid-level complexity include the Delft hand. Mechanical grippers can come in various types, including friction and encompassing jaws. Friction jaws use all the force of the gripper to hold the object in place using friction. Encompassing jaws cradle the object in place, using less friction.

(2) Suction end-effectors. Suction end-effectors, powered by vacuum generators, are very simple astrictive devices that can hold very large loads provided the prehension surface is smooth enough to ensure suction.

Pick and place robots for electronic components and for large objects like car windscreens, often use very simple vacuum end-effectors.

Suction is a highly used type of end-effector in industry, in part because the natural compliance of soft suction end-effectors can enable a robot to be more robust in the presence of imperfect robotic perception. As an example: consider the case of a robot vision system estimates the position of a water bottle, but has 1cm of error. While this may cause a rigid mechanical gripper to puncture the water bottle, the soft suction end-effector may just bend slightly and conform to the shape of the water bottle surface.

(3) General purpose effectors. Some advanced robots are beginning to use fully humanoid hands, like the Shadow hand, Manus, and the Schunk hand[15]. These are highly dexterous manipulators, with as many as 20 degrees of freedom and hundreds of tactile sensors.

5.4.5 Locomotion

5.4.5.1 Rolling Robots

For simplicity, most mobile robots have four wheels or a number of continuous tracks. Some researchers have tried to create more complex wheeled robots with only one or two wheels. These can have certain advantages such as greater efficiency and reduced parts, as well as allowing a robot to navigate in confined places that a four-wheeled robot would not be able to.

(1) Two-wheeled balancing robots. Balancing robots generally use a gyroscope to detect how much a robot is falling and then drive the wheels proportionally in the same direction, to counterbalance the fall at hundreds of times per second, based on the dynamics of an inverted pendulum. Many different balancing robots have been designed. While the Segway is not commonly thought of as a robot, it can be thought of as a component of a robot, when used as such Segway refer to

them as RMP (robotic mobility platform). An example of this use has been as NASA's Robonaut that has been mounted on a Segway.

(2) One-wheeled balancing robots. A one-wheeled balancing robot is an extension of a two-wheeled balancing robot so that it can move in any 2D direction using a round ball as its only wheel. Several one-wheeled balancing robots have been designed recently, such as Carnegie Mellon University's "Ballbot" that is the approximate height and width of a person, and Tohoku Gakuin University's "BallIP". Because of the long, thin shape and ability to maneuver in tight spaces, they have the potential to function better than other robots in environments with people.

(3) Spherical orb robots. Several attempts have been made in robots that are completely inside a spherical ball, either by spinning a weight inside the ball, or by rotating the outer shells of the sphere. These have also been referred to as an orb bot or a ball bot.

(4) Six-wheeled robots. Using six wheels instead of four wheels can give better traction or grip in outdoor terrain such as on rocky dirt or grass.

(5) Tracked robots. Tank tracks provide even more traction than a six-wheeled robot. Tracked wheels behave as if they were made of hundreds of wheels, therefore are very common for outdoor and military robots, where the robot must drive on very rough terrain. However, they are difficult to use indoors such as on carpets and smooth floors. Examples include NASA's Urban Robot "Urbie".

5. 4. 5. 2 Walking Applied to Robots

Walking is a difficult and dynamic problem to solve. Several robots have been made which can walk reliably on two legs, however, none have yet been made which are as robust as a human. There has been much study on human inspired walking, such as AMBER Lab which was established in 2008 by the Mechanical Engineering Department at Texas A&M University. Many other robots have been built that walk on more than two legs, due to these robots being significantly easier to construct. Walking robots can be used for uneven terrains, which would provide better mobility and energy efficiency than other locomotion methods. Typically, robots on two legs can walk well on flat floors and can occasionally walk up stairs. None can walk over rocky, uneven terrain. Some of the methods which have been tried are:

(1) ZMP technique. The zero moment point (ZMP) is the algorithm used by robots such as Honda's ASIMO. The robot's onboard computer tries to keep the total inertial forces (the combination of Earth's gravity and the acceleration and deceleration of walking), exactly opposed by the floor reaction force (the force of the floor pushing back on the robot's foot). In this way, the two forces cancel out, leaving no moment (force causing the robot to rotate and fall over). However, this is not exactly how a human walks, and the difference is obvious to human observers, some of whom have pointed out that ASIMO walks as if it needs the lavatory. ASIMO's walking algorithm is not static, and some dynamic balancing is used (see below). However, it still requires a smooth surface to walk on.

(2) Hopping. Several robots, built in the 1980s by Marc Raibert at the MIT Leg Laboratory, successfully demonstrated very dynamic walking. Initially, a robot with only one leg, and a very small foot could stay upright simply by hopping. The movement is the same as that of a person on a pogo stick. As the robot falls to one side, it would jump slightly in that direction, in order to catch itself. Soon, the algorithm was generalised to two and four legs. A bipedal robot was de-

monstrated running and even performing somersaults. A quadruped was also demonstrated which could trot, run, pace and bound.

(3) Dynamic balancing (controlled falling). A more advanced way for a robot to walk is by using a dynamic balancing algorithm, which is potentially more robust than the Zero Moment Point technique, as it constantly monitors the robot's motion, and places the feet in order to maintain stability. This technique was recently demonstrated by Anybots' Dexter Robot, which is so stable, and can even jump. Another example is the TU Delft Flame.

(4) Passive dynamics. Perhaps the most promising approach utilizes passive dynamics where the momentum of swinging limbs is used for greater efficiency. It has been shown that totally unpowered humanoid mechanisms can walk down a gentle slope, using only gravity to propel themselves. Using this technique, a robot need only supply a small amount of motor power to walk along a flat surface or a little more to walk up a hill. This technique promises to make walking robots at least ten times more efficient than ZMP walkers, like ASIMO[16].

5.4.5.3 Other Methods of Locomotion

(1) Flying. A modern passenger airliner is essentially a flying robot, with two humans to manage it. The autopilot can control the plane for each stage of the journey, including takeoff, normal flight and even landing. Other flying robots are uninhabited and are known as unmanned aerial vehicles (UAVs). They can be smaller and lighter without a human pilot on board, and fly into dangerous territory for military surveillance missions. Some can even fire on targets under command. UAVs are also being developed which can fire on targets automatically, without the need for a command from a human. Other flying robots include cruise missiles, the Entomopter, and the Epson micro helicopter robot. Robots such as the Air Penguin, Air Ray and Air Jelly have lighter-than-air bodies, propelled by paddles and guided by sonar.

(2) Snaking. Two robot snakes. Left one has 64 motors (with 2 degrees of freedom per segment), the right one 10. Several snake robots have been successfully developed. Mimicking the way real snakes move, these robots can navigate very confined spaces, meaning they may one day be used to search for people trapped in collapsed buildings. The Japanese ACM-R5 snake robot can even navigate both on land and in water.

(3) Skating. A small number of skating robots have been developed, one of which is a multi-mode walking and skating device. It has four legs, with unpowered wheels, which can either step or roll. Another robot, Plen, can use a miniature skateboard or roller-skates, and skate across a desktop.

(4) Climbing. Several different approaches have been used to develop robots that have the ability to climb vertical surfaces. One approach mimics the movements of a human climber on a wall with protrusions; adjusting the center of mass and moving each limb in turn to gain leverage. An example of this is Capuchin, built by Dr. Ruixiang Zhang at Stanford University, California. Another approach uses the specialized toe pad method of wall-climbing geckoes, which can run on smooth surfaces such as vertical glass. Examples of this approach include Wallbot and Stickybot.

China's Technology Daily reported on 15 November 2008, that Dr. Yeung and his research group of New Concept Aircraft (Zhuhai) Co., Ltd. had successfully developed a bionic gecko robot named "Speedy Freelander". According to Dr. Yeung, the gecko robot could rapidly climb up and down a variety of building walls, navigate through ground and wall fissures, and walk upside-down on the ceiling. It was also able to adapt to the surfaces of smooth glass, rough, sticky or dusty walls as well as various types of metallic materials. It could also identify and circumvent obstacles automatically. Its flexibility and speed were comparable to a natural gecko. A third approach is to mimic the motion of a snake climbing a pole.

(5) Swimming (piscine). It is calculated that when swimming some fish can achieve a propulsive efficiency greater than 90%. Furthermore, they can accelerate and maneuver far better than any man-made boat or submarine, and produce less noise and water disturbance. Therefore, many researchers studying underwater robots would like to copy this type of locomotion. Notable examples are the Essex University Computer Science Robotic Fish G9, and the Robot Tuna built by the Institute of Field Robotics, to analyze and mathematically model thunniform motion. The Aqua Penguin, designed and built by Festo of Germany, copies the streamlined shape and propulsion by front "flippers" of penguins. Festo have also built the Aqua Ray and Aqua Jelly, which emulate the locomotion of manta ray, and jellyfish, respectively.

In 2014 iSplash-II was developed by Ph. D student Richard James Clapham and Prof. Huosheng Hu at Essex University. It was the first robotic fish capable of outperforming real carangiform fish in terms of average maximum velocity (measured in body lengths/second) and endurance, the duration that top speed is maintained. This build attained swimming speeds of 11. 6BL/s (i. e. 3. 7 m/s). The first build, iSplash-I (2014) was the first robotic platform to apply a full-body length carangiform swimming motion which was found to increase swimming speed by 27% over the traditional approach of a posterior confined waveform.

(6) Sailing. Sailboat robots have also been developed in order to make measurements at the surface of the ocean. A typical sailboat robot is Vaimos[17] built by IFREMER and ENSTA-Bretagne. Since the propulsion of sailboat robots uses the wind, the energy of the batteries is only used for the computer, for the communication and for the actuators (to tune the rudder and the sail). If the robot is equipped with solar panels, the robot could theoretically navigate forever. The two main competitions of sailboat robots are WRSC, which takes place every year in Europe, and Sailbot.

5.4.6 Environmental Interaction and Navigation

Though a significant percentage of robots in commission today are either human controlled or operate in a static environment, there is an increasing interest in robots that can operate autonomously in a dynamic environment. These robots require some combination of navigation hardware and software in order to traverse their environment. In particular, unforeseen events (e. g. people and other obstacles that are not stationary) can cause problems or collisions. Some highly advanced robots such as ASIMO and Meinürobot have particularly good robot navigation hardware and software. Also, self-controlled cars, Ernst Dickmanns' driverless car, and the entries in the DARPA Grand Challenge, are capable of sensing the environment well and subsequently making navigational decisions based on this information, including by a swarm of autonomous robots. Most of these robots employ a GPS navigation device with waypoints, along with radar, sometimes combined with other sensory data such as lidar, video cameras, and inertial guidance systems for better navigation between waypoints.

5.4.7 Human-Robot Interaction

The state of the art in sensory intelligence for robots will have to progress through several orders of magnitude if we want the robots working in our homes to go beyond vacuum-cleaning the floors. If robots are to work effectively in homes and other non-industrial environments, the way they are instructed to perform their jobs, and especially how they will be told to stop will be of critical importance. The people who interact with them may have little or no training in robotics, and so any interface will need to be extremely intuitive. Science fiction authors also typically assume that robots will eventually be capable of communicating with humans through speech, gestures, and facial expressions, rather than a command-line interface. Although speech would be the most natural way for the human to communicate, it is unnatural for the robot. It will probably be a long time before robots interact as naturally as the fictional C-3PO, or Data of Star Trek, Next Generation.

(1) Speech recognition. Interpreting the continuous flow of sounds coming from a human, in real time, is a difficult task for a computer, mostly because of the great variability of speech. The same word, spoken by the same person may sound different depending on local acoustics, volume, the previous word, whether or not the speaker has a cold, etc. It becomes even harder when the speaker has a different accent. Nevertheless, great strides have been made in the field since Davis, Biddulph, and Balashek designed the first "voice input system" which recognized "ten digits spoken by a single user with 100% accuracy" in 1952[18]. Currently, the best systems can recognize continuous, natural speech, up to 160 words per minute, with an accuracy of 95%. With the help of artificial intelligence, machines nowadays can use people's voice to identify their emotions such as satisfied or angry[19].

(2) Robotic voice. Other hurdles exist when allowing the robot to use voice for interacting with humans. For social reasons, synthetic voice proves suboptimal as a communication medium, making it necessary to develop the emotional component of robotic voice through various techniques. An advantage of diphonic branching is the emotion that the robot is programmed to project, can be carried on the voice tape, or phoneme, already pre-programmed onto the voice media. One of the earliest examples is a teaching robot named Leachim developed in 1974 by Michael J. Freeman. Leachim was able to convert digital memory to rudimentary verbal speech on pre-recorded computer discs. It was programmed to teach students in the Bronx, New York.

(3) Gestures. One can imagine, in the future, explaining to a robot chef how to make a pastry, or asking directions from a robot police officer. In both of these cases, making hand gestures would aid the verbal descriptions. In the first case, the robot would be recognizing gestures made by the human, and perhaps repeating them for confirmation. In the second case, the robot police officer would gesture to indicate "go down the road, then turn right". It is likely that gestures will make up a part of the interaction between humans and robots. A great many systems have been developed to recognize human hand gestures.

(4) Facial expression. Facial expressions can provide rapid feedback on the progress of a dialog between two humans, and soon may be able to do the same for humans and robots. Robotic faces have been constructed by Hanson Robotics using their elastic polymer called frubber, allowing a large number of facial expressions due to the elasticity of the rubber facial coating and embedded subsurface motors (servos). The coating and servos are built on a metal skull. A robot should

know how to approach a human, judging by their facial expression and body language. Whether the person is happy, frightened or crazy-looking affects the type of interaction expected of the robot. Likewise, robots like Kismet and the more recent addition, Nexi can produce a range of facial expressions, allowing it to have meaningful social exchanges with humans.

(5) Artificial emotions. Artificial emotions can also be generated, composed of a sequence of facial expressions and/or gestures. As can be seen from the movie "Final Fantasy: The Spirits Within", the programming of these artificial emotions is complex and requires a large amount of human observation. To simplify this programming in the movie, presets were created together with a special software program. This decreased the amount of time needed to make the film. These presets could possibly be transferred for use in real-life robots (Figure 5-1).

(6) Personality. Many of the robots of science fiction have a personality, something which may or may not be desirable in the commercial robots of the future. Nevertheless, researchers are trying to create robots which appear to have a personality: i. e. they use sounds, facial expressions and body language to try to convey an internal state, which may be joy, sadness or fear. One commercial example is Pleo, a toy robot dinosaur, which can exhibit several apparent emotions (Figure 5-2).

Figure 5-1 The picture of an ideal robot

Figure 5-2 A picture of humanoid robot

(7) Social intelligence. The Socially Intelligent Machines Lab of the Georgia Institute of Technology researches new concepts of guided teaching interaction with robots. The aim of the projects is a social robot that learns task and goals from human demonstrations without prior knowledge of high-level concepts. These new concepts are grounded from low-level continuous sensor data through unsupervised learning, and task goals are subsequently learned using a Bayesian approach. These concepts can be used to transfer knowledge to future tasks, resulting in faster learning of those tasks. The results are demonstrated by the robot Curi who can scoop some pasta from a pot onto a plate and serve the sauce on top.

5. 5 Control

The mechanical structure of a robot must be controlled to perform tasks. The control of a robot involves three distinct phases-perception, processing and action (robotic paradigms). Sensors give information about the environment or the robot itself (e. g. the position of its joints or its end effector). This information is then processed to be stored or transmitted and to calculate the appropriate signals to the actuators (motors) which move the mechanical.

The processing phase can range in complexity. At a reactive level, it may translate raw sensor information directly into actuator commands. Sensor fusion may first be used to estimate parameters of interest (e. g. the position of the robot's gripper) from noisy sensor data. An immediate task (such as moving the gripper in a certain direction) is inferred from these estimates. Techniques from control theory convert the task into commands that drive the actuators.

At longer time scales or with more sophisticated tasks, the robot may need to build and reason with a "cognitive" model. Cognitive models try to represent the robot, the world, and how they interact. Pattern recognition and computer vision can be used to track objects. Mapping techniques can be used to build maps of the world. Finally, motion planning and other artificial intelligence techniques may be used to figure out how to act. For example, a planner may figure out how to achieve a task without hitting obstacles, falling over, etc.

Control systems may also have varying levels of autonomy.

(1) Direct interaction is used for haptic or teleoperated devices, and the human has nearly complete control over the robot's motion.

(2) Operator-assist modes have the operator commanding medium-to-high-level tasks, with the robot automatically figuring out how to achieve them.

(3) An autonomous robot may go without human interaction for extended periods of time. Higher levels of autonomy do not necessarily require more complex cognitive capabilities. For example, robots in assembly plants are completely autonomous but operate in a fixed pattern.

Another classification takes into account the interaction between human control and the machine motions.

(1) Teleoperation. A human controls each movement, each machine actuator change is specified by the operator.

(2) Supervisory. A human specifies general moves or position changes and the machine decides specific movements of its actuators.

(3) Task-level autonomy. The operator specifies only the task and the robot manages itself to complete it.

(4) Full autonomy. The machine will create and complete all its tasks without human interaction.

5.6 Research

Much of the research in robotics focuses not on specific industrial tasks, but on investigations into new types of robots, alternative ways to think about or design robots, and new ways to manufacture them. Other investigations, such as MIT's cyberflora project, are almost wholly academic.

A first particular new innovation in robot design is the open sourcing of robot-projects. To describe the level of advancement of a robot, the term "Generation Robots" can be used. This term is coined by Professor Hans Moravec, principal research scientist at the Carnegie Mellon University

Robotics Institute in describing the near future evolution of robot technology. First generation robots, Moravec predicted in 1997, should have an intellectual capacity comparable to perhaps a lizard and should become available by 2010. Because the first generation robot would be incapable of learning, however, Moravec predicts that the second generation robot would be an improvement over the first and become available by 2020, with the intelligence maybe comparable to that of a mouse. The third generation robot should have the intelligence comparable to that of a monkey. Though fourth generation robots, robots with human intelligence, professor Moravec predicts, would become possible, he does not predict this happening before around 2040 or 2050.

The second is evolutionary robots. This is a methodology that uses evolutionary computation to help design robots, especially the body form, or motion and behavior controllers. In a similar way to natural evolution, a large population of robots is allowed to compete in some way, or their ability to perform a task is measured using a fitness function. Those that perform worst are removed from the population and replaced by a new set, which have new behaviors based on those of the winners. Over time the population improves, and eventually a satisfactory robot may appear. This happens without any direct programming of the robots by the researchers. Researchers use this method both to create better robots, and to explore the nature of evolution. Because the process often requires many generations of robots to besimulated[20], this technique may be run entirely or mostly in simulation, using a robot simulator software package, then tested on real robots once the evolved algorithms are good enough. Currently, there are about 10 million industrial robots toiling around the world, and Japan is the top country having high density of utilizing robots in its manufacturing industry.

The study of motion can be divided into kinematics and dynamics. Direct kinematics or forward kinematics refers to the calculation of end effector position, orientation, velocity, and acceleration when the corresponding joint values are known. Inverse kinematics refers to the opposite case in which required joint values are calculated for given end effector values, as done in path planning. Some special aspects of kinematics include handling of redundancy (different possibilities of performing the same movement), collision avoidance, and singularity avoidance. Once all relevant positions, velocities, and accelerations have been calculated using kinematics, methods from the field of dynamics are used to study the effect of forces upon these movements. Direct dynamics refers to the calculation of accelerations in the robot once the applied forces are known. Direct dynamics is used in computer simulations of the robot. Inverse dynamics refers to the calculation of the actuator forces necessary to create prescribed end-effector acceleration. This information can be used to improve the control algorithms of a robot.

In each area mentioned above, researchers strive to develop new concepts and strategies, improve existing ones, and improve the interaction between these areas. To do this, criteria for "optimal" performance and ways to optimize design, structure, and control of robots must be developed and implemented.

Bionics and biomimetics apply the physiology and methods of locomotion of animals to the design of robots. For example, the design of Bionic Kangaroo was based on the way kangaroos jump.

There has been some research into whether robotics algorithms can be run more quickly on quantum computers than they can be run on digital computers. This area has been referred to as quantum robotics.

5.7 Education and Training

Robotics engineers design robots, maintain them, develop new applications for them, and conduct research to expand the potential of robotics. Robots have become a popular educational tool in some middle and high schools, particularly in parts of the USA[21], as well as in numerous youth summer camps, raising interest in programming, artificial intelligence, and robotics among students.

5.7.1 Career Training

Universities like Worcester Polytechnic Institute (WPI) offer bachelors, masters, and doctoral degrees in the field of robotics. Vocational schools offer robotics training aimed at careers in robotics.

5.7.2 Certification

The Robotics Certification Standards Alliance (RCSA) is an international robotics certification authority that confers various industry- and educational-related robotics certifications.

5.7.3 Summer Robotics Camp

Several national summer camp programs include robotics as part of their core curriculum. In addition, youth summer robotics programs are frequently offered by celebrated museums and institutions.

5.7.4 Robotics Competitions

There are many competitions around the globe. The Sea Perch curriculum is aimed as students of all ages.

(1) Competitions for younger children. The FIRST organization offers the FIRST Lego League Jr. competitions for younger children. This competition's goal is to offer younger children an opportunity to start learning about science and technology. Children in this competition build Lego models and have the option of using the Lego WeDo robotics kit.

(2) Competitions for children ages 9-14. One of the most important competitions is the FLL or FIRST Lego League. The idea of this specific competition is that kids start developing knowledge and getting into robotics while playing with Lego since they are nine years old. This competition is associated with National Instruments. Children use Lego Mindstorms to solve autonomous robotics challenges in this competition.

(3) Competitions for teenagers. The FIRST Tech Challenge is designed for intermediate students, as a transition from the FIRST Lego League to the FIRST Robotics Competition. The FIRST Robotics Competition focuses more on mechanical design, with a specific game being played each year. Robots are built specifically for that year's game. In match play, the robot moves autonomously during the first 15 seconds of the game (although certain years such as 2019s Deep Space change this rule), and is manually operated for the rest of the match.

(4) Competitions for older students. The various RoboCup competitions include teams of teenagers and university students. These competitions focus on soccer competitions with different types of robots, dance competitions, and urban search and rescue competitions. All of the robots in these competitions must be autonomous. Some of these competitions focus on simulated robots.

AUVSI runs competitions for flying robots, robot boats, and underwater robots. The Student AUV Competition Europe (SAUC-E) mainly attracts undergraduate and graduate student teams. As in the AUVSI competitions, the robots must be fully autonomous while they are participating in the competition. The Microtransat Challenge is a competition to sail a boat across the Atlantic Ocean.

(5) Competitions open to anyone. RoboGames is open to anyone wishing to compete in their over 50 categories of robot competitions. Federation of International Robot-Soccer Association holds the FIRA World Cup competitions. There are flying robot competitions, robot soccer competitions, and other challenges, including weightlifting barbells made from dowels and CDs.

5.7.5 Robotics Afterschool Programs

Many schools across the country are beginning to add robotics programs to their after school curriculum. Some major programs for afterschool robotics include FIRST Robotics Competition, Botball and B. E. S. T. Robotics. Robotics competitions often include aspects of business and marketing as well as engineering and design. The Lego Company began a program for children to learn and get excited about robotics at a young age.

5.7.6 Decolonial Educational Robotics

Decolonial Educational Robotics is a branch of Decolonial Technology, and Decolonial AI[22], practiced in various places around the world. This methodology is summarized in pedagogical theories and practices such as Pedagogy of the Oppressed and Montessori methods. And it aims at teaching robotics from the local culture, to pluralize and mix technological knowledge.

5.8 Employment

Robotics is an essential component in many modern manufacturing environments. As factories increase their use of robots, the number of robotics-related jobs grows and has been observed to be steadily rising. The employment of robots in industries has increased productivity and efficiency savings and is typically seen as a long-term investment for benefactors. A paper by Michael Osborne and Carl Benedikt Frey found that 47 percent of US jobs are at risk to automation "over some unspecified number of years"[23]. These claims have been criticized on the ground that social policy, not AI, causes unemployment. In a 2016 article in The Guardian, Stephen Hawking stated "The automation of factories has already decimated jobs in traditional manufacturing, and the rise of artificial intelligence is likely to extend this job destruction deep into the middle classes, with only the most caring, creative or supervisory roles remaining".

5.9 Occupational Safety and Health Implications

A discussion paper has drawn up by EU-OSHA highlights how the spread of robotics presents both opportunities and challenges for occupational safety and health (OSH).

The greatest OSH benefits stemming from the wider use of robotics should be substitution for people working in unhealthy or dangerous environments. In space, defence, security, or the nuclear industry, but also in logistics, maintenance, and inspection, autonomous robots are particularly useful in replacing human workers performing dirty, dull or unsafe tasks, thus avoiding workers' exposures to hazardous agents and conditions and reducing physical, ergonomic and psychosocial risks. For example, robots are already used to perform repetitive and monotonous tasks, to handle radioactive material or to work in explosive atmospheres. In the future, many other highly repetitive, risky or unpleasant tasks will be performed by robots in a variety of sectors like agriculture, construction, transport, healthcare, firefighting or cleaning services.

Despite these advances, there are certain skills to which humans will be better suited than machines for some time to come and the question is how to achieve the best combination of human and robot skills. The advantages of robotics include heavy-duty jobs with precision and repeatability, whereas the advantages of humans include creativity, decision-making, flexibility, and adaptability. This need to combine optimal skills has resulted in collaborative robots and humans sharing a common workspace more closely and led to the development of new approaches and standards to guarantee the safety of the "man-robot merger". Some European countries are including robotics in their national programmes and trying to promote a safe and flexible co-operation between robots and operators to achieve better productivity. For example, the German Federal Institute for Occupational Safety and Health organises annual workshops on the topic "human-robot collaboration".

In the future, co-operation between robots and humans will be diversified, with robots increasing their autonomy and human-robot collaboration reaching completely new forms. Current approaches and technical standards aiming to protect employees from the risk of working with collaborative robots will have to be revised.

Notes and References

[1] Cheng K. Machining dynamics: fundamentals, applications and practices. Springer Science & Business Media, 2008.

[2] Altintas Y. Manufacturing automation: metal cutting mechanics, machine tool vibrations, and CNC design. Cambridge university press, 2012.

[3] Appleton E, Williams D J. Industrial robot applications. Halsted Press, New York, 1987.

[4] Milutinovic D, Glavonjic M, Slavkovic N, et al. Reconfigurable robotic machining system controlled and programmed in a machine tool manner. Internaticnal Journal of Advanced Manufacturing Technology, 2011, 53: 1217-1229.

[5] Roozing W, Li Z, Caldwell D G, et al. Design optimisation and control of compliant actuation arrangements in articulated robots for improved energy efficiency. IEEE Robotics & Automation Letters, 2017, 1(2): 1110-1117.

[6] Kim M, et al. Online walking pattern generation for humanoid robot with compliant motion control. International conference on Robotics & Automation, 2019.

[7] Pratt J E, Krupp B T. Series elastic actuators for legged robots. Proceedings of SPIE-The International Society for Optical Engineering, 2004, 29 (3) : 234-241.

[8] Li Z, Tsagarakis N G, Caldwell D G. Walking pattern generation for a humanoid robot with compliant joints. Autonomous Robots, 2013, 35(1): 1-14.

[9] Andrea, et al. Impedance control of series elastic actuators: passivity and acceleration-based control. Mechatronics, 2017, 47: 37-48.

[10] Tosun F E, Patoglu V. Necessary and sufficient conditions for the passivity of impedance rendering with velocity-sourced series elastic actuation. IEEE Transactions on Robotics, 2020, (99): 1-16.

[11] Tondu B. Modelling of the McKibben artificial muscle: A review. Journal of Intelligent Material Systems & Structures, 2012, 23(3): 225-253.

[12] Nishibori K, Okuma S, Obata H, et al. Robot hand with fingers using vibration-type ultrasonic motors (driving characteristics). International Conference on Industrial Electronics, 2002.

[13] Wettels N, Santos V J, Johansson R S, et al. Biomimetic tactile sensor array. Advanced Robotics, 2008, 22 (8), 829-849.

[14] CD C Iii, Duffy J. Kinematic analysis of robot manipulators. Cambridge University Press, 1998.

[15] Xiong C H, Chen W R, et al. Design and implementation of an anthropomorphic hand for replicating human grasping functions. IEEE Transactions on Robotics, 2016, 32(3), 652-671.

[16] Collins S, Ruina A, Tedrake R, et al. Efficient bipedal robots based on passive-dynamic walkers. Science, 2005, 307(5712): 1082-1085.

[17] Jaulin L, Le Bars F. An interval approach for stability analysis: application to sailboat robotics. IEEE Transactions on Robotics, 2013, 29(1): 282-287.

[18] Fournier R S, Schmidt B J. Voice input technology: learning style and attitude toward its use. Delta Pi Epsilon Journal, 1995, 37: 1-12.

[19] Huang C L, et al. Facial emotion recognition towards affective computing-based learning. Library Hi Tech. 2013, 31 (2) : 294-307.

[20] Žlajpah L. Simulation in robotics. Mathematics & Computers in Simulation, 2008, 79(4): 879-897.

[21] Saad A, Kroutil R M. Hands-on learning of programming concepts using robotics for middle and high school students. In Proceedings of the 50th Annual Southeast Regional Conference, 2012: 361-362.

[22] Mohamed S, Png M T, et al. Decolonial AI: decolonial theory as sociotechnical foresight in artificial intelligence. Philosophy & Technology, 2020, 33(4), 659-684.

[23] Frey C B, et al. The future of employment: How susceptible are jobs to computerisation. Technological Forecasting and Social Change, 2017, 114: 254-280.

Chapter 6 Electric Car

An electric car is a car that is propelled by one or more electric motors, using energy stored in rechargeable batteries. Compared to internal combustion engine (ICE) vehicles, electric cars are quieter, have no exhaust emissions, and lower emissions overall. In the United States, as of 2020, the total cost of ownership of recent EVs is cheaper than that of equivalent ICE cars, due to lower fueling and maintenance costs. Charging an electric car can be done at a variety of charging stations, and these charging stations can be installed in both houses and public areas.

Several countries have established government incentives for plug-in electric vehicles, tax credits, subsidies, and other non-monetary incentives. Several countries have established a phase-out of fossil fuel vehicles, and California, which is one of the largest vehicle markets, has an executive order to ban sales of new gasoline powered vehicles by 2035[1,2].

The Tesla Model 3, which has a maximum range of 570km (353miles) according to the EPA, has been the world's best-selling electric vehicle (EV) on an annual basis since 2018, and became the world's all-time best-selling electric car in early 2020.

As of December 2019, the global stock of pure electric passenger cars totaled 4.8 million units, representing two-thirds of all plug-in passenger cars in use. In 2019, over half (54%) of the world's all-electric car fleet was in China. Despite rapid growth, the global stock of fully electric and plug-in hybrid cars represented about 1 out of every 200 vehicles (0.48%) on the world's roads by the end of 2019, of which pure electrics comprised 0.32%.

6.1 Terminology

Electric cars are a type of electric vehicle (EV). The term "electric vehicle" refers to any vehicle that uses electric motors for propulsion, while "electric car" generally refers to highway-capable automobiles. Low-speed electric vehicles, classified as neighborhood electric vehicles (NEVs) in the United States, and as electric motorised quadricycles in Europe, are plug-in electric-powered microcars or city cars with limitations in terms of weight, power and maximum speed that are allowed to travel on public roads and city streets up to a certain posted speed limit, which varies by country.

While an electric car's power source is not explicitly an on-board battery, electric cars with motors powered by other energy sources are typically referred to by a different name. An electric car using solar panels as a power source is a solar car, and an electric car powered by a gasoline generator is a form of hybrid car. Thus, an electric car that derives its power from an on-board battery pack is a form of battery electric vehicle (BEV). Most often, the term "electric car" is used to refer to battery electric vehicles, but may also refer to plug-in hybrid electric vehicles (PHEV) (Figure 6-1).

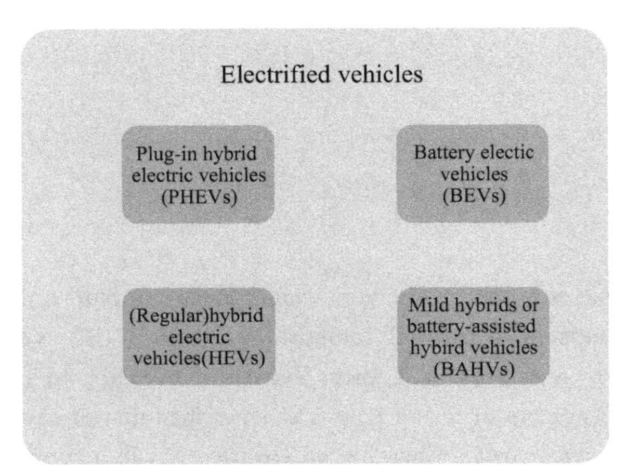

Figure 6-1 Venn diagram of electrified vehicles

6.2 History

The first practical electric cars were produced in the1880s[3]. In November 1881, Gustave Trouvé presented an electric car at the Exposition International d'Électricitéde Paris[4]. In 1884, over 20 years before the Ford Model T, Thomas Parker built a practical production electric car in Wolverhampton using his own specially designed high-capacity rechargeable batteries, although the only documentation is a photograph from 1895 (see below)[5]. The Flocken Elektrowagen of 1888 was designed by German inventor Andreas Flocken and is regarded as the first real electric car[6,7]. The schematic of the electric car can be found in Figure 6-2.

Electric cars were among the preferred methods for automobile propulsion in the late 19th and early 20th century, providing a level of comfort and ease of operation that could not be achieved by the gasoline cars of the time. The electric vehicle stock peaked at approximately 30000 vehicles at the turn of the 20th century.

In 1897, electric cars found their first commercial use as taxis in Britain and the US. In London, Walter Bersey's electric cabs were the first self-propelled vehicles for hire at a time when cabs were horse-drawn[8]. In New York City, a fleet of twelve hansom cabs and one brougham, based

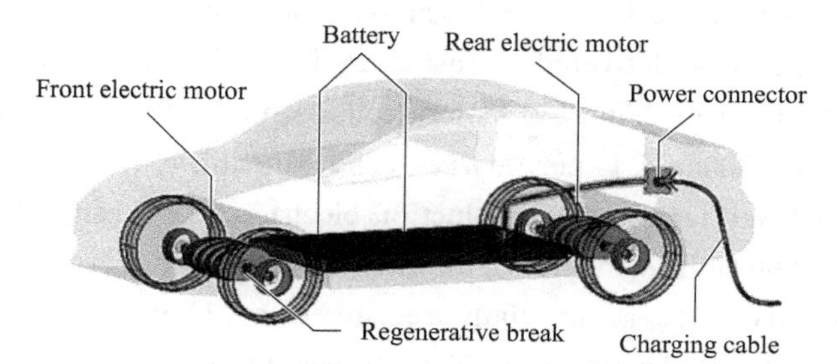

Figure 6-2 The schematic of the electric car.

on the design of the Electrobat Ⅱ, were part of a project funded in part by the Electric Storage Battery Company of Philadelphia[9]. During the 20th century, the main manufacturers of electric vehicles in the US were Anthony Electric, Baker, Columbia, Anderson, Edison, Riker, Milburn, Bailey Electric, Detroit Electric and others. Unlike gasoline-powered vehicles, the electric ones were less noisy, and did not require gear changes[10].

Six electric cars held the land speed record in the 19th century. The last of them was the rocket-shaped La Jamais Contente, driven by Camille Jenatzy, which broke the 100km/h (62mph) speed barrier by reaching a top speed of 105. 88km/h (65. 79 mph) on 29 April 1899.

Electric cars were popular until advances in internal combustion engine (ICE) cars (electric starters in particular) and mass production of cheaper petrol (gasoline) and diesel vehicles led to a decline. ICE cars' much quicker refueling times and cheaper production costs made them more popular. However, a decisive moment was the introduction in 1912 of the electric starter motor that replaced other, often laborious, methods of starting the ICE, such as hand-cranking.

The emergence of metal-oxide -semiconductor (MOS) technology led to the development of modern electric road vehicles. The MOSFET (MOS field-effect transistor, or MOS transistor), invented by Mohamed M. Atalla and Dawon Kahng at Bell Labs in 1959, led to the development of the power MOSFET by Hitachi in 1969, and the single-chip microprocessor by Federico Faggin, Marcian Hoff, Masatoshi Shima and Stanley Mazor at Intel in 1971. The power MOSFET and the microcontroller, a type of single-chip microprocessor, led to significant advances in electric automobile technology. MOSFET power converters allowed operation at much higher switching frequencies, made it easier to drive, reduced power losses, and significantly reduced prices, while single-chip microcontrollers could manage all aspects of the drive control and had the capacity for battery management. Another important technology that enabled modern highway-capable electric cars is the lithium-ion battery, invented by John Goodenough, Rachid Yazami and Akira Yoshino in the 1980s, which was responsible for the development of electric cars capable of long-distance travel.

In the early 1990s, the California Air Resources Board (CARB) began a push for more fuel-efficient, lower-emissions vehicles, with the ultimate goal being a move to zero-emissions vehicles such as electricvehicles[11]. In response, automakers developed electric models, including the Chrysler TEVan, Ford Ranger EV pickup truck, GM EV1, and S10 EV pickup, Honda EV Plus hatchback, Nissan Altra EV miniwagon, and Toyota RAV4 EV. Both US Electricar and Solectria produced 3 -phase AC Geo-bodied electric cars with the support of GM, Hughes, and Delco. These early cars were eventually withdrawn from the U S market.

California electric automaker Tesla Motors began development in 2004 of what would become the Tesla Roadster, which was first delivered to customers in 2008. The Roadster was the first highway legal all-electric car to use lithium-ion battery cells, and the first production all-electric car to travel more than 320km (200miles) per charge. The Mitsubishi i-MiEV, launched in 2009 in Japan, was the first highway legal series production electric car[12], and also the first all-electric car to sell more than 10000 units (including the models badged in Europe as Citroën C-Zero and Peugeot iOn) in February 2011 as officially registered by Guinness World Records. Several months later, the Nissan Leaf, launched in 2010, surpassed the i-MiEV as the all-time best selling all-electric car.

Starting in 2008, a renaissance in electric vehicle manufacturing occurred due to advances in batteries, and the desire to reduce greenhouse gas emissions and improve urban air quality.

In July 2019, US-based "Motor Trend" magazine awarded the fully electric Tesla Model S the title "ultimate car of the year". In March 2020, the Tesla Model 3 passed the Nissan Leaf to become the world's all-time best-selling electric car, with more than 500000 units delivered. The Leaf passed the 500000 unit mark in December 2020.

In November 2020, GM announced it plans to spend more on electric car development over next 5 years than it spends on gas and diesel vehicles.

6.3 Economics

6.3.1 Total Cost of Ownership

As of 2020 in the United States, the total cost of ownership of electric cars is less than comparable ICE cars, due to the lower cost of fueling and maintenance[13], more than making up for the higher initial cost.

The greater the distance driven per year, the more likely the total cost of ownership for an electric car will be less than for an equivalent ICE car. The break even distance varies by country depending on the taxes, subsidies, and different costs of energy. In some countries the comparison may vary by city, as a type of car may have different charges to enter different cities. For example, the UK city of London charges ICE cars more than the UK city of Birmingham does.

6.3.2 Purchase Cost

Several national and local governments have established EV incentives to reduce the purchase price of electric cars and other plug-ins[14].

When designing an electric vehicle, manufacturers may find that for low production, converting existing platforms may be cheaper, as development cost is lower. However, for higher production, a dedicated platform may be preferred to optimize design and cost. As of 2020 the electric vehicle battery is more than a quarter of the total cost of the car. Purchase prices are expected to drop below those of new ICE cars when battery costs fall below US $100 per kW·h, which is forecast to be in the mid-2020s.

Leasing or subscriptions are popular in some countries[15], depending somewhat on national taxes and subsidies, and end of lease cars are expanding the second hand market.

6.3.3 Operating Cost

The examples and perspective in this section may not represent a worldwide view of the subject. You may improve this section, discuss the issue on the talk page, or create a new section, as appropriate.

According to a study done in 2018, examining only fuel costs, the average fueling cost of an electric vehicle in the United States is $485 per year, as opposed to an ICE cars' $1117 per year. Estimated gasoline costs varied from $993 in Alabama to $1509 in Hawaii. Electric costs varied from $372 in Washington to $1106 in Hawaii.

6.3.4 Manufacturing Cost

The main cost driver of an electric car is its battery. The price decreased from €600 per kW·h in 2010, to €170 in 2017, to €100 in 2019.

6.4 Environmental Aspects

Electric cars have several benefits over ICE cars, including a significant reduction of local air pollution, as they do not directly emit pollutants such as volatile organic compounds, hydrocarbons, carbon monoxide, ozone, lead, and various oxides ofnitrogen[16].

Depending on the production process and the source of the electricity to charge the vehicle, emissions may be partly shifted from cities to the plants that generate electricity and produce the car as well as to the transportation of material. The amount of carbon dioxide emitted depends on the emissions of the electricity source and the efficiency of the vehicle. For electricity from the grid, the emissions vary significantly depending on the region, the availability of renewable sources and the efficiency of the fossil fuel-based generation used[17]. Considering the average electricity mix in the EU, driving electric cars emits 44%-56% less greenhouse gas than driving conventional cars. Including the energy intensive production of batteries in the analysis results in 31%-46% less greenhouse gas emissions than conventional cars. For context, 94% of EU transport depended on oil in 2017.

Similar to ICE vehicles, electric cars emit particulates from tyre and brake wear, although regenerative braking in electric cars means less brake dust. The sourcing of fossil fuels (oil well to gasoline tank) causes further damage as well as use of resources during the extraction and refinement processes, including high amounts of electricity.

The cost of installing charging infrastructure has been estimated to be repaid by health cost savings in less than 3 years.

According to a 2020 study, balancing lithium supply and demand for the rest of the century will require good recycling systems, vehicle-to-grid integration, and lower lithium intensity of transportation.

6.5 Performance

Electric motors can provide high power-to-weight ratios. Batteries can be designed to supply the electrical current needed to support these motors. Electric motors have a flat torque curve down to zero speed. For simplicity and reliability, most electric cars use fixed-ratio gearboxes and have no clutch.

Many electric cars have faster acceleration than average ICE cars, largely due to reduced drivetrain frictional losses and the more quickly-available torque of an electric motor. However NEVs may have a low acceleration due to their relatively weak motors.

Electric vehicles can also use a direct motor-to-wheel configuration that increases the available power. Having motors connected directly to each wheel allows the use of the motor for both propulsion and braking, increasing traction. Electric vehicles that lack an axle, differential, or transmission can have less drive-train inertia.

For example, the Venturi Fetish delivers supercar acceleration despite a relatively modest 220kW (300hp) motor and top speed of around 160km/h (100mph). Some direct current motor-equipped drag racer EVs have simple two-speed manual transmissions to improve top speed. The 2008 Tesla Roadster 2.5 Sport can accelerate from 0 to 97km/h (0 to 60mph) in 3.7 seconds with a motor rated at 215kW (288hp). Tesla Model S P100D (performance/100kW • h/4-wheel drive) is capable of 2.28 seconds for 0 to 60mph at a price of $140000. As of May 2017, the P100D is the second quickest production car ever built, taking only 0.08 seconds longer for 0−97km/h (0−60mph), compared to a $847975 Porsche 918 Spyder. The concept electric supercar Rimac Concept One claims it can go from 0−97km/h (0−60mph) in 2.5 seconds. Tesla claims the upcoming Tesla Roadster will go 0−60mph (0−97km/h) in 1.9 seconds.

6.6 Energy Efficiency

Internal combustion engines have thermodynamic limits on efficiency, expressed as fraction of energy used to propel the vehicle compared to energy produced by burning fuel. Gasoline engines effectively use only 15% of the fuel energy content to move the vehicle or to power accessories; diesel engines can reach on-board efficiency of 20%; electric vehicles have efficiencies of 69%−72%, when counted against stored chemical energy, or around 59%−62%, when counted against required energy to recharge.

Electric motors are more efficient than internal combustion engines in converting stored energy into driving a vehicle. However, they are not equally efficient at all speeds. To allow for this, some cars with dual electric motors have one electric motor with a gear optimised for city speeds and the second electric motor with a gear optimised for highway speeds. The electronics select the motor that has the best efficiency for the current speed and acceleration. Regenerative braking, which is most common in electric vehicles, can recover as much as one fifth of the energy normally lost during braking.

While heating can be provided with an electric resistance heater, higher efficiency and integral cooling can be obtained with a reversible heat pump, such as on the Nissan Leaf. PTC junction cooling is also attractive for its simplicity—this kind of system is used, for example, in the 2008 Tesla Roadster.

To avoid using part of the battery's energy for heating and thus reducing the range, some models allow the cabin to be heated while the car is plugged in. For example, the Nissan Leaf, the Mitsubishi i-MiEV, Renault Zoe and Tesla cars can be pre-heated while the vehicle is plugged in.

Some electric cars (for example, the Citroën Berlingo Electrique) use an auxiliary heating system (for example gasoline-fueled units manufactured by Webasto or Eberspächer) but sacrifice "green" and "zero emissions" credentials. Cabin cooling can be augmented with solar power external batteries and USB fans or coolers, or by automatically allowing outside air to flow through the car when parked, and two models of the 2010 Toyota Prius include this feature as an option.

6.7 Safety

The safety issues of BEVs are largely dealt with by the international standard ISO 6469. This document is divided in three parts dealing with specific issues:

(1) On-board electrical energy storage, i. e. the battery.

(2) Functional safety means and protection against failures.

(3) Protection of persons against electrical hazards.

6.7.1 Risk of Fire

Like their ICE counterparts, electric vehicle batteries can catch fire after a crash or mechanical failure. Plug-in electric vehicle fire incidents have occurred, albeit less per distance travelled than ICE vehicles. The first modern crash-related fire was reported in China in May 2012, after a high-speed car crashed into a BYD e6 taxi in Shenzhen.

In the United States, General Motors ran a training program in several cities for firefighters and first responders to demonstrate how to safely disable the Chevrolet Volt's powertrain and its 12 volt electrical system. The Volt's high-voltage system is designed to shut down automatically in the event of an airbag deployment, and to detect a loss of communication from an airbag control module. GM made available an "Emergency Response Guide" for the 2011 volt for use by emergency responders. The guide describes methods of disabling the high voltage system and identifies cut zone information. Nissan also published a guide for first responders that details procedures for handling a damaged 2011 Leaf at the scene of an accident, including a manual high-voltage system shutdown, rather than the automatic process built in the car's safety systems.

6.7.2 Vehicle Safety

The weight of the batteries themselves usually makes an EV heavier than a comparable gasoline vehicle. In a collision, the occupants of a heavy vehicle will, on average, suffer fewer and less serious injuries than the occupants of a lighter vehicle. Therefore, the additional weight brings safety benefits (to the occupant). Depending on where the battery is located, it may lower the center of gravity, increasing driving stability, lowering the risk of an accident through loss of control. An accident will, on average, cause about 50% more injuries to the occupants of a 2000lb (900kg) vehicle than those in a 3000lb (1400kg) vehicle.

Some electric cars use low rolling resistance tires, which typically offer less grip than normal tires. The Insurance Institute for Highway Safety in America had condemned the use of low speed vehicles and "mini trucks", called NEVs when powered by electric motors, on public roads.

Mindful of this, several companies (Tesla Motors, BMW, and Uniti) have succeeded in keeping the body light, while making it very strong.

6.8 Controls

As of 2018, most electric cars have similar driving controls to that of a car with a conventional automatic transmission. Even though the motor may be permanently connected to the wheels through a fixed-ratio gear, and no parking pawl may be present, the modes "P" and "N" are often still provided on the selector. In this case, the motor is disabled in "N" and an electrically actuated hand brake provides the "P" mode.

In some cars, the motor will spin slowly to provide a small amount of creep in "D", similar to a traditional automatic transmission car.

When an internal combustion vehicle's accelerator is released, it may slow by engine braking, depending on the type of transmission and mode. EVs are usually equipped with regenerative braking that slows the vehicle and recharges the battery somewhat. Regenerative braking systems also decrease the use of the conventional brakes (similar to engine braking in an ICE vehicle), reducing brake wear and maintenance costs.

6.9 Batteries

Lithium-ion-based batteries are often used for their high power and energy density. Other battery types are cheaper, such as nickel metal hydride (NiMH), but have a poorer power-to-weight ratio than lithium-ion. Batteries with different chemical compositions are in development such as zinc-air battery that could be much lighter.

6.9.1 Range

The range of an electric car depends on the number and type of batteries used, and (as with all vehicles), the aerodynamics, weight and type of vehicle, performance requirements, and the weather.

The EPA range of production electric vehicles in 2017 ranged from 100km (60miles) in the Renault Twizy to 540km (340miles) in the Tesla Model S 100D. Real-world range tests conducted by "What Car" in early 2019 found that the highest real-world range was 417km (259miles) in the Hyundai Kona.

The majority of electric cars are fitted with a display of the expected range. This may take into account how the vehicle is being used and what the battery is powering. However, since factors can vary over the route, the estimate can vary from the actual range. The display allows the driver to make informed choices about driving speed and whether to stop at a charging point enroute. Some roadside assistance organizations offer charge trucks to recharge electric cars in case of emergency.

A study in 2016 stated that 87% of US vehicle-days could be met by the then-current affordable electric cars.

6.9.2 Charging

Electric cars are typically charged overnight from a charging station installed in the owner's house, or from faster charging stations found in businesses and public areas.

Compared to fossil fuel vehicles, the need for charging using public infrastructure is diminished because of the opportunities for home charging. Vehicles can be plugged in and begin each day with a full charge, assuming the home charging station can charge quickly enough. An overnight charge of 8 hours using a 120V AC outlet will provide around 65km (40miles) of range, while a 240V AC outlet will provide approximately 290km (180miles).

Charging an electric vehicle using public charging stations takes longer than refueling a fossil fuel vehicle. The speed at which a vehicle can recharge depends on the charging station's charging speed and the vehicle's own capacity to receive a charge. Connecting a vehicle that can accommodate very fast charging to a charging station with a very high rate of charge can refill the vehicle's battery to 80% in 15 minutes. Vehicles and charging stations with slower charging speeds may take as long as 2 hours to refill a battery to 80%. As with a mobile phone, the final 20% takes longer because the systems slow down to fill the battery safely and avoid damaging it.

Some companies have been experimenting with battery swapping to substantially reduce the effective time to recharge.

Electric vehicle charging plugs are not yet universal throughout the world. Europe uses the CCS standard, while CHAdeMO is used in Japan, and a GB/T standard is used in China. The United States has no de facto standard, with a mix of CCS, Tesla Superchargers, and CHAdeMO charging stations. However vehicles using one type of plug are generally able to charge at other types of charging stations through the use of plug adapters.

Some electric cars (for example, the BMW i3) have an optional gasoline range extender. The system is intended as an emergency backup to extend range to the next recharging location, and not for long-distance travel.

The range-extender option of the BMW i3 was designed to meet the CARB regulation for an auxiliary power unit (APU) called REx. According to rules adopted in March 2012 by CARB, the 2014 BMW i3 with a REx unit fitted was the first car ever to qualify as a range-extended battery-electric vehicle or "BEVx".

6.9.3 Lifespan

As with all lithium-ion batteries, electric vehicle batteries may degrade over long periods of time, especially if they are frequently charged to 100%, however, this may take at least several years before being noticeable.

Nissan stated in 2015 that at that point only 0.01 percent of batteries had to be replaced because of failures or problems, and then only because of externally inflicted damage. Vehicles that had already covered more than 200000km had no problems with the battery.

6.9.4 Future

(1) Autonomous park-and-charge. Volkswagen, in collaboration with six partners, is developing an EU research project that is focused on automating the parking and charging of electric vehi-

cles. The objective of this project is to develop a smart car system that allows for autonomous driving in designated areas (e. g. valet parking, park and ride) and can offer advanced driver support in urban environments. Tesla has shown a prototype of a robot arm that automatically charges their vehicles.

(2) Other methods of energy storage. Experimental supercapacitors and flywheel energy storage devices offer comparable storage capacity, faster charging, and lower volatility. In 2010, they were considered to have the potential to overtake batteries as the preferred rechargeable storage for EVs[18, 19]. The FIA included their use in its sporting regulations of energy systems for Formula One race vehicles in 2007 (for supercapacitors) and 2009 (for flywheel energy storage devices).

(3) Solar cars. Solar cars are electric vehicles powered completely or significantly by direct solar energy, usually through photovoltaic (PV) cells contained in solar panels that convert the sun's energy directly into electric energy, usually to charge a battery.

(4) Dynamic charging. Dynamic charging allows electric vehicles to charge while driving on roads or highways. Sweden is testing four different dynamic charging technologies, three of which are suitable for passenger cars.

6.10 Electric Vehicle Charging Patents

Qualcomm, Hyundai, Ford and Mitsubishi are the top patent holders of the close to 800 electric vehicle charging patents filed between 2014 and 2017. A majority of patents filed between 2014 and 2017 on electric vehicle charging were filed in Japan, followed by the US and then China.

6.11 Infrastructure

6.11.1 Charging Station

Battery electric vehicles are most commonly charged from the power grid overnight at the owner's house. The electricity on the grid is in turn generated from a variety of sources; such as coal, hydroelectricity, nuclear and others. Power sources such as photovoltaic solar cell panels, micro hydro or wind may also be used and are promoted because of concerns regarding global warming.

Charging stations can have a variety of different speeds of charging, with slower charging being more common for houses, and more powerful charging stations on public roads and areas for trips. The BMW i3 can charge 0–80% of the battery in under 30 minutes in rapid charging mode. The superchargers developed by Tesla Motors provide up to 250kW of charging, allowing a 250-mile charge in 30 minutes.

Most electric cars use conductive coupling to supply electricity for recharging after CARB settled on the "SAE J1772-2001" standard as the charging interface for electric vehicles in California in June 2001. In Europe, the ACEA has decided to use the Type 2 connector from the range of IEC-62196 plug types for conductive charging of electric vehicles in the European Union, as the Type 1 connector ("SAE J1772-2009") does not provide for three-phase charging.

Another approach is inductive charging using a non-conducting "paddle" inserted into a slot in the car. Delco Electronics developed the Magne Charge inductive charging system in 1998 for the General Motors EV1 that was also used for the Chevrolet S-10 EV and Toyota RAV4 EV vehicles.

6.11.2 Vehicle-to-Grid: Uploading and Grid Buffering

During peak load periods, when the cost of generation can be very high, electric vehicles with vehicle-to-grid capabilities could contribute energy to the grid. These vehicles can then be recharged during off-peak hours at cheaper rates while helping to absorb excess night time generation. The batteries in the vehicles serve as a distributed storage system to buffer power.

6.12 Currently Available Electric Cars

6.12.1 Highway Capable

According to "Bloomberg New Energy Finance", as of December 2018, there were almost 180 models of highway-capable all-electric passenger cars and utility vans available for retail sales globally[20].

Tesla became the world's leading electric vehicle manufacturer in December 2019, with cumulative global sales of over 900000 all-electric cars since 2008[21]. Its Model S was the world's top selling plug-in electric car in 2015 and 2016, and its Model 3 has been the world's best selling plug-in electric car for three years in a row, from 2018 to 2020. The Tesla Model 3 surpassed the Leaf in early 2020 to become the world's cumulative best selling electric car, with more than 500000 sold by March 2020. Tesla produced its 1 millionth electric car in March 2020, becoming the first auto manufacturer to do so[22]. Tesla has listed as the world's top selling plug-in electric car manufacturer, both as a brand and by automotive group for three years running, from 2018 to 2020.

As of December 2019, the Renault-Nissan-Mitsubishi Alliance is one of the world's leading all-electric vehicle manufacturer. Since 2010, the Alliance's global all-electric vehicle sales totaled over 800000 light-duty electric vehicles through December 2019, including those manufactured by Mitsubishi Motors, now part of the Alliance. Nissan leads global sales within the Alliance, with about 500000 cars and vans sold by April 2020, followed by the Groupe Renault with more than 273550 electric vehicles sold worldwide through December 2019, including its Twizy heavy quadricycle. Mitsubishi's only all-electric vehicle is the i-MiEV, with global sales of over 50000 units by March 2015, accounting for all variants of the i-MiEV, including the two minicab versions sold in Japan. The Alliance's best-selling Nissan Leaf was the world's top-selling plug-in electric car in 2013 and 2014. Through 2019, the Nissan Leaf was the world's all-time top-selling highway-legal electric car with global sales of almost 450000 units. The Renault Kangoo Z. E. utility van is the European leader of the light-duty all-electric segment with sales of 57595 units through November 2020.

Other leading electric vehicles manufacturers are BAIC Motor, with 480000 units sold, SAIC Motor with 314000 units, and Geely with 228700, all cumulative sales in China as of December 2019. BMW is also a leading plug-in car manufacturer, with over 500000 plug-in electric cars sold globally by December 2019, but its electrified vehicle lineup only includes one all-electric model, the BMW i3, with 200000 units produced up to October 2020, including the REx variant.

6.12.2 Retrofitted Electric Vehicles

Any car can be converted to an electric vehicle using plug-and-play custom solution kits. The car resulting from a conversion of an ICE car to an electric car is called an Retrofitted Electric Vehicle.

6.12.3 Electric Cars by Country

Global sales of highway legal plug-in electric passenger cars and light utility vehicles achieved the one million milestone in September 2015, almost twice as fast as hybrid electric vehicles (HEV). Cumulative global sales of light-duty all-electric vehicles reached one million units in September 2016. Global sales of plug-in passenger cars passed 2 million in December 2016, the 3 million mark in November 2017, the 5 million milestone in December 2018, and totaled 7.2 million units in December 2019. Despite rapid growth, the global stock of plug-in electric cars represented about 1 out of every 250 vehicles (0.40%) on the world's roads by the end of 2018.

As of December 2019, the global stock of pure electric passenger cars totaled 4.79 million units, representing two-thirds of all plug-in passenger cars on the world's roads. China has the largest all-electric car fleet in use, with 2.58 million at the end of 2019, more than half (53.9%) of the world's electric car stock. In addition, there were almost 378000 electric light commercial vehicles in use by the end of 2019, mainly in China and Europe.

All-electric cars have oversold plug-in hybrids for several years, and by the end of 2019, the plug-in market continues to shift towards fully electric battery vehicles. The global ratio between annual sales of battery BEVs and PHEVs went from 56:44 in 2012 to 74:26 in 2019.

6.13 Government Policies and Incentives

Several national, provincial, and local governments around the world have introduced policies to support the mass-market adoption of plug-in electric vehicles. A variety of policies have been established to provide: financial support to consumers and manufacturers, non-monetary incentives, subsidies for the deployment of charging infrastructure, and long-term regulations with specific targets.

Financial incentives for consumers are aiming to make electric car purchase price competitive with conventional cars due to the higher upfront cost of electric vehicles. Depending on battery size, there are one-time purchase incentives such as grants and tax credits, exemptions from import duties, exemptions from road tolls and congestion charges, and exemption of registration and annual fees.

The U S offers a federal income tax credit up to US $7500. The UK offers a Plug-in Car Grant up to GB £4500 (US $5929). France introduced a bonus-malus CO_2-based tax that penalizes fossil-fuel vehicle sales. As of 2020, monetary incentives are available in several European Union member states, China, Norway, some provinces in Canada, South Korea, India, Among the non-monetary incentives, there are several perks such allowing plug-in vehicles access to bus lanes and high-occupancy vehicle lanes, free parking and free charging. Some countries or cities that restrict private car ownership (for example, a purchase quota system for new vehicles), or have implemented permanent driving restrictions (for example, no-drive days), have these schemes exclude electric vehicles to promote their adoption. and other countries.

Among the non-monetary incentives, there are several perks such allowing plug-in vehicles access to bus lanes and high-occupancy vehicle lanes, free parking and free charging. Some countries or cities that restrict private car ownership (for example, a purchase quota system for new vehicles), or have implemented permanent driving restrictions (for example, no-drive days), have these schemes exclude electric vehicles to promote their adoption.

Some government have also established long term regulatory signals with specific targets such as zero-emissions vehicle (ZEV) mandates, national or regional CO_2 emission regulations, stringent fuel economy standards, and the phase out of internal combustion engine vehicle sales. For example, Norway set a national goal that by 2025 all new car sales should be ZEVs (battery electric or hydrogen).

6.14 EV Plans From Major Manufacturers

Plans 27 electric vehicles by 2022, on a dedicated EV platform dubbed "Modular Electric Toolkit" and initialed as MEB. In November 2020, it announced the intention to invest $86 billion in the following 5 years, aimed at developing EVs and increasing its share in the EV market. Total capital expenditure will include "digital factories", automotive software and self-driving cars.

6.15 Psychological Barriers to Adoption

For the past century, most people have driven ICE cars, making them feel common, familiar, and low risk. Even though EV technology has been around for over a century and modern EVs have been on the market for decades, multiple studies show that various psychological factors impair EV adoption.

6.15.1 Range Anxiety

A 2019 study found that the dominant fear hindering EV adoption was range anxiety. ICE car drivers are accustomed to going on trips without having to plan refueling stops, and may worry that an EV will lack the range to reach their destination or the closest charging station. Range anxiety has been shown to diminish among drivers who have gained familiarity and experience with EVs.

6.15.2 Identity Concerns

This same study also found that people view driving an EV as an action taken by those with "stronger attitudes in favor of environmental and energy security" or by those that are "attracted to the novelty and status associated with being among the first to adopt new technology". Thus, people may be resistant to EV ownership if they do not consider themselves environmentalists or early adopters of new technology, or do not want others to think of themselves in this way.

The perceived value associated with driving an EV can also differ by gender. A 2019 survey conducted in Norway found that people believe women drive EVs for sustainability purposes, while men drive EVs for the new technology. The thought process behind this stereotype is that "big and expensive cars are driven by men, while women drive smaller, less valuable cars". Since the reasons behind adopting EVs have a gender component, it can be argued that some fear driving an EV will result ina disconnect between their gender identity and how they are perceived by others.

Notes and References

[1] Collantes G, et al. The origin of California's zero emission vehicle mandate. Transportation Research Part A: Policy and Practice, 2008, 42 (10), 1302-1313.

[2] Parkhurst E L, Conner L B, Ferraro J C, et al. Heuristic evaluation of a Tesla Model 3 interface. In Proceedings of the Human Factors and Ergonomics Society Annual Meeting, 2019, 63 (1): 1515-1519.

[3] Worsnop R L. Electric cars. CQ Researcher, 1993, 3: 25.

[4] Wakefield E H, et al. History of the electric automobile. Warrendale, PA: SAE International, 1993.

[5] Guarnieri M. Looking back to electric cars. 2012 Third IEEE History of ELectro-Technology Conference, 2012: 1-6.

[6] Boyle D. 30-second great inventions. Ivy Press, 2018.

[7] Denton T. Electric and hybrid vehicles. Routledge, 2016.

[8] Høyer K G. The history of alternative fuels in transportation: the case of electric and hybrid cars. Utilities Policy, 2008, 16(2) : 63-71.

[9] Burton N. History of electric cars. Crowood, 2013.

[10] Perujo A, et al. The introduction of electric vehicles in the private fleet: potential impact on the electric supply system and on the environment. A case study for the Province of Milan, Italy. Energy Policy, 2010, 38 (8) : 4549-4561.

[11] Boschert S. Plug-in hybrids: the cars that will recharge America. New Society Publishers, 2006.

[12] Larson P D, Viáfara J, Parsons R V, et al. Consumer attitudes about electric cars: pricing analysis and policy implications. Transportation Research Part A: policy and Practice, 2014, 69: 299-314.

[13] Milev G, Hastings A, et al. The environmental and financial implications of expanding the use of electric cars-a case study of Scotland. Energy and Built Environment, 2021, 2(2) : 204-213.

[14] Zheng J, Mehndiratta S, Guo J Y, et al. Strategic policies and demonstration program of electric vehicle in China. Transport Policy, 2012, 19 (1), 17-25.

[15] Huang Y, Qian L, Soopramanien D, et al. Buy, lease, or share? Consumer preferences for innovative business models in the market for electric vehicles. Technological Forecasting and Social Change, 2021, 166: 120639.

[16] Gould J, et al. Clean air forever? A longitudinal analysis of opinions about air pollution and electric vehicles. Transportation Research Part D: Transport and Environment, 1998, 3(3), 157-169.

[17] Naira V R, Das D, et al. Real time light intensity based carbon dioxide feeding for high cell-density microalgae cultivation and biodiesel production in a bubble column photobioreactor under outdoor natural sunlight. Bioresource technology, 2019, 284, 43-55.

[18] Hilton J. Flywheel hybrid as an alternative to electric vehicles. ATZ Worldwide, 2012, 114(11), 26-30.

[19] Shukla A K, et al. Electrochemical supercapacitors: energy storage beyond batteries. Current Science, 2000, 79(12), 1656-1661.
[20] Harding G G. Electric vehicles in the next millennium. Journal of Power Sources, 1999, 78(1-2), 193-198.
[21] Ahmad S, Khan M. Tesla: Disruptor or Sustaining Innovator. Journal of Case Research, 2019, 10(1).
[22] Wu L, Yang W. Tackling supply chain challenges of Tesla Model 3. Neilson Journals Publishing, 2017.

Chapter 7
Electric Power Transmission

Electric power transmission is the bulk movement of electrical energy from a generating site, such as a power plant, to an electrical substation. The interconnected lines which facilitate this movement are known as a transmission network. This is distinct from the local wiring between high-voltage substations and customers, which is typically referred to as electric power distribution. The combined transmission and distribution network is part of electricity delivery, known as the electrical grid.

Efficient long-distance transmission of electric power requires high voltages. This reduces the losses produced by heavy current. Transmission lines mostly use high-voltage AC, but an important class of transmission line uses high voltage direct current. The voltage level is changed with transformers, stepping up the voltage for transmission, and then reducing voltage for local distribution and then use by customers. An example of 500kV electric power transmission line can be found in Figure 7-1.

Figure 7-1 500kV three-phase electric power transmission lines at Grand Coulee Dam

A wide area synchronous grid, also known as an "interconnection" in North America, directly connects many generators delivering AC power with the same relative frequency to many consumers. For example, there are four major interconnections in North America (the Western Interconnection, the Eastern Interconnection, the Quebec Interconnection and the Electric Reliability Council of Texas (ERCOT) grid). In Europe one large grid connects most of continental Europe.

Historically, transmission and distribution lines were often owned by the same company, but starting in the 1990s, many countries have liberalized the regulation of the electricity market in ways that have led to the separation of the electricity transmission business from the distribution business.

7.1 System

Most transmission lines are high-voltage three-phase alternating current (AC), although single phase AC is sometimes used in railway electrification systems. High-voltage direct-current (HVDC) technology is used for greater efficiency over very long distances (typically hundreds of miles). HVDC technology is also used in submarine power cables (typically longer than 30 miles), and in the interchange of power between grids that are not mutually synchronized. HVDC links are used to stabilize large power distribution networks where sudden new loads, or blackouts, in one part of a network can result in synchronization problems and cascading failures.

Electricity is transmitted at high voltages (66kV or above) to reduce the energy loss which occurs in long-distance transmission. Power is usually transmitted through overhead power lines. Underground power transmission has a significantly higher installation cost and greater operational limitations, but reduced maintenance costs. Underground transmission is sometimes used in urban areas or environmentally sensitive locations.

A lack of electrical energy storage facilities in transmission systems leads to a key limitation. Electrical energy must be generated at the same rate at which it is consumed. A sophisticated control system is required to ensure that the power generation very closely matches the demand. If the demand for power exceeds supply, the imbalance can cause generation plant and transmission equipment to automatically disconnect or shut down to prevent damage. In the worst case, this may lead to a cascading series of shut downs and a major regional blackout. Examples include the US Northeast blackouts of 1965, 1977, 2003, and major blackouts in other US regions in 1996 and 2011. Electric transmission networks are interconnected into regional, national, and even continent wide networks to reduce the risk of such a failure by providing multiple redundant, alternative routes for power to flow should such shut downs occur. Transmission companies determine the maximum reliable capacity of each line (ordinarily less than its physical or thermal limit) to ensure that spare capacity is available in the event of a failure in another part of the network.

7.2 Overhead Transmission

High-voltage overhead conductors are not covered by insulation. The conductor material is nearly always an aluminum alloy, made into several strands and possibly reinforced with steel strands. Copper was sometimes used for overhead transmission, but aluminum is lighter, yields only marginally reduced performance and costs much less. Overhead conductors are a commodity supplied by several companies worldwide. Improved conductor material and shapes are regularly used to allow increased capacity and modernize transmission circuits. Conductor sizes range from $12mm^2$

(#6 American wire gauge) to 750mm^2 (1590000 circular mils area), with varying resistance and current-carrying capacity. For large conductors (more than a few centimetres in diameter) at power frequency, much of the current flow is concentrated near the surface due to the skin effect. The center part of the conductor carries little current, but contributes weight and cost to the conductor. Because of this current limitation, multiple parallel cables (called bundle conductors) are used when higher capacity is needed. Bundle conductors are also used at high voltages to reduce energy loss caused by corona discharge.

Today, transmission-level voltages are usually considered to be 110kV and above. Lower voltages, such as 66kV and 33kV, are usually considered subtransmission voltages, but are occasionally used on long lines with light loads. Voltages less than 33kV are usually used for distribution. Voltages above 765kV are considered extra high voltage and require different designs compared to equipment used at lower voltages.

Since overhead transmission wires depend on air for insulation, the design of these lines requires minimum clearances to be observed to maintain safety. Adverse weather conditions, such as high winds and low temperatures, can lead to power outages. Wind speeds as low as 23 knots (43km/h) can permit conductors to encroach operating clearances, resulting in a flashover and loss of supply[1]. Oscillatory motion of the physical line can be termed conductor gallop or flutter depending on the frequency and amplitude of oscillation.

7.3 Underground Transmission

Electric power can also be transmitted by underground power cables instead of overhead power lines. Underground cables take up less right-of-way than overhead lines, have lower visibility, and are less affected by bad weather. However, costs of insulated cable and excavation are much higher than overhead construction. Faults in buried transmission lines take longer to locate and repair.

In some metropolitan areas, underground transmission cables are enclosed by metal pipe and insulated with dielectric fluid (usually oil) that is either static or circulated via pumps. If an electric fault damages the pipe and produces a dielectric leak into the surrounding soil, liquid nitrogen trucks are mobilized to freeze portions of the pipe to enable the draining and repair of the damaged pipe location. This type of underground transmission cable can prolong the repair period and increase repair costs. The temperature of the pipe and soil are usually monitored constantly throughout the repair period[2].

Underground lines are strictly limited by their thermal capacity, which permits fewer overloads or re-rating than overhead lines. Long underground AC cables have significant capacitance, which may reduce their ability to provide useful power to loads beyond 50 miles (80 kilometres). DC cables are not limited in length by their capacitance, however, they do require HVDC converter stations at both ends of the line to convert from DC to AC before being interconnected with the transmission network.

7.4 History

In the early days of commercial electric power, transmission of electric power at the same voltage as used by lighting and mechanical loads restricted the distance between generating plant and consumers. In 1882, generation was with direct current (DC), which could not easily be increased in voltage for long-distance transmission. Different classes of loads (for example, lighting, fixed motors and traction/railway systems) required different voltages, and so used different generators andcircuits[3, 4].

Due to this specialization of lines and because transmission was inefficient for low-voltage high-current circuits, generators needed to be near their loads. It seemed, at the time, that the industry would develop into what is now known as a distributed generation system with large numbers of small generators located near their loads.

The transmission of electric power with alternating current (AC) became possible after Lucien Gaulard and John Dixon Gibbs built what they called the secondary generator, an early transformer provided with 1:1 turn ratio and open magnetic circuit, in 1881.

The first long distance AC line was 34 kilometres (21 miles) long, built for the 1884 International Exhibition of Turin, Italy. It was powered by a 2kV, 130Hz Siemens&Halske alternator and featured several Gaulard "secondary generators" (transformers) with their primary windings connected in series, which fed incandescent lamps. The system proved the feasibility of AC electric power transmission over long distances.

The very first AC distribution system to operate was in service in 1885 in via dei Cerchi, Rome, Italy, for public lighting. It was powered by two Siemens&Halske alternators rated 30hp (22kW), 2kV at 120Hz and used 19km of cables and 200 parallel-connected 2kV to 20V step-down transformers provided with a closed magnetic circuit, one for each lamp. A few months later it was followed by the first British AC system, which was put into service at the Grosvenor Gallery, London. It also featured Siemens alternators and 2.4kV to 100V step-down transformers-one per user-with shunt-connected primaries[5].

Working from what he considered an impractical Gaulard-Gibbs design, electrical engineer William Stanley, Jr. developed what is considered the first practical series AC transformer in 1885. Working with the support of George Westinghouse, in 1886 he demonstrated a transformer based alternating current lighting system in Great Barrington, Massachusetts. Powered by a steam engine driven 500V Siemens generator, voltage was stepped down to 100V using the new Stanley transformer to power incandescent lamps at 23 businesses along main street with very little power loss over 4000 feet (1200m). This practical demonstration of a transformer and alternating current lighting system would lead Westinghouse to begin installing AC based systems later that year.

1888 saw designs for a functional AC motor, something these systems had lacked up till then. These were induction motors running on polyphase current, independently invented by Galileo Ferraris and Nikola Tesla (with Tesla's design being licensed by Westinghouse in the US). This design was further developed into the modern practical three-phase form by Mikhail Dolivo-Dobrovolsky and Charles Eugene Lancelot Brown. Practical use of these types of motors would be delayed many years by development problems and the scarcity of poly-phase power systems needed to power them[6].

The late 1880s and early 1890s would see the financial merger of smaller electric companies into a few larger corporations such as Ganz and AEG in Europe and General Electric and Westinghouse Electric in the US. These companies continued to develop AC systems but the technical difference between direct and alternating current systems would follow a much longer technical merger. Due to innovation in the US and Europe, alternating current's economy of scale with very large generating plants linked to loads via long-distance transmission was slowly being combined with the ability to link it up with all of the existing systems that needed to be supplied. These included single phase AC systems, poly-phase AC systems, low voltage incandescent lighting, high voltage arc lighting, and existing DC motors in factories and street cars. In what was becoming a universal system, these technological differences were temporarily being bridged via the development of rotary converters and motor-generators that would allow the large number of legacy systems to be connected to the AC grid. These stopgaps would slowly be replaced as older systems were retired or upgraded.

Westinghouse alternating current polyphase generators on display at the 1893 World's Fair in Chicago, part of their "Tesla Poly-phase System". Such polyphase innovations revolutionized transmission

The first transmission of single-phase alternating current using high voltage took place in Oregon in 1890 when power was delivered from a hydroelectric plant at Willamette Falls to the city of Portland 14 miles (23km) downriver. The first three-phase alternating current using high voltage took place in 1891 during the international electricity exhibition in Frankfurt. A 15kV transmission line, approximately 175km long, connected Lauffen on the Neckar and Frankfurt.

Voltages used for electric power transmission increased throughout the 20th century. By 1914, fifty-five transmission systems each operating at more than 70kV were in service. The highest voltage then used was 150kV. By allowing multiple generating plants to be interconnected over a wide area, electricity production cost was reduced. The most efficient available plants could be used to supply the varying loads during the day. Reliability was improved and capital investment cost was reduced, since stand-by generating capacity could be shared over many more customers and a wider geographic area. Remote and low-cost sources of energy, such as hydroelectric power or mine-mouth coal, could be exploited to lower energy production cost.

The rapid industrialization in the 20th century made electrical transmission lines and grids critical infrastructure items in most industrialized nations. The interconnection of local generation plants and small distribution networks was spurred by the requirements of World War I, with large electrical generating plants built by governments to provide power to munitions factories. Later these generating plants were connected to supply civil loads through long-distance transmission.

7.5 Bulk Power Transmission

Engineers design transmission networks to transport the energy as efficiently as possible, while at the same time taking into account the economic factors, network safety and redundancy. These networks use components such as power lines, cables, circuit breakers, switches and transformers. The transmission network is usually administered on a regional basis by an entity such as a regional transmission organization or transmission system operator.

Transmission efficiency is greatly improved by devices that increase the voltage (and thereby proportionately reduce the current), in the line conductors, thus allowing power to be transmitted with acceptable losses. The reduced current flowing through the line reduces the heating losses in the conductors. According to Joule's Law, energy losses are directly proportional to the square of the current. Thus, reducing the current by a factor of two will lower the energy lost to conductor resistance by a factor of four for any given size of conductor.

The optimum size of a conductor for a given voltage and current can be estimated by Kelvin's law for conductor size, which states that the size is at its optimum when the annual cost of energy wasted in the resistance is equal to the annual capital charges of providing the conductor. At times of lower interest rates, Kelvin's law indicates that thicker wires are optimal; while, when metals are expensive, thinner conductors are indicated: however, power lines are designed for long-term use, so Kelvin's law has to be used in conjunction with long-term estimates of the price of copper and aluminum as well as interest rates for capital.

The increase in voltage is achieved in AC circuits by using a step-up transformer. HVDC systems require relatively costly conversion equipment which may be economically justified for particular projects such as submarine cables and longer distance high capacity point-to-point transmission. HVDC is necessary for the import and export of energy between grid systems that are not synchronized with each other.

A transmission grid is a network of power stations (Figure 7-2), transmission lines and substations. Energy is usually transmitted within a grid with three-phase AC. Single-phase AC is used only for distribution to end users since it is not usable for large polyphase induction motors. In the 19th century, two-phase transmission was used but required either four wires or three wires with unequal currents. Higher order phase systems require more than three wires, but deliver little or no benefit.

Figure 7-2 The PacifiCorp Hale Substation, Orem, Utah, USA

(A transmission substation decreases the voltage of incoming electricity, allowing it to connect from long-distance high voltage transmission, to local lower voltage distribution. It also reroutes power to other transmission lines that serve local markets)

The price of electric power station capacity is high, and electric demand is variable, so it is often cheaper to import some portion of the needed power than to generate it locally. Because loads are often regionally correlated (hot weather in the Southwest portion of the US might cause many people to use air conditioners), electric power often comes from distant sources. Because of the economic benefits of load sharing between regions, wide area transmission grids now span countries and even continents. The web of interconnections between power producers and consumers should enable power to flow, even if some links are inoperative.

The unvarying (or slowly varying over many hours) portion of the electric demand is known as the base load and is generally served by large facilities (which are more efficient due to economies of scale) with fixed costs for fuel and operation. Such facilities are nuclear, coal-fired or hydroelectric, while other energy sources such as concentrated solar thermal and geothermal power have the potential to provide base load power. Renewable energy sources, such as solar photovoltaics, wind, wave and tidal, are, due to their intermittency, not considered as supplying "base load" but will still add power to the grid. The remaining or "peak" power demand, is supplied by peaking power plants, which are typically smaller, faster-responding, and higher cost sources, such as combined cycle or combustion turbine plants fueled by natural gas.

Long-distance transmission of electricity (hundreds of kilometers) is cheap and efficient, with costs of US $0.005-0.02 per kW•h (compared to annual averaged large producer costs of US $0.01-0.025 per kW•h, retail rates upwards of US $0.10 per kW•h, and multiples of retail for instantaneous suppliers at unpredicted highest demand moments). Thus distant suppliers can be cheaper than local sources (e.g., New York often buys over 1000MW of electricity from Canada). Multiple local sources (even if more expensive and infrequently used) can make the transmission grid more fault tolerant to weather and other disasters that can disconnect distant suppliers.

Long-distance transmission allows remote renewable energy resources to be used to displace fossil fuel consumption. Hydro and wind sources cannot be moved closer to populous cities, and solar costs are lowest in remote areas where local power needs are minimal. Connection costs alone can determine whether any particular renewable alternative is economically sensible. Costs can be prohibitive for transmission lines, but various proposals for massive infrastructure investment in high capacity, very long distance super grid transmission networks could be recovered with modest usage fees.

7.5.1 Grid Input

At the power stations, the power is produced at a relatively low voltage between about 2.3kV and 30kV, depending on the size of the unit. The generator terminal voltage is then stepped up by the power station transformer to a higher voltage (115kV to 765kV AC, varying by the transmission system and by the country) for transmission over long distances.

In the United States, power transmission is, variously, 230kV to 500kV, with less than 230kV or more than 500kV being local exceptions. For example, the Western Interconnection has two primary interchange voltages: 500kV AC at 60Hz, and ±500kV (1000kV net) DC from North to South (Columbia River to Southern California) and Northeast to Southwest (Utah to Southern California). The 287.5kV (Hoover Dam to Los Angeles line, via Victorville) and 345kV (Arizona Public Service (APS) line) being local standards, both of which were implemented before 500kV became practical, and thereafter the Western Interconnection standard for long distance AC power transmission.

7.5.2 Losses

Transmitting electricity at high voltage reduces the fraction of energy lost to resistance, which varies depending on the specific conductors, the current flowing and the length of the transmission line. For example, a 100m (160km) span at 765kV carrying 1000MW of power can have losses of 1.1% to 0.5%. A 345kV line carrying the same load across the same distance has losses of 4.2%. For a given amount of power, a higher voltage reduces the current and thus the resistive losses in the conductor. For example, raising the voltage by a factor of 10 reduces the current by a corresponding factor of 10 and therefore the losses by a factor of 100, provided the same sized conductors are used in both cases. Even if the conductor size (cross-sectional area) is decreased tenfold to match the lower current, the losses are still reduced ten-fold. Long-distance transmission is typically done with overhead lines at voltages of 115kV to 1200kV. At extremely high voltages, where more than 2000kV exists between conductor and ground, corona discharge losses are so large that they can offset the lower resistive losses in the line conductors. Measures to reduce corona losses include conductors having larger diameters; often hollow to save weight, or bundles of two or more conductors.

Factors that affect the resistance, and thus loss, of conductors used in transmission and distribution lines include temperature, spiraling and the skin effect. The resistance of a conductor increases with its temperature. Temperature changes in electric power lines can have a significant effect on power losses in the line. Spiraling, which refers to the way stranded conductors spiral about the center, also contributes to increases in conductor resistance. The skin effect causes the effective resistance of a conductor to increase at higher alternating current frequencies. Corona and resistive losses can be estimated using a mathematical model.

Transmission and distribution losses in the United States were estimated at 6.6% in 1997, 6.5% in 2007 and 5% from 2013 to 2019. In general, losses are estimated from the discrepancy between power produced (as reported by power plants) and power sold to the end customers. The difference between what is produced and what is consumed constitute transmission and distribution losses, assuming no utility theft occurs.

As of 1980, the longest cost-effective distance for direct-current transmission was determined to be 7000 kilometres (4300 miles). For alternating current it was 4000 kilometres (2500 miles), though all transmission lines in use today are substantially shorter than this.

In any alternating current transmission line, the inductance and capacitance of the conductors can be significant. Currents that flow solely in "reaction" to these properties of the circuit, (which together with the resistance define the impedance) constitute reactive power flow, which transmits no "real" power to the load. These reactive currents, however, are very real and cause extra heating losses in the transmission circuit. The ratio of "real" power (transmitted to the load) to "apparent" power (the product of a circuit's voltage and current, without reference to phase angle) is the power factor. As reactive current increases, the reactive power increases and the power factor decreases. For transmission systems with low power factor, losses are higher than for systems with high power factor. Utilities add capacitor banks, reactors and other components (such as phase-shifting transformers, static VAR compensators and flexible AC transmission systems) throughout the system help to compensate for the reactive power flow, reduce the losses in power transmission and stabilize system voltages. These measures are collectively called "reactive support".

7.5.3 Transposition

Current flowing through transmission lines induces a magnetic field that surrounds the lines of each phase and affects the inductance of the surrounding conductors of other phases. The mutual inductance of the conductors is partially dependent on the physical orientation of the lines with respect to each other. Three-phase power transmission lines are conventionally strung with phases separated on different vertical levels. The mutual inductance seen by a conductor of the phase in the middle of the other two phases will be different than the inductance seen by the conductors on the top or bottom. An imbalanced inductance among the three conductors is problematic because it may result in the middle line carrying a disproportionate amount of the total power transmitted.

Similarly, an imbalanced load may occur if one line is consistently closest to the ground and operating at lower impedance. Because of this phenomenon, conductors must be periodically transposed along the length of the transmission line so that each phase sees equal time in each relative position to balance out the mutual inductance seen by all three phases. To accomplish this, line position is swapped at specially designed transposition towers at regular intervals along the length of the transmission line in various transposition schemes.

7.5.4 Subtransmission

Subtransmission is part of an electric power transmission system that runs at relatively lower voltage. It is uneconomical to connect all distribution substations to the high main transmission voltage, because the equipment is larger and more expensive. Typically, only larger substations connect with this high voltage. It is stepped down and sent to smaller substations in towns and neighborhoods. Subtransmission circuits are usually arranged in loops so that a single line failure does not cut off service to many customers for more than a short time. Loops can be "normally closed", where loss of one circuit should result in no interruption, or "normally open" where substations can switch to a backup supply. While subtransmission circuits are usually carried on overhead lines, in urban areas buried cable may be used. The lower-voltage subtransmission lines use less right-of-way and simpler structures, and it is much more feasible to put them underground where needed. Higher-voltage lines require more space and are usually above-ground since putting them underground is very expensive.

There is no fixed cutoff between subtransmission and transmission, or subtransmission and distribution. The voltage ranges overlap somewhat. Voltages of 69kV, 115kV, and 138kV are often used for subtransmission in North America. As power systems evolved, voltages formerly used for transmission were used for subtransmission, and subtransmission voltages became distribution voltages. Like transmission, subtransmission moves relatively large amounts of power, and like distribution, subtransmission covers an area instead of just point-to-point[7].

7.5.5 Transmission Grid Exit

At the substations, transformers reduce the voltage to a lower level for distribution to commercial and residential users. This distribution is accomplished with a combination of sub-transmission (33kV to 132kV) and distribution (3.3kV to 25kV). Finally, at the point of use, the energy is transformed to low voltage (varying by country and customer requirements).

7.6 Advantage of High-Voltage Power Transmission

High-voltage power transmission allows for lesser resistive losses over long distances in the wiring. This efficiency of high voltage transmission allows for the transmission of a larger proportion of the generated power to the substations and in turn to the loads, translating to operational cost savings.

7.7 High-Voltage Direct Current

High-voltage direct current (HVDC) is used to transmit large amounts of power over long distances or for interconnections between asynchronous grids. When electrical energy is to be transmitted over very long distances, the power lost in AC transmission becomes appreciable and it is less expensive to use direct current instead of alternating current. For a very long transmission line, these lower losses (and reduced construction cost of a DC line) can offset the additional cost of the required converter stations at each end.

HVDC is also used for long submarine cables where AC cannot be used because of the cable capacitance[8]. In these cases special high-voltage cables for DC are used. Submarine HVDC systems are often used to connect the electricity grids of islands, for example, between Great Britain and continental Europe, between Great Britain and Ireland, between Tasmania and the Australian mainland, between the North and South Islands of New Zealand, between New Jersey and New York City, and between New Jersey and Long Island. Submarine connections up to 600 kilometres in length are presently in use[9].

HVDC links can be used to control problems in the grid with AC electricity flow. The power transmitted by an AC line increases as the phase angle between source end voltage and destination ends increases, but too large a phase angle will allow the systems at either end of the line to fall out of step. Since the power flow in a DC link is controlled independently of the phases of the AC networks at either end of the link, this phase angle limit does not exist, and a DC link is always able to transfer its full rated power. A DC link therefore stabilizes the AC grid at either end, since power flow and phase angle can then be controlled independently.

As an example, to adjust the flow of AC power on a hypothetical line between Seattle and Boston would require adjustment of the relative phase of the two regional electrical grids. This is an everyday occurrence in AC systems, but one that can become disrupted when AC system components fail and place unexpected loads on the remaining working grid system. With an HVDC line instead, such an interconnection would:

(1) Convert AC in Seattle into HVDC.

(2) Use HVDC for the 3000 miles (4800 kilometres) of cross-country transmission.

(3) Convert the HVDC to locally synchronized AC in Boston (and possibly in other cooperating cities along the transmission route). Such a system could be less prone to failure if parts of it were suddenly shut down. One example of a long DC transmission line is the Pacific DC Intertie located in the Western United States.

7.8 Capacity

The amount of power that can be sent over a transmission line is limited. The origins of the limits vary depending on the length of the line. For a short line, the heating of conductors due to line losses sets a thermal limit. If too much current is drawn, conductors may sag too close to the ground, or conductors and equipment may be damaged by overheating. For intermediate-length lines on the order of 100 kilometres (62 miles), the limit is set by the voltage drop in the line. For longer AC lines, system stability sets the limit to the power that can be transferred. Approximately, the power flowing over an AC line is proportional to the cosine of the phase angle of the voltage and current at the receiving and transmitting ends. This angle varies depending on system loading and generation. It is undesirable for the angle to approach 90 degrees, as the power flowing decreases but the resistive losses remain. Very approximately, the allowable product of line length and maximum load is proportional to the square of the system voltage. Series capacitors or phase-shifting transformers are used on long lines to improve stability. High-voltage direct current lines are restricted only by thermal and voltage drop limits, since the phase angle is not material to their operation.

Up to now, it has been almost impossible to foresee the temperature distribution along the cable route, so that the maximum applicable current load was usually set as a compromise between understanding of operation conditions and risk minimization. The availability of industrial distributed temperature sensing (DTS) systems that measure in real time temperatures all along the cable is a first step in monitoring the transmission system capacity. This monitoring solution is based on using passive optical fibers as temperature sensors, either integrated directly inside a high voltage cable or mounted externally on the cable insulation. A solution for overhead lines is also available. In this case, the optical fiber is integrated into the core of a phase wire of overhead transmission lines (OPPC). The integrated dynamic cable rating (DCR) or also called real time thermal rating (RTTR) solution enables not only to continuously monitor the temperature of a high voltage cable circuit in real time, but to safely utilize the existing network capacity to its maximum. Furthermore, it provides the ability to the operator to predict the behavior of the transmission system upon major changes made to its initial operating conditions.

7.9 Control

To ensure safe and predictable operation, the components of the transmission system are controlled with generators, switches, circuit breakers and loads. The voltage, power, frequency, load factor and reliability capabilities of the transmission system are designed to provide cost effective performance for the customers.

7.9.1 Load Balancing

The transmission system provides for base load and peak load capability, with safety and fault tolerance margins. The peak load times vary by region largely due to the industry mix. In very hot and very cold climates home air conditioning and heating loads have an effect on the overall load. They are typically highest in the late afternoon in the hottest part of the year and in mid-morning and mid-evening in the coldest part of the year. This makes the power requirements vary by the season and the time of day. Distribution system designs always take the base load and the peak load into consideration.

The transmission system usually does not have a large buffering capability to match the loads with the generation. Thus generation has to be kept matched to the load, to prevent overloading failures of the generation equipment.

Multiple sources and loads can be connected to the transmission system and they must be controlled to provide orderly transfer of power. In centralized power generation, only local control of generation is necessary, and it involves synchronization of the generation units, to prevent large transients and overload conditions.

In distributed power generation the generators are geographically distributed and the process to bring them online and offline must be carefully controlled. The load control signals can either be sent on separate lines or on the power lines themselves. Voltage and frequency can be used as signaling mechanisms to balance the loads.

In voltage signaling, the variation of voltage is used to increase generation. The power added by any system increases as the line voltage decreases. This arrangement is stable in principle. Voltage-based regulation is complex to use in mesh networks, since the individual components and setpoints would need to be reconfigured every time a new generator is added to the mesh.

In frequency signaling, the generating units match the frequency of the power transmission system. In droop speed control, if the frequency decreases, the power is increased (the drop in line frequency is an indication that the increased load is causing the generators to slow down).

Wind turbines, vehicle-to-grid and other locally distributed storage and generation systems can be connected to the power grid, and interact with it to improve system operation. Internationally, the trend has been a slow move from a heavily centralized power system to a decentralized power system. The main draw of locally distributed generation systems which involve a number of new and innovative solutions is that they reduce transmission losses by leading to consumption of electricity closer to where it was produced.

7.9.2 Failure Protection

Under excess load conditions, the system can be designed to fail gracefully rather than all at once. Brownouts occur when the supply power drops below the demand. Blackouts occur when the supply fails completely.

Rolling blackouts (also called load shedding) are intentionally engineered electrical power outages, used to distribute insufficient power when the demand for electricity exceeds the supply.

7.10 Communications

Operators of long transmission lines require reliable communications for control of the power grid and, often, associated generation and distribution facilities. Fault-sensing protective relays at each end of the line must communicate to monitor the flow of power into and out of the protected line section so that faulted conductors or equipment can be quickly de-energized and the balance of the system restored. Protection of the transmission line from short circuits and other faults is usually so critical that common carrier telecommunications are insufficiently reliable, and in remote areas a common carrier may not be available. Communication systems associated with a transmission project may use microwaves, power-line communication, or optical fibers.

Rarely, and for short distances, a utility will use pilot-wires strung along the transmission line path. Leased circuits from common carriers are not preferred since availability is not under control of the electric power transmission organization.

Transmission lines can also be used to carry data: this is called power-line carrier, or power line communication (PLC). PLC signals can be easily received with a radio for the long wave range.

Optical fibers can be included in the stranded conductors of a transmission line, in the overhead shield wires. These cables are known as optical ground wire (OPGW). Sometimes a standalone cable is used, all-dielectric self-supporting (ADSS) cable, attached to the transmission line cross arms.

Some jurisdictions, such as Minnesota, prohibit energy transmission companies from selling surplus communication bandwidth or acting as a telecommunications common carrier. Where the regulatory structure permits, the utility can sell capacity in extra dark fibers to a common carrier, providing another revenue stream.

7.11 Electricity Market Reform

Some regulators regard electric transmission to be a natural monopoly[10,11] and there are moves in many countries to separately regulate transmission (see electricity market).

Spain was the first country to establish a regional transmission organization. In that country, transmission operations and market operations are controlled by separate companies. The transmission system operator is Red Eléctrica de España (REE) and the wholesale electricity market operator is Operador del Mercado Ibérico de Energía-Polo Español, S. A. Omel Holding. Spain's transmission system is interconnected with those of France, Portugal, and Morocco.

The establishment of RTOs in the United States was spurred by the "FERC's Order 888": "Promoting Wholesale Competition Through Open Access Non-discriminatory Transmission Services by Public Utilities", and "Recovery of Stranded Costs by Public Utilities and Transmitting Utilities" issued in 1996[12]. In the United States and parts of Canada, several electric transmission companies operate independently of generation companies, but there are still regions—the Southern United States—where vertical integration of the electric system is intact. In regions of separation, transmission owners and generation owners continue to interact with each other as market participants with voting rights within their RTO. RTOs in the United States are regulated by the Federal Energy Regulatory Commission.

7.12 Cost of Electric Power Transmission

The cost of high voltage electricity transmission (as opposed to the costs of electric power distribution) is comparatively low, compared to all other costs arising in a consumer's electricity bill. In the UK, transmission costs are about 0.2 pennies per kW•h compared to a delivered domestic price of around 10 pennies per kW•h.

Research evaluates the level of capital expenditure in the electric power T&D equipment market will be worth $128.9 billion in 2011[13].

7.13 Merchant Transmission

Merchant transmission is an arrangement where a third party constructs and operates electric transmission lines through the franchise area of an unrelated incumbent utility.

Operating merchant transmission projects in the United States include the Cross Sound Cable from Shoreham, New York to New Haven, Connecticut, Neptune RTS Transmission Line from Sayreville, New Jersey to New Bridge, New York, and Path 15 in California. Additional projects are in development or have been proposed throughout the United States, including the Lake Erie

Connector, an underwater transmission line proposed by ITC Holdings Corp., connecting Ontario to load serving entities in the PJM Interconnection region.

There is only one unregulated or market interconnector in Australia: Basslink between Tasmania and Victoria. Two DC links originally implemented as market interconnectors, Directlink and Murraylink, have been converted to regulated interconnectors.

A major barrier to wider adoption of merchant transmission is the difficulty in identifying who benefits from the facility so that the beneficiaries will pay the toll. Also, it is difficult for a merchant transmission line to compete when the alternative transmission lines are subsidized by incumbent utility businesses with a monopolized and regulated rate base[14]. In the United States, the "FERC's Order 1000", issued in 2010, attempts to reduce barriers to third party investment and creation of merchant transmission lines where a public policy need is found.

7.14 Health Concerns

Some large studies, including a large study in the United States, have failed to find any link between living near power lines and developing any sickness or diseases, such as cancer. A 1997 study found that it did not matter how close one was to a power line or a sub-station, there was no increased risk of cancer or illness.

The mainstream scientific evidence suggests that low-power, low-frequency, electromagnetic radiation associated with household currents and high transmission power lines does not constitute a short or long-term health hazard. Some studies, however, have found statistical correlations between various diseases and living or working near power lines. No adverse health effects have been substantiated for people not living close to power lines.

The New York State Public Service Commission conducted a study, documented in "Opinion No. 78-13" (issued on June 19, 1978), to evaluate potential health effects of electric fields. The study's case number is too old to be listed as a case number in the commission's online database, DMM, and so the original study can be difficult to find. The study chose to utilize the electric field strength that was measured at the edge of an existing (but newly built) right-of-way on a 765kV transmission line from New York to Canada, 1.6kV/m, as the interim standard maximum electric field at the edge of any new transmission line right-of-way built in New York State after issuance of the order. The opinion also limited the voltage of all new transmission lines built in New York to 345kV. On September 11, 1990, after a similar study of magnetic field strengths, the NYSPSC issued their "Interim Policy Statement on Magnetic Fields". This study established a magnetic field interim standard of 200mG at the edge of the right-of-way using the winter-normal conductor rating. This later document can also be difficult to find on the NYSPSC's online database, since it predates the online database system. As a comparison with everyday items, a hair dryer or electric blanket produces a 100-500mG magnetic field. An electric razor can produce 2.6kV/m. Whereas electric fields can be shielded, magnetic fields cannot be shielded, but are usually minimized by optimizing the location of each phase of a circuit in cross-section[15].

When a new transmission line is proposed, within the application to the applicable regulatory body (usually a public utility commission), there is often an analysis of electric and magnetic field levels at the edge of rights-of-way. These analyses are performed by a utility or by an electrical engineering consultant using modelling software. At least one state public utility commission has access to software developed by an engineer or engineers at the Bonneville Power Administration to analyze electric and magnetic fields at edge of rights-of-way for proposed transmission lines. Often, public utility commissions will not comment on any health impacts due to electric and magnetic fields and will refer information seekers to the state's affiliated department of health.

There are established biological effects for acute high level exposure to magnetic fields well above 100μT (1G) (1000mG). In a residential setting, there is "limited evidence of carcinogenicity in humans and less than sufficient evidence for carcinogenicity in experimental animals", in particular, childhood leukemia, associated with average exposure to residential power-frequency magnetic field above 0.3μT (3mG) to 0.4μT (4mG). These levels exceed average residential power-frequency magnetic fields in homes, which are about 0.07μT (0.7mG) in Europe and 0.11μT (1.1mG) in North America.

The Earth's natural geomagnetic field strength varies over the surface of the planet between 0.035mT and 0.07mT (35-70μT or 350-700mG) while the International Standard for the continuous exposure limit is set at 40mT (400000mG or 400G) for the general public.

7.15 Policy by Country

The Federal Energy Regulatory Commission (FERC) is the primary regulatory agency of electric power transmission and wholesale electricity sales within the United States. It was originally established by Congress in 1920 as the Federal Power Commission and has since undergone multiple name and responsibility modifications. That which is not regulated by FERC, primarily electric power distribution and the retail sale of power, is under the jurisdiction of state authority.

Two of the more notable US energy policies impacting electricity transmission are "Order No. 888" and the "Energy Policy Act of 2005."

"Order No. 888" adopted by FERC on 24 April 1996, was designed to remove impediments to competition in the wholesale bulk power marketplace and to bring more efficient, lower cost power to the nation's electricity consumers. The legal and policy cornerstone of these rules is to remedy undue discrimination in access to the monopoly owned transmission wires that control whether and to whom electricity can be transported in interstate commerce. "Order No. 888" required all public utilities that own, control, or operate facilities used for transmitting electric energy in interstate commerce, to have open access non-discriminatory transmission tariffs. These tariffs allow any electricity generator to utilize the already existing power lines for the transmission of the power that they generate. "Order No. 888" also permits public utilities to recover the costs associated with providing their power lines as an open access service.

The "Energy Policy Act of 2005" (EPAct) signed into law by congress on August 8, 2005, further expanded the federal authority of regulating power transmission. EPAct gave FERC significant new responsibilities including but not limited to the enforcement of electric transmission reliability standards and the establishment of rate incentives to encourage investment in electric transmission[16].

Historically, local governments have exercised authority over the grid and have significant disincentives to encourage actions that would benefit states other than their own. Localities with cheap electricity have a disincentive to encourage making interstate commerce in electricity trading easier, since other regions will be able to compete for local energy and drive up rates. For example, some regulators in Maine do not wish to address congestion problems because the congestion serves to keep Maine rates low. Further, local constituencies can block or slow permitting by pointing to visual impact, environmental and perceived health concerns. In the US, generation is growing four times faster than transmission, but big transmission upgrades require the coordination of multiple states, a multitude of interlocking permits, and cooperation between a significant portion of the 500 companies that own the grid. From a policy perspective, the control of the grid is balkanized, and even former energy secretary Bill Richardson refers to it as a third world grid. There have been efforts in the EU and US to confront the problem. The US national security interest in significantly growing transmission capacity drove passage of the 2005 energy act giving the "Department of Energy" the authority to approve transmission if states refuse to act. However, soon after the "Department of Energy" used its power to designate two "National Interest Electric Transmission Corridors", 14 senators signed a letter stating the "Department of Energy" was being too aggressive[17].

7.16 Special Transmission

7.16.1 Grids for Railways

In some countries where electric locomotives or electric multiple units run on low frequency AC power, there are separate single phase traction power networks operated by the railways. Prime examples are countries in Europe (including Austria, Germany and Switzerland) which utilize the older AC technology based on (16+2/3) Hz (Norway and Sweden also use this frequency but use conversion from the 50Hz public supply; Sweden has a (16+2/3) Hz traction grid but only for part of the system).

7.16.2 Superconducting Cables

The Federal Government of the United States admits that the power grid is susceptible to cyberwarfare. The United States Department of Homeland Security works with industry to identify vulnerabilities and to help industry enhance the security of control system networks, the federal government is also working to ensure that security is built in as the US develops the next generation of "smart grid" networks.

In June 2019, Russia has conceded that it is "possible" its electrical grid is under cyber-attack by the United States. The New York Times reported that American hackers from the United States Cyber Command planted malware potentially capable of disrupting the Russian electrical grid.

Notes and References

[1] Betz H D, et al. Lightning: principles, instruments and applications. Springer, 2009.

[2] Dominianni C, Lane K, Johnson S, et al. Health impacts of citywide and localized power outages in New York City. Environmental Health Perspectives, 2018, 126(6): 067003.

[3] Thomas P. Networks of power: electrification in western society, 1880-1930. Baltimore: Johns Hopkins University Press, 1993: 119-122.

[4] Guarnieri M. The beginning of electric energy transmission: part one. IEEE Industrial Electronics Magazine. 2013, 7(1): 57-60.

[5] Guarnieri M. The beginning of electric energy transmission: part two. IEEE Industrial Electronics Magazine. 2013, 7(2): 52-59.

[6] Jonnes J. Empires of light: Edison, Tesla, Westinghouse, and the race to electrify the world. Random House Trade Paperbacks, 2004.

[7] Donald G, et al. Standard handbook for electrical engineers. 15th edition, New York: McGraw-Hill, 2007.

[8] Donald G, et al. Standard handbook for electrical engineers. 11th edition, McGraw Hill, 1978.

[9] Guarnieri M. The alternating evolution of DC power transmission. IEEE Industrial Electronics Magazine. 2013, 7(3): 60-63.

[10] Hogan W W. Contract networks for electric power transmission. Journal of Regulatory Economics, 1992, 4 (3): 211-242.

[11] Ngan H W. Electricity regulation and electricity market reforms in China. Energy Policy, 2010, 38 (5): 2142-2148.

[12] Ghesu F C, et al. Multi-scale deep reinforcement learning for real-time 3D-landmark detection in CT scans. IEEE Transactions on Pattern Analysis and Machine Intelligence, 2017, 41(1): 176-189.

[13] Carreras B A, et al. Critical points and transitions in an electric power transmission model for cascading failure blackouts. Chaos: An Interdisciplinary Journal of Nonlinear Science, 2002, 12 (4): 985-994.

[14] Fiona W. Global transmission expansion, 2003.

[15] Murphy J C, Kaden D A, et al. Power frequency electric and magnetic fields: a review of genetic toxicology. Mutation Research/Reviews in Genetic Toxicology, 1993, 296(3): 221-240.

[16] Malmedal K, Kroposki B, et al. Energy policy act of 2005 and its impact on renewable energy applications in USA. 2007 IEEE Power Engineering Society General Meeting, 2007: 1-8.

[17] Shafiullah G M, et al. Potential challenges of integrating large-scale wind energy into the power grid. Renewable and Sustainable Energy Reviews. 2013, 20: 306-321.

Chapter 8
China Ultrahigh Voltage Project

Over the past 100 years, the development of electric power transmission always focused on the theme of improving transfer capacity and reducing transmission cost. Raising voltage level is the most efficient way to improve transmission power. The electric power transmission at 1000kV and above AC (alternating current) voltages is known as UHV (ultra high voltage) AC transmission, and the voltages at above 600kV DC (direct current) are known as UHV DC transmission. Especially, 1000kV AC and 800kV DC are UHV AC and UHV DC transmission voltages in China, respectively.

In order to meet the growing of power load, to carry out long distance and bulk capacity power transmission, Russia (the former USSR), Japan, the United States of America (USA), Italy, Canada, Brazil began study on the UHV transmission relevant technologies in 1960s and 1970s[1-5]. Wuhan High Voltage Research Institute (WHVRI) of State Grid Corporation of China (SGCC), China Electric Power Research Institute (CEPRI), Beijing Electric Power Construction Research Institute (EPCRI) of SGCC and some universities in China began study on the UHV AC transmission technologies in 1986. China, India and Southern Africa[6,7] began study on the UHV DC transmission technologies in recent years. Many research works such as line parameters (conductors, towers, insulators), effects on environment, lightning performance, over-voltage and insulation coordination, live-line working, equipment manufacture etc., have been carried out to discuss UHV AC transmission in the range of 1000-1500kV and UHV DC transmission at the voltages of 800kV and above.

The research results indicate that UHV transmission lines can transmit large block of electric energy over a long distance, reduce number of circuit lines and right-of-way, lower electric energy loss, etc. Especially, the UHV AC transmission suits to interconnect large power grid. The UHV DC alternative of voltages above 600kV (UHV DC) can be economically attractive for very long distance and high capacity transmission. The obtained results demonstrate that there is no insurmountable technical obstacle in the design and construction of UHV transmission, and the UHV transmission is presently available and awaiting commercial applications. CIGRE working group 38-04 evaluated the UHV technique and conclusions were drawn that the application of UHV AC transmission technology was fully developed and 800kV UHV DC was technically feasible.

The UHV transmission research activities and achievements of China in the past twenty years, including the UHV AC, UHV DC test and demonstration lines, test bases, etc. , the future prospects of UHV transmission in China are introduced in this paper.

8.1 Research History and Background of UHV Transmission

The UHV AC and DC transmission and transformation development history and research background in Russia (the former USSR), Japan, USA, Italy, Canada, Brazil, China and other countries are introduced in this section.

Rapid load growth in the 1960s and the prospects of continued load growth in future several decades were the driving forces for research and development of UHV AC power transmission lines at voltages above 1000kV. Even as the first transmission lines at 500kV and 750kV were being built and operated in the 1960s, there was a heightened interest in developing the next higher transmission voltages in the so called UHV range of 1000kV to 1500kV AC and above 600kV DC. In order to gather the vast amount of technical information necessary to design transmission lines above 1000kV AC and above 600kV, research and test facilities were built in several countries in the 1970s and the following about thirty years. Information on the progress of UHV technology research works with focus on UHV system planning, performance and reliability aspects, UHV transmission lines, UHV substations and equipment, UHV testing facilities and new technologies that carried out in Russia (The Former USSR), Japan, the USA, Italy, Canada, Brazil, India, France were presented in three excellent CIGRE Working Group (WG) reports— "WG 31. 04 1983""WG 38. 04 1988" "WG 38. 04 1994" and two overview papers.

8.1.1 Russia (The Former USSR)

In the 1970s, in order to satisfy the need of strengthening the electrical links between integrated power systems, as well as the need for transfer large quantities of power overlong distances, USSR had an in-depth study on the insulation system, line and equipment of UHV AC transmission at voltages in the range of 1150 to 1500 kV.

A circuit of three phases 1.17km long 1150kV test line was constructed at the Bely Rast Substation. Test data were obtained on the corona performance of conductor bundles. Tests of the air insulation, insulation of equipment, studies of switching over-voltage, audible noise (AN), radio interference (RI), electric fields in a substation, and installation, operation and maintenance of equipment were also carried out.

2362km 1150kV AC transmission lines were constructed successively in USSR from the end of the 1980s to the beginning of the 1990s. Three substations and two segments were put into service, but the line, after a few years of operation at the design voltage of 1150kV, has been operated at the lower level of 500kV. In the end of the 1970s, USSR also practiced 750kV DC project (6000MW and 2414km). The main equipment passed the type test and 1090 km transmission lines were constructed. Most earthwork and equipment installation in converter stations at the both ends of the line were completed.

8.1.2 Japan

Japan began study on UHV transmission technology in 1973. The need for overcoming stability problems of the existing 500kV network and obviating the problems of excessive short-circuit currents led to the consideration of transmission above 1000kV to overlay the existing network. And UHV research was carried out at Central Research Institute of Electric Power Industry (CRIEPI), Tokyo Electric Power Company (TEPCO), and the NGK Insulator Company.

Testing facilities of CRIEPI have a UHV fog chamber for testing of polluted insulators, a facility for insulator testing under continuous energized with phase-to-ground voltages up to 900kV, a corona cage used for AN test, and a double-circuit 600m long test line of voltage 1000kV AC (convertible to a ±500 to ±650kV DC). On the UHV test line, the behavior of 8-, 10-, and 12-conductor bundles and towers under strong wind and earthquake were investigated. Construction and maintenance techniques, AN, RI and television interference (TVI), as well as studies of the effects of electric fields were also investigated.

Takaishiyama test line of TEPCO has two spans with 10 aluminium conductor steel reinforced (ACSR) conductor bundles. Research and development work for mechanical performance of bundled conductor and insulator assemblies, such as galloping and icing, were carried out on the test line.

RI and AN tests on insulator assemblies under polluted conditions were performed with corona testing equipment and the 1000kV pollution testing equipment constructed at NGK high voltage laboratory. A significant amount of information was obtained on the withstand voltages of contaminated and snow-covered insulator strings.

TEPCO began construct the 1000kV transmission project in 1988, and Sin-Haruna UHV equipment test field was constructed in 1996, the construction of 427km, 1000kV double-circuit transmission line on the same tower was completed in 1999, but the line has been operated at 500kV since it was energized, and is planned to be upgraded to 1000kV AC around 2015.

8.1.3 The USA

The USA began study on UHV transmission technologies in 1967, the purpose of the new transmission systems was to transmit large blocks of power, improve system stability, and reduce environmental impact.

In the USA, UHV studies were conducted at the General Electric Company (GE), the Electric Power Research Institute (EPRI), the American Electric Power Company (AEP), and the Bonneville Power Administration (BPA). Research works of AC and DC environment tests, AC line and DC line tests in the same corridor, etc., were carried out.

Three separate research and test facilities were built to evaluate the technical feasibility of transmission lines above 1000kV:

(1) The GE/EPRI Project UHV comprises a three-phase experimental line, a test cage and a pollution chamber. The facilities have the capability of testing the corona performance of conductor bundles, with stand strength of air clearances and the pollution performance of line and station insulators.

(2) The AEP/ASEA test station, jointly operated by AEP and the ASEA (combined to a part of ABB since 1988) company of Sweden, locates near South Bend, Indiana, has the capability of testing single-phase conductor bundles at voltages corresponding to transmission system voltages up to, and even beyond 1500kV, but now this UHV AC test facility has been decommissioned.

(3) At BPA, a full-scale three-phase, 1200kV prototype test line, near Lyons, Oregon, was used to evaluate the long-term corona performance of an 8-conductor bundle. In addition, the facility at Carey High Voltage Laboratory was used for studies on air insulation, while conductor vibration and galloping studies were carried out at the Moro mechanical test line.

In 1967, a research program to study on overhead transmission lines with voltages of 1000 to 1500kV was initiated at GE Project UHV research facility located in Lenox, Massachusetts. A single-phase experimental line consisting of three spans each 305m long, a station with a UHV transformer manufactured by ASEA (rated voltage at 420/835/1785kV, three-phase equivalents, and 333MVA) and two test cages, each cage is 30.5m long. The cages have a square section and the dimensions could vary between 6.1m × 6.1m and 9.1m × 9.1m, were used to evaluate the corona performance of large conductor bundles. In 1974, a new three-year program to construct and operate a three-phase test line in the range of 1000kV to 1500kV was started by the EPRI. The single-phase test line was expanded to three-phase operation with the addition of two UHV transformers, surge arresters, coupling capacitors, and associated equipment. The three-phase UHV test line was 523m long and test voltages up to 1500kV phase-to-phase were utilized.

At Project UHV, extensive switching impulse tests on many different types of line and substation equipment were performed and power frequency tests on contaminated insulators were performed at UHV voltages. The AN, RI, corona loss, TVI, electric field at ground level, and ozone generation of 11 different conductor bundles with subconductor diameters in the range of 33-56mm and the number of subconductors in the range of 6-16 were measured. The facilities at Project UHV were also used for HVDC research later. New equipment had been installed as part of this project to make possible a comprehensive research program with test voltage up to 1500kV.

Data on the corona performance, including AN, RI, TVI and corona loss of several bundles, with up to 18 subconductors, were obtained at the AEP/ASEA UHV Project test station. In addition, ASEA had performed developmental testing to determine and verify the insulation design of UHV equipment. Research on the GIS equipment had been conducted for support insulators, entrance bushings, and on the effects of varying gas quality and pressure.

At BPA, extensive research and development on UHV transmission have been conducted at the test facility at Lyons, Oregon, and the mechanical test line at Moro, Oregon. The Lyons UHV test facility consists of a 2.1km three-phase, 1200kV line, has been used for electrical studies. The test line at Moro has been used for structural and mechanical studies without voltage. Investigations at Lyons and Moro have been supported with tests and studies in the BPA laboratories.

Studies of corona performance on conductors, insulators, and hardware fittings had been carried out both in the BPA's Carey Laboratory and on the Lyons 1200kV test line. Long term AN, RI,

TVI, corona loss, and ozone generation for 7-and 8-conductor bundles of 41mm diameter subconductor have been investigated. Mechanical and structural tests including studies of line loadings (wind and ice load), conductor motion (aeolian vibration, subconductor oscillation, and galloping), switching surge with stand strength of air gaps, pollution performance characteristics of ceramic and nonceramic insulator strings, have been performed in the BPA's Mangan mechanical-electrical laboratory and at the Moro mechanical test line.

In addition, other studies include studies of substation noise and electric fields, and evaluation of the performance of transformers, arresters, and SF_6 equipment.

Between 1982 and 1985, EPRI (USA) with Cepel/Eletrobras (Brazil) studied the critical problems in developing HVDC converter station equipment for voltages in the range of±600 to±1200kV. The conclusion was that converter stations at voltage of 800kV DC was technically feasible.

8.1.4 Italy

In the middle of the 1970s, Italy began study on UHV transmission technology, and the purpose was to transmit large blocks of power from large power generation facilities to the load centers far away. The UHV transmission studies were carried out in Italy at several testing stations and laboratories.

At the Suvereto 1000kV Project, a 1km long test line was used for air insulation and corona studies, and a 40m outdoor test cage was also used for corona studies. Switching impulse behavior of air clearances, behavior of surface insulation of UHV system in polluted atmosphere, performance of SF_6 insulation, and development of nonconventional insulators were carried out. Studies on the interference levels produced by UHV insulators and fittings were also carried out.

A test line at Pradarena Pass was used for icing and wind loading studies in winter and vibration, sub-span galloping, and spacer performance studies in summer. Studies on air insulation and performance of polluted insulators were carried out at the CESI laboratories in Milan.

The researches generated a large amount of data for determining phase-to-ground and phase-to-phase air clearances, selecting ceramic and non-ceramic insulator strings, and selecting conductor bundles for a 1050kV prototype transmission line. The test data were also used in the development of vibration dampers, spacers, and non conventional tower structures and foundations for 1050kV transmission lines.

A±700kV generator was used for the dielectric tests of UHV DC insulation. A test plant was also used for the functional tests of thyristor modules of converter valves.

8.1.5 Canada

In Canada, the need for transmission systems above 1000kV was foreseen in the provinces of British Columbia and Quebec to bring large blocks of power from remotehydro-electric projects to the load centers.

The main research and test facilities for studies at system voltages up to 1500kV were located at the HV laboratory of Hydro-Quebec Institute of Research (IREQ). The test facilities at IREQ comprising a large indoor high voltage laboratory, with capabilities for air insulation studies on tower window mock ups for system voltages up to 1500kV, a large pollution chamber for studies on insulators, and an outdoor experimental line and test cages were used for the corona test of conductor bundles for AC systems up to 1500kV and DC systems up to 1200kV. IREQ also studied the corona, electric field, and ion current performance of DC transmission lines in the range of 600kV to 1200kV. Phase-to-ground and phase-to-phase air insulation tests on line and substation configurations at IREQ provided a large amount of data necessary for determining air clearances for transmission lines and substations at system voltages of 1200kV and 1500kV.

A test line was also built at Magdalen Islands to study vibration performance of 6 and 12 conductor bundles and development of spacer dampers.

8.1.6 Brazil

The purpose for research on transmission lines above 1000kV in Brazil was the need for transmitting a block of power on the order of 20000MW from the Amazon Basin to the load centers at distances in the range of 1500-2500km.

Research and test facilities were built at the research institute CEPEL in Adrianopolis, Brazil. A 360m long test line and a 7.5m×7.5m test cage were installed and used to conduct research for AC transmission systems up to1500kV and DC systems up to ±1000kV. In addition to a large indoor high voltage laboratory for tests on equipment, the facilities at CEPEL include an outdoor area where full-scale or mock up transmission towers can be tested for air insulation clearances and an outdoor experimental line and test cages for corona studies.

Since 1978, Brazil has been associated with ENEL, in Italy, B. C. HYFRO, in Canada, for a joint UHV AC research and development program.

Basic research in HVDC 800kV systems in 1987-1995 was carried out and some equipment were designed and manufactured in Brazil. Since then design work has continued within ABB (combined of ASEA/Sweden and BBC Brown Boveri/Switzerland since 1988). Several studies and meetings confirmed that 800kV HVDC is a feasible voltage level.

The Itaipu 600kV, 6300MW transmission line is operating at 600kV since 1984, Brazil, which is the highest voltage and capacity DC transmission system in the world, and the design and implementation of this project was a joint effort of ASEA/Sweden and ASEA/Brazil.

8.1.7 China

WHVRI, CEPRI, EPCRI of SGCC and some universities began study on the UHV transmission technology in 1986. Since 1986, the UHV transmission research was included in the mega-projects of Scientific Research for 7th Five-Year Plan, 8th Five-Year Plan and 10th Five-Year Plan in China. Some subject researches had been developed, including the prophase research of UHV AC transmission (1986-1990) structured by Importance Project Ministry of State Council; demonstration on long distance transmission and voltage level (1990-1995) structured by Importance Project Ministry of State Council; the feasibility study on 1000kV AC transmission (1990-1995) and the prophase demonstration on UHV AC transmission (1997-1999) structured by Ministry of Science and Technology of China; UHV AC test line (1994-1996), the effects on environment of UHV AC transmission line (1997-1999) and the generation of long front switching wave by using power frequency test equipment (1997-1999) structured by Ministry of Power Industry of China; the background factor of UHV AC transmission technology development (1998-2000) and external insulation characteristics of UHV AC transmission line (1999-2001) structured by State Power Corporation; higher voltage level application in Southern Power Grid (2003-2004) structured by China Southern Power Grid Corporation (CSG); the economical feasibility study on 1000kV AC transmission (2003-2004) structured by SGCC.

Test facilities of WHVRI of SGCC comprising a 450m×120m outdoor test yard, a 5.4MV, 530kJ impulse generator, a 24m×24m×26m artificial pollution chamber with a 800kV (phase-earth) rated voltage wall bushing, a 3×750kV, 4A transformer cascade (Figure 8-1), 2250kVA regulator (Figure 8-2), 7500kVA synchronous generation voltage regulation unit; and a 1000kV 200m long test line that was constructed at WHVRI of SGCC in 1994.

Figure 8-1 2250kV power frequency AC test transformer at WHVRI (3×750kV, 4A transformer cascade)

Figure 8-2 UHV AC test line section at WHVRI

Test facilities of CEPRI comprising a 6MV, 300kJ impulse generators and a transformer cascade. A tower test site was constructed at ECPRI of SGCC in 2004. China, Brazil, India and South-Africa began study on the UHV DC transmission technologies at the voltage of 800kV and above in recent years.

From the above review of UHV AC and DC development in China and other countries, the conclusion can be drawn that, although the technical feasibility is approved, there is no practical transmission systems at the voltages of 1000kV AC, 800kV DC and above being operated at the present time.

8.2 Target Design and Research Background of China's UHV System

8.2.1 The Demand and Goals Analysis of UHV Transmission in China

In the past 20 years, the power industry in China has been developing very fast. Both the installed capacity and the total power consumptions in China has been the second largest one in the world since 1996. By 2005, the installed capacity reached 512GW, and the annual growth rate is higher than 10%. China is now building a well-off society in an all-round way, and the GDP should be increased from USD 1653 billion by 2004 to USD 4000 billion by 2020. Strengthening the power supply is one of the reliable guarantees to attain the above economic objective. A huge installed capacity is indispensable to meet the rapid load growth in the coming years. By 2010, it is expected that the total installed capacity will be 900GW, the power consumption be 4200TW • h per year. By 2020, it is expected that the total installed capacity will be 1400-1600GW, and 1300GW at least, the power consumption be 7000-8000TW • h per year. Because the present power system could not meet the future power transmission needs of China, developing a 1000kV AC network supported by a series of ±800kV DC transmission projects is needed urgently.

(1) The excellent characteristics of UHV transmission: The UHV transmission has obvious advantages of improving transmission capacity, increasing power transmission distance, reducing line losses, lowering project investment, saving line corridors.

Increase transmission capacity: The UHV transmission can increase the transmission capacity. The natural transmission capacity of a 1000kV AC circuit is about 5GW, and that is approximately 4-5 times that of a 500kV AC transmission line. A circuit ±800kV DC transmission line has the capacity of 6.4GW, which is 2.1 times that of a ±500kV DC power line.

Increase transmission distance: The UHV transmission could increase the economic power transmission distance. A 1000kV AC line can economically transmit power distances of 1000km to 2000km. A ±800kV DC power line can economically transmit power over distances of 2000km to 3000km. A ±800kV DC power line is economical than 1000kV AC line when the transmission distance is longer than 1200km.

Reduce transmission loss: If the conductor cross section and transmission power are regard as constant, the resistance losses of a 1000kV AC power line is 25% that of the 500kV AC power line. The resistance loss of ±800kV DC transmission line is about 39% that of a comparable ±500kV DC line.

Reduce cost: At the same conditions, the resistance loss of the 1000kV AC line is only 1/4 of the 500kV AC line, and the project investment, etc. can be saved. The cost per unit of transmission capacity of 1000kV AC and ±800kV DC transmission scheme are 73% and 72% that of 500kV AC and ±500kV DC schemes, respectively.

Reduce land requirements: UHV transmission has obvious advantages in reducing the land occupation of the line. A 1000kV AC power line saves 50% to 66% of the corridor area required by 500kV AC lines in transmitting the same capacity. A ±800kV DC line would save 23% of the corridor area required by a 500kV DC lines in transmitting the same capacity.

(2) The need of bulk power transmission.

Power demands increase rapidly with fast economic growth: At present, China is in a critical period to build the well-off society. The industrialization and urbanization keep speeding up and the demand for electric power keeps growing. The whole society power consumption has been annually increasing at more than 10% in the past four years. According to the national economic and social development plan, by 2020, the total installed capacity is predicted to be 1300GW at least, the power consumption be 7000-8000TW • h per year. Hence, the national power grid faces a big challenge to ensure the safe and reliable supply of the bulk electric power transmission.

Imbalance distribution of energy resources and loads: The distribution of generation energy resources and power demands in China differs sharply from place to place. In China, the resources of hydro-power and coal are the main power generation resources. The proven amount of the coal resource is over 1000 billion tons among which more than 2/3 are in north and northwest China including Shanxi province and the Inner Mongolian Autonomous Region. The exploitable capacity of hydro resource is over 400GW among which more than 3/4, such as Jinsha River, Yalong River, Daduhe River and Lancang River are in southwest region including Sichuan province, Yunnan province and the Tibet Autonomous Region. However, more than 2/3 energy demand is in the relative developed central and eastern region.

The important energy bases are always 800-3000km away from the load centers. Because of the high pressure on environment protection, high cost of transportation and limited land resources, the east region is no longer suitable for the building of large-scale coal power plant. To meet the continuous increasing demand for electricity, it is necessary to optimize the energy resources allocation nationwide and transmit power in a trans-region, cross drainage area, long distance and large scale manner by the construction of strong electric grid.

The power grid backbones consisting of 500kV AC and ±500kV DC now in China are difficult to overcome the problems of imbalance distribution of energy sources and loads for the following reasons: without enough substation sites and line corridors; short circuit currents beyond standards at heavy loaded areas; difficulties in coal transportation; weak basis for power grids security; heavy pressure on environmental protection.

Because the ability in power transmission is insufficient and short circuit current exceeds the breaking capability of circuit breaker, the present 500kV AC and ±500kV DC grid is hard to meet the future development needs. Especially, 12 hydro-power plants were planned to be built in the trunk stream of the Jinsha River, including the hydropower plants designed in Yalong River and Daduhe River. The area of the Jinsha River will have 100GW installed capacity, which accounts for 25% of all exploitable hydropower in China. The installed capacity of prophase I project for the lower reaches of the Jinsha River will be about 18.6GW, even which is 0.4GW more than that of the Three Gorges project. These hydro-power bases with large installed capacity scale are concentrated and remote from load centers with distance of more than 1000km with limited line corridors. Hence, UHV DC transmission is required to transmit hydro-power out. Therefore it is urgent to develop a UHV electric grid with strong ability for resources allocation and build an electric power highway.

(3) The need of national power grid development: China has six regional grids—the North China Grid, Northeast Grid, Central China Grid, East China Grid, Northwest Grid and South China Grid. For the most part, inter connection between these grids has been accomplished. Because of the insufficient long-term investment, power grid development in China lagged behind, and resulted in a very weak grid structure. The ability of grid to optimize the resources allocation cannot be brought into play. Its ability to resist accidents and risks is not strong; and the risk of large area blackout always exists.

The construction of the UHV grid, namely 1000kV AC and 800kV DC, can effectively solve the safety and stability problems caused by the present insufficient ability of 500kV AC and ±500kV DC grid, optimize the layout of electric power and obviously improve the safe and reliable operation.

UHV AC transmission system is flexible for transmission, interchange and distribution of power on the strong power grids. The UHV AC transmission is oriented to network configuration of higher voltage level and bulk power transmission between regions. While ±800kV DC transmission is oriented to long distance electric power send-out from large hydro-power bases.

For the proposed UHV synchronized network connecting North, East and Central China, the total installed capacity will exceed 700GW by 2020. Power system simulation shows that the stability level will be high enough to transmit bulk power while protecting the system from high current faults. In case of a bipolar failure of ±800kV or a single transmission corridor failure of UHV AC, the system will be capable of maintaining system stability without experiencing serious low-frequency oscillation.

8.2.2 Important Innovations and Progress

In order to build a strong and reliable national grid and meet the load growth, construction of UHV backbone transmission network comprising 1000kV AC and ±800kV DC transmission projects was proposed by SGCC and CSG.

Although a large amount of research and test data are obtained from the different facilities around the world, the situation of China is different from other countries, especially high altitude and heavy pollution, hence, many technical problems relevant UHV transmission and transformation need to be discussed.

The SGCC started study on key technologies of UHV including power system analysis of UHV AC and DC transmission, construction and test, engineering design, manufacture of main equipment, etc., at the end of 2004. The CSG started that at 2003. Many other enterprises concern UHV transmission also research, design and manufacture the UHV equipment actively. The UHV research works are promoted greatly and many valuable achievements for the development of the UHV transmission have been achieved.

Important innovations and progress have been achieved in the following aspects:

(1) Made a systematic demonstration of the necessity and feasibility of UHV transmission and revealed the objective necessity of transformation for Chinese grid development pattern and large scale development of the UHV transmission.

(2) The key technologies like voltage standards, over voltage and insulation coordination, electromagnetic environment, live-line working, etc., have been carefully and systematically studied based on the 1000kV AC and ±800kV DC power transmission and transformation projects. The recent research developments of 1000kV AC and ±800kV DC transmission and transformation key technologies in China will be introduced in detail in the following sections.

(3) The 1000kV AC and ±800kV DC tests and demonstration project were approved and the project design were completed, and are under construction at the present.

(4) The UHV AC and DC test bases were energized, and the state grid simulation center is under construction.

(5) The UHV equipment research was fully promoted. A whole technology specification was formed and conceptual design of the equipment was completed, and some equipment was successfully developed, this will be introduced in the following section of equipment manufacturing.

(6) The UHV grid plan was formed and a scheme was proposed to build the north, central and east China UHV AC synchronous electric grid and realize large capacity transmission with super long distance through UHV DC.

8.2.3 UHV AC and DC Key Technologies Researches and Achievements

More than 2000 academicians, experts and engineers from various consulting organizations, scientific research institutions, universities and engineering and equipment manufacturing organizations to make a in depth research and repeated evaluations on more than 100 key technical problems of UHV transmission. Lots of researches have been carried out for over-voltage and insulation coordination, live-line working, lightning performance, electromagnetic environment (AN, RI, and TVI, effects of electric field and magnetic field on human bodies under UHV AC and DC transmission lines) of 1000kV AC and ±800kV DC transmission and transformation projects.

(1) Voltage standards: 1000kV AC and 800kV DC are the voltage level of UHV transmission in China.

(2) Over-voltage and insulation coordination: The external insulation discharge characteristics of power frequency, switching impulse and lighting impulse on AC transmission and substation equipment air clearances, heat stable extending radius flexible bundle conductor (bus), bundle conductor, tubular bus were carried out by mimic real configuration test on UHV transmission line. The relationship curves between air clearances and discharge voltage had been obtained by switching impulse, lightning impulse and power frequency voltage tests.

The front time of switching impulse test voltage is 1000μs, which is close to the switching over-voltage of real transmission line. The selection of air clearance under operation voltage considers the maximum operating voltage; the maximum wind speed happened once in 100 years; the rate of flash over is 0.13%. The minimum operating frequency voltage air clearance for altitudes 500m, 1000m, and 2000m requires 2.7m, 2.9m, and 3.1m respectively. The selection of air clearance under switching voltage considers the 1.7p.u. of maximum statistic 2% over-voltage level along the line; the influence of the multiply clearance connected in parallel on flash over voltage; the ratio of flash over is 0.13%. The minimum switching voltage air clearance for altitude 500m, 1000m, and 2000m requires 6.7m, 7.2m, and 7.7m for middle phase, respectively. For side phase it requires 5.9m, 6.2m, and 6.4m, respectively. The air clearance of side phase is controlled by operating voltage and the air clearance of middle phase is controlled by switching voltage impulse voltage. The lightning impulse voltage does not control the distance of tower air clearance. The requirement of air clearance under lightning impulse voltage can not be specified.

The switching impulse discharge characteristic curve of 1000kV AC substation equipment air clearance was studied at WHVRI of the SGCC. Two sets of impulse voltage generators, whose voltages are 5400kV and 3000kV respectively, are used conjointly in the experiments. By adjusting the time delay unit to make the two impulse generators in synchronism, The test used up and down method to obtain the discharge characteristic curve under switching impulse between the ring phases, the tubular bus phases and the 4 multiple conductor phases which have 5-9m distance.

Over-voltage and insulation coordination of 1000kV AC transmission line, substation (or switch station) and the equipment were studied. High voltage shunt reactor configuration was proposed when avoidance of non complete phases' power frequency resonant over-voltage, limitation of over-voltage level and reduction of number of spare high voltage shunt reactor were considered. Power frequency temporary over-voltage (TOV), secondary arc current and recovery voltage, parameter choice of MOA, switching over-voltage (including energized and single phase energized unloaded line over-voltage, ground fault over-voltage and clearing short-circuit faults switching over-voltage), gas insulated switchgear (GIS) isolator switching over-voltage, circuit breaker transient recovery voltage (TRV), DC component decayed time constant of short-circuit current cleared by circuit breaker were studied. Air clearances choice of UHV transmission line tower under working voltage, lightning impulse and switching impulse voltage, air clearances distance choice of substation and switch station as well as the choice of UHV equipment insulation level of Jindongnan-Nanyang-Jingmen transmission line had been carried out. The very fast front over-voltage (VFTO) of GIS and hybrid gas Insulated switchgear (HGIS) substation of Jindongnan-Nanyang-Jingmen UHV AC demonstration line was calculated. The Jindongnan GIS VFTO produced by disconnector switching on and off can be up to 2795kV, and 500 switching on and off parallel resistance installed on the disconnector can limit the VFTO to about 1008kV. The VFTO of a transformer in GIS substation is not high when the transformer is connected to GIS by overhead line.

The UHV transmission research results in China indicate that long air clearance switching impulse discharge voltage is influenced greatly by the shape ofelectrodes[8]. In all kinds of shape electrodes, the discharge voltage of pole-plate air clearance is the lowest, and its saturation trend is obvious. Whereas, the switching impulse discharge characteristics of conductors to tower air clearance in transmission line are influenced by the type of conductors, hardware fittings configuration, type of insulators, framework of tower, width, etc.; and its clearance coefficient is obvious bigger than that of pole-plate air clearance. Hence, the switching impulse discharge characteristics of UHV AC transmission line have no obvious saturation. The tower head dimension of UHV AC transmission line is proposed based on large amount of 1:1 true type testing in China.

Power frequency voltage withstands performance tests were carried out on artificially polluted insulator string with real model arrangement. Influence of high altitude on insulator pollution withstands voltage, influence of NSDD on insulator, influence of different pollution accumulation on pollution withstand voltage and the correction coefficient of different pollution accumulation had been obtained by tests and studies. The results show that $U_{50\%}$ of insulator string rise nonlinearly with string length. The linearly analogous $U_{50\%}$ is 1.6%–10.2% higher than tested one. The $U_{50\%}$ of different type insulator string is correlated to ESDD with negative exponential power of −0.202–−0.195. The $U_{50\%}$ values are also correlated with NSDD with negative exponential power of about −0.1341. Under the same test conditions, the $U_{50\%}$ of single string of double-shed insulator is about 5% higher than that of normal insulator, the $U_{50\%}$ of double string of normal insulators is about 6% lower than that of single string, and that of V string is 4%−13% higher than that of single string[9].

The effects of high altitude, contamination, ice covered, snow-covered, acid rain and acid fog on discharge characteristics of insulators for UHV DC transmission lines as well as the discharge characteristics of long air clearance under high altitude were expounded and discussed in China[10]. Icing flash over performance tests of short samples of two different type of silicone rubber (SIR) composite long rod insulators intended for UHV AC transmission lines were carried out. Ice thickness, pollution severity on the surface of insulators before ice accretion, atmospheric pressure, and shed profiles were also considered.

The discussion meeting for the first draft of 1000kV AC transmission system over-voltage and insulation coordination guide was hold on November 10, 2007 at Wuhan, China, and the amendatory advices were proposed.

The causes and characteristics of the over-voltage of ±800kV UHV DC transmission system was investigated under operation and fault conditions by a complete simulation model of PSCAD/EMT-DC, and the factors which influence the over-voltage level.

±800kV DC transmission tower head air clearance switching impulse and 50% lightning impulse breakdown voltage tests were carried out at outdoor test yard of CEPRI of SGCC. It was found that the air clearance of ±800kV DC transmission tower head should not less than 6.1m; the 2000m altitude correction coefficient for air clearance switching impulse breakdown voltage was 1.13; and the influence of barrier that shorter than 2m can be in the DC field air clearance design of converter station. ±800kV DC and the insulation coordination were also studied and the values of AC and DC equipment switching impulse insulation level and lightning insulation level were proposed.

Preliminary recommendations on the decisive parameters of suitable shed profiles and a test method for station composite insulators were discussed at Tsinghua University on November, 22, 2005, with a group of invited experts on external insulation from Beijing Wanglian HVDC Engineering Technology of SGCC, Technology Research Center of CSG, CEPRI, Xi'an Electro-Ceramic Research Institute (XECRI), Tsinghua University, ABBHVDC Sweden, and ABB Corporate Research in China. Flash over performances of three kinds of composite insulators were studied at the high voltage test base of Yunnan Electric Power Research Institute (YEPRI) by Graduate School at Shenzhen, Tsinghua University. It was found that the DC flashover voltage was influenced by shed profiles, and the flashover voltage can be increased 20% by reasonable optimization of shed profiles.

The artificial pollution flashover performance of the short samples of one kind FXBW-±800/400 DC SIR composite long rod insulator was investigated in the multifunction artificial climate chamber (a diameter of 7.8m and a height of 11.6m) in the High Voltage and Insulation Technological Laboratory of Chongqing University. The effects of pollution and high altitude on the flashover performance were analyzed. The exponent characterizing the influence of Equivalent Salt Deposit Density (ESDD) on the flashover voltage was related with the profile and the material of the insulator shed. The values of the samples'exponents vary between 0.24 and 0.30, which were smaller than those of porcelain or glass cap and pin insulators, namely, the influence of the pollution on the composite long rod insulators was less, relatively. The best ratio of the leakage distance to the arcing distance was about 3.35. The exponent characterizing the influence of air pressure on the flashover voltage is related with the profile and the material of the insulator shed and the pollution severity; the values of the samples'exponents vary between 0.6 and 0.8, which are larger than those of porcelain or glass cap-and-pin insulators. Therefore, the DC composite insulator used in high altitude regions should have enough arcing distance. If FXBW-±800/400 DC SIR composite long rod insulator is selected for the ±800kV UHV DC transmission lines, the basic arcing distance should be no less than 8.16m and the basic leakage distance no less than 30.2m.

It was reported in that 64 units 210kN cap and pin insulators can be used for ±800kV I insulator string in light pollution level according to the principle of same creepage distance, and 67 units 210kN cap and pin insulators can be used according to principle of the same spacing height. If composite insulator is used, the spacing height and creepage distance are suggested to be 80%

length of 210kN cap and pin insulators. The external insulation arrangement design in much light pollution level should be same as light pollution class.

The configuration scheme of lightning arrester protection for ±800kV converter station, the principle of insulation coordination and over-voltage protection strategy for converter station were analyzed; the parameters and characteristic of lightning arresters were calculated; after analyzing the over-voltage protection for equipment and insulation level of equipment in detail, the discharge voltage of air clearances in converter station was given preliminarily.

(3) Live line working: The live line working research works for 1000kV AC transmission line were firstly and systematically carried out in China at WHVRI of the SGCC. The research results indicate that the live line working of 1000kV AC transmission line in China is feasible and safe.

The minimum approach distance and combined clearance at different system over-voltage level, altitude above sea level, working conditions, on side phase, middle phase and tension string were studied respectively.

Power frequency breakdown test, power frequency withstand voltage test and switching impulse flashover test of portable protective clearances were carried out. The maximum portable protective clearances correlated to different altitude were calculated according to the test results. Various actual work conditions were simulated in 1:1 tower window, and the switching impulse discharge test concerning insulation matching between portable protective clearances and working clearance was carried out by up and down method. The research results indicate that insulated tools made in China can satisfy the requirement of 1000kV AC transmission line live working[11].

Suitable full set shielding clothes were also developed, and then characteristics of material that the shielding clothes made of and the ready-to-wear were tested in accordance with national standards of China live line working. Measuring of electrical field intensity in and outside of shielding clothes at different part of body while climbing tower and during equal-potential process, the electric field intensity inside the screening shielding clothes were 0.4−10kV/m, 8.4−137kV/m inside the mask, and the current through the body of equipotential operator was 32μA[12]. The arc test and impulse current measuring while during equal-potential process were also carried out. The tests results indicate that the shielding clothes developed have good performances on electrical field shielding, current splitting and voltage sharing, thus meets the requirements for the safety protection[13].

(4) Lightning performance: Lightning performance of UHVAC transmission line and lightning invaded wave over voltage of UHV substations were studied. The measures to improve shielding performance were recommended[14].

The shielding performance of ground lines and back flashover were studied at CEPRI of SGCC. The shielding angles of typical UHV AC towers in China were recommended[15].

Some special design problems on direct lightning stroke shielding of 1000kV AC substation were calculated and discussed. Lighting invaded wave protection of typical substations was also simulated to suggest the number and location of MOA in substation. For different station operating conditions, the safety operation guideline was calculated. According to the calculation results, the MOA installation schemes were put forward for ensuring safety operation of every station in 1500−2000 years.

The striking distance factor for high tower was studied. The results showed that the striking distance factor (β) would reduce with the height (H) of tower increasing, and the value of lightning current did not affect the striking distance factor. An empirical formula was proposed (H: $\beta = 1.18 - H/108.69$), and then the striking distance factor was introduced to the improved electric-geometry model (EGM) to analyze the lightning protection performance of shielding failure for UHV transmission line. The lightning performance of Yun-Guang ±800kV UHV DC transmission line was also analyzed.

(5) Electromagnetic environment: The electromagnetic environment for 1000kV AC power transmission line such as power frequency electric field, power frequency magnetic field, RI and AN were studied. ±800kV DC ion flow density, combined field strength at ground surface, RI and AN were also studied. And the recommended control criteria were proposed. The control criteria can be satisfied by using multi-bundled and large section conductors and increasing the lowest phase conductor to the ground surface. Based on the control criteria, the $6\times720mm^2$ conductors can be used when the transmission capacity is 6400MW, 45cm bundle spacing can be used; $6\times800mm^2$ conductors can be used when the altitude is above 2600m, however, the $6\times720mm^2$ conductors can still be used when the distance between the two poles increased properly[16].

The tests for critical corona onset voltage of the heat resistant diameter expanded flexible conductor, the bundle conductor and the tubular bus bar, the visible corona onset voltage of typical grading ring for transmission line and insulator strings were carried out by using UHV test line section of WHVRI. The relationship between corona voltage and the height of conductors, and the minimum of the grading ring's diameter were also achieved.

(6) Relay protection: The development of a test environment based on a real-time digital simulator (RTDS) of a UHV power system model and the results of testing a distance relay using the model was presented[17]. The conclusion that the protective zone will be enlarged in the UHV system with shunt reactor compensation and reduced without the compensation was drawn. The suggestion that new protective algorithms should be developed and examined was proposed.

The control and protection system differences between UHV DC project and conventional HVDC project as well as the special requirements of UHV DC project were analyzed. The integral structure, control strategy, hierarchy of the structure and redundancy, distribution of control functions and configuration of the protection, etc. of the control and protection system, were carried out. Then a possible integral scheme of control and protection system for UHV DC project was put forward. Simulation results show that the proposed control strategy can completely satisfy the requirement of the design for HVDC power transmission system, the faulty 12-pulse converters can be reliably deblocked to ensure continuous operation of UHV DC system[18-20].

A novel transient based protection for ±800kV DC transmission lines was proposed based on a ±800kV DC bipolar model that was built with the PSCAD/EMTDC software. Wavelet-multiresolution signal decomposition technique was applied to analyze the transient voltages. Based on spectral energy distribution of transient voltages, the criteria were presented to distinguish the ±800kV line faults from other transient phenomena.

Some new protection principles and actions, which include a criteria of differential current between bi-pole under bipolar operation mode and an action of switching to metallic return mode under unipolar operation mode, in the bi-pole area of ±800kV DC system was put forward.

(7) Sharing earth electrodes: Sharing earth electrode was firstly proposed in China to make it

easier to choose the electrode sites for the close distance between multi-converter stations, the technical and economic feasibility were studied thoroughly and approved.

The different operation modes of the UHV DC system sharing earth electrode from the aspect of power system stable operation were discussed. Two UHV DC system rectifiers and inverters sharing earth electrodes were studied respectively by electromagnetic transient analysis software EMTP-RV. The influence on the normal operation of UHV DC system sharing earth electrode was simulated and analyzed. The merits and demerits of sharing earth electrode were also summarized in detail.

The analysis of the choice of burial depth and the effects on step voltage, earth resistance and current density caused by multi-ring electrodes with equal or unequal depth were studied. The conclusions was drawn that the maximum depth of electrodes buried in soil should be controlled less than 4m; the position of inner rings has little influence on the running parameters; the model of triple concentric-ring electrodes laid with different depth in the upper soils with a smaller resistivity could be more economical in the investments on the premise that all parameters are within their permissible running limits. The feasibility of sharing earth electrodes by two or three converter stations was also validated.

(8) Tower and truss tests: EPCRI of SGCC has built the aeolian vibration laboratory, stranded wire fatigue laboratory, heavy current laboratory, mechanical property laboratory and transmission line galloping laboratory, electrical and mechanical performance tests of conductor and fittings for 1000kV transmission line can be carried out with the testing facilities.

Tower test station of EPCRI of SGCC has the test capability of 1000kV single circuit tower, 750kV double circuit tower and 500kV multi-circuit towers. The maximum height of test tower is 100m; the maximum span of tested tower foot is 30m; the maximum uplifting force per tower leg is 10000kN; the over-turn torque is bigger than 240000kN • m.

ZM2 straight tower strength tests: Strength tests of the first UHV true type ZM2 straight tower in China under fourteen working conditions were successfully carried out from September 28 to October 2, 2006, in the tower test station of EPCRI.

The head of the ZM2 straight tower is cat head shape, the nominal tower height is 59m, the total tower height is 79.3m, the root span is 16.66m, and the weight is 59t. High strength Q420 steel is used in the main part of tower body and lower bent arm.

The tests under fourteen working conditions, including the wire breakage fault working condition, installing working condition, normal operation working condition, etc., were carried to test the stress characteristics of truss member, stress transfer relationship between truss members and the reasonability of joint structure.

The 60° strong wind overload test, testing the ultimate bearing capacity of tower under strong wind, was carried out on October 2, 2006. The testing load was increased from 0 to 100% design load, and then was increased step by step according to 5% design load. The fracture phenomena was occurred on the tower when the testing load was increased from 120% to 125% design load, then the tower fell down. This was identical with the design and calculated results.

Transformation truss 1∶2 model strength tests: Transformation truss 1∶2 model (Figure 8-3) truetypestrengthtests of 1000kV AC test and demonstration project were successfully carried out from April 15 to April 18, 2007. The height of the outlet line beam is 27.5m, the height of busbar beam is 19m, and the horizontal span is 7.5m. The testing transformation truss is made up of steel pipe, and the maximum design with stand wind velocity is 25m/s.

SZ1 straight tower strength tests: UHV true type SZ1straight tower strength tests (Figure 8-4) were successfully carried out on August 17, 2007, in the tower test stationof EPCRI. This SZ1 tower is made up of Q460 high strength steel, and this is the first time use of Q460 high strength steel in power transmission tower.

Figure 8-3 Transformation truss strength tests of 1000kV AC test and demonstration project (1 : 2 model)

Figure 8-4 SZ1 straight tower strength tests

8.3 Equipment Manufacturing

Research institutions and equipment manufacturers conduct in-depth investigations on the design and manufacture of UHV equipment. Over the past several years, great breakthroughs have been made in the development and manufacture of UHV equipment. Performance of equipment test, equipment research, development and manufacture of 1000kV AC and 800kV DC transmission and transformation projects have been carried out.

8.3.1 UHV AC Equipment Manufacture

(1) Transformer: One type of UHV AC oil-immersed 250MVA/1200kV testing power transformer which can be used in UHV test was developed; another type of UHV AC oil-immersed 610MVA/1700kV testing power transformer passed all the type tests successfully on July 8, 2007.

The first unit of 1000kV/1000MVA UHV AC transformer in the world passed the long time induced voltage (with partial discharge measurement) test on June 30, 2008. And the normal test, switching impulse test and lightning impulse test of this type transformer had been passed before.

(2) Reactor: The 1100kV/240MVar reactor passed all the test items on February 13, 2008. The 1100kV/320MVar reactor, the biggest capacity reactor in the world, passed all the test items on March 9, 2008. Configuration of twin columns for improving the magnetic leakage distribution was adopted.

BKD-200MVar/1100kV UHV shunt reactor passed all the test items on May 18, 2008. The reliability of insulation was improved by application of unique connection and winding insulation configuration.

Especially, the partial discharge, temperature rising, noise and vibration of these three types reactor all reached the advanced level in the world.

(3) Circuit breaker: The technique of 550kV single-break arc chamber with independent intellectual property right had been achieved. Based on two 550kV single-break arc chamber series, internationally advanced 1100kV double break circuit breaker can be developed.

All the type tests of one typical type of 1100kV circuit breaker, which has double-break chamber, closing resistor and capacitor are paralleled for each break chamber, each phase has a hydraulic operating mechanism, will be finished in 2008. LW10-1100kV SF_6 circuit breaker was also developed.

(4) Switch: Prototypes of 1100kV disconnecting switch (DS) and earthing switch (ES) in China were assembled and tested.

The insulation level of one typical type 1100kV/63kA ES developed in China has the technical characteristics that the lightning impulse voltage is 2680kV; the switching impulse voltage is 1860kV; and the power frequency withstand voltage is 1230kV.

(5) Bushing: The bushing of 1100kV capacitive transformer/reactor were successfully developed in February, 2008.

The technical parameters of one typical type 1100kV bushing developed in China are: the rated voltage is 1100kV; the rated current is 2000A; the lightning impulse withstand voltage is 2400kV; the switching impulse withstand voltage is 1960/1800kV; and the power frequency withstand voltage is 1200kV (5min).

1100kV OIP type condenser bushing developed in China is with porcelain insulator, and the tan $\delta \leqslant 0.4\%$; the partial discharge$<$10pC under 1100kV voltage.

(6) Insulators: Normal type porcelain insulator, normal type glass insulator, double-shed porcelain insulator and trished porcelain insulator were developed and manufactured. 1000kV composite insulators were successfully developed.

The technical parameters of typical 1100kV post insulators made in China are: the lightning impulse withstand voltage is 2550kV; the switching impulse withstand voltage is 1800kV; the power frequency wet withstand voltage is 1100kV; and the bend failing load is 12.5kN and 16kN respectively.

(7) Arrester: The technical parameters of one typical 1000kV arrester are: the maximum current non-uniform coefficient in multi-columns of disks is 5%; the maximum potential distribution non-uniform coefficient is 1.17 (with grading capacitor configuration); the energy absorption capability is 40MJ; and the 4/10μs withstands impulse current is 4$\times$100kA.

The developed high capacity resistance piece can completely satisfy the requirements with enough margin for 1000kV UHV AC GIS and porcelain-clad metal oxide arrester. The porcelain-clad arrester was the initiate product on the world. Three GIS and twelve porcelain-clad metal oxide arresters had been assembled and tested on August 8, 2008 and before.

(8) Current transformer, voltage transformer: Current transformer (CT) and voltage transformer (VT) were also developed.

The developed typical 1000kV capacitor voltage transformer (CVT) has the technical characteristics of that the tan $\delta \leqslant 0.07\%$; the partial discharge level$\leqslant$3pC; and the anti-corrosion performance is excellent. All the test items had been passed before December 2006.

Eighteen 1000kV CVTs of 1000kV Jindongnan-Nanyang-Jingmen test and demonstration line were successfully passed the on-site experiments from September 22 to October 8, 2008.

(9) Fittings: The UHV AC transmission line hardware fittings include grading ring, yoke plate, spacer, suspension hardware fittings, tension hardware fittings, jumper line hardware fittings, protection hardware fittings, antigalloping device, vibration damper etc. developed by some corporations in China. The performance testing of some hardware fittings had been carried out at WHVRI of SGCC (Figure 8-5).

Figure 8-5 Performance test of some UHV hardware fittings at WHVRI outdoor test yard

(10) HGIS: The first 1100kV HGIS, developed with the cooperation of ABB had been successfully manufactured in June, 2008.

8.3.2 UHV DC Equipment Manufacture

(1) Converter transformer: The parameters of developed ZZDFPZ-321100/500kV converter transformer are: ±(400+400) kV DC rectifier, the rated power is 321.1MVA; the impedance is 18%.

Line side insulation level: the lightning impulse withstand voltage (full wave) is 1550kV; the switching impulse withstand voltage is 1175kV; the power frequency withstand voltage is 680kV. Valve side insulation level: the lightning impulse withstand voltage (full wave) is 1800kV; the switching impulse withstand voltage is 1600kV; the power frequency withstand voltage is 921kV; the applied DC withstand voltage is 1271kV (120min).

(2) Converter valve: The preliminary research and development of ±800kV converter valves had been carried out actively based on the ±500kV converter valve technologies and has obtained the intermediate achievements as follows: the electrical design of the valve section; the mechanical components design of the valve section; the thermal design of the valve section; the technical specification of valve-based electronic equipment (VBE).

(3) Smoothing reactor: Technical parameters of manufactured ±800kV dry type smoothing reactor are: the rated voltage is 800kV; the rated current is 4000A; the inductance is 75mH; the lightning impulse withstand voltage (full wave) is 2100kV; and the applied DC withstand voltage is 960kV (120min).

Technical parameters of another type ±800kV/4000A smoothing reactor primary layout are: the rated DC current is 4296A; the rated inductance is 75mH; the insulating thermal endurance grade of turn insulation is H, and holistic insulation is F; the temperature rise of average is 79K, and the hotspot is 90K; the dry-arcing distance on coil surface is 4300mm; the lightning impulse withstand voltage is 1225kV (between terminals); the switching impulse withstand voltage is 1005kV (between terminals); the diameter is 5000mm.

(4) Thyristor: Through technology import and domestic innovation, the prototype of 6inch ultra high power electric triggered thyristors for UHV DC applications had been independently developed, which is under perfection through tests and studies.

The developed prototype of 6 inches thyristor (KPE4000-80), has the excellent technical characteristics: outstanding dynamic performance; advanced electron irradiation technology; evaporation of thick aluminum layer technology; unique chip packaging technology; the design of explosion-proof packages. 5 inches thyristor module for UHV DC transmission was also developed, and its rated current is 3125A. The thyristor module passed the routine test according to the test specification of the thyristor valve. The test results meet the requirements of ±800kV UHV DC thyristor valve.

(5) Insulators: ±800kV DC composite insulators were also successfully developed. 16 kinds of different shed structure ±800kV DC composite insulators sample were manufactured and large amount of optimal DC pollution flashover tests were carried out by Shenzhen Graduate School of Tsinghua University in the extra high voltage (EHV) test base of Yunnan Electric Power Research Institute (YEPRI). The most optimal structural parameters of DC composite insulator were achieved and the pollution withstand characteristic of composite insulator was improved.

Notes and References

[1] C B. Electric power transmission at voltages of 1000kV and above: Plans for future AC and DC transmission, data on technical and economic feasibility and on general design, information on testing facilities and the research in progress. Electra, 1983, 91: 83-133.

[2] Scherer H N, Vassell G S. Transmission of electric power at ultra-high voltages: Current status and future prospects. Proc. IEEE, 1985, 73(8): 1252-1278.

[3] C B. Electric power transmission at voltages of 1000kV AC or 600kV DC and above: Network problems and solutions peculiar to UHV AC transmission. Electra, 1998, 122: 41-75.

[4] Cigre B. Ultra high voltage technology. Electra, 1994.

[5] Lings R, Chartier V, Maruvada P S, et al. Overview of transmission lines above 700kV. Inaugural IEEE PES 2005 Conference and Exposition in Africa, Durban, South Africa, 2005, 33-43.

[6] Guan Z C, Wang G L. The projects and related key techniques of ultra high voltage transmission in China. China Southern Power Grid Technology Research, 2005, 1(6): 12-18.

[7] Bisewski B, Rao A. Considerations for the application of 800kV HVDC transmission from a system perspective. International Workshop 800kV HVDC Systems, New Delhi, 2005: 1-7.

[8] Shu Y B, Hu Y. Research and application of the key technologies of UHV AC transmission line. Proc. CSEE, 2007, 27(36): 1-7.

[9] Fang K F, Wu G Y, Zhang R. Study on pollution performance of insulator string for AC transmission line of 1000kV. Insulators and Surge Arresters, 2007, 2: 7-11.

[10] Jiang X L, Yuan J H, Sun C X, et al. External insulation of 800kV UHV DC power transmission lines in China. Power System Technology, 2006, 30(9): 1-9.

[11] Liu K, Hu Y, Wang L N, et al. Research on portable protective gaps for live working on 1000kV AC transmission line. High Voltage Engineering, 2006, 32(12): 83-88.

[12] Shao G W, Hu Y, Wang L N, et al. Safty protection tools and measures of live working on 1000 kV AC transmission lines. High Voltage Engineering, 2007, 33(10): 46-50.

[13] Liu K, Wang L, Hu Y, et al. Research of safety protection for live working on 1000 kV ultra high voltage transmission line. High Voltage Engineering, 2006, 32(12): 74-77.

[14] Gu D X, Zhou P H, Dai M, et al. Study on lightning performance of 1000kV AC transmission project. High Voltage Engineering, 2006, 32(12): 40-44.

[15] Ge D, Du S C, Zhang C X. Lightning protection of AC 1000kV UHV transmission line. Electric Power, 2006, 39 (10): 24-28.

[16] Zhang W L, Lu J Y, Ju Y, et al. Design consideration of conductor bundles of 800kV DC transmission lines. Proc. CSEE, 2007, 27(27): 1-6.

[17] Wang B, Dong X, Bo Z, et al. RTDS environment development of ultra-high-voltage power system and relay protection test. 2007 IEEE Power Engineering Society General Meeting, Tampa, Florida, 2007, 1-7.

[18] Zhang M, Shi Y, Han W. Research on action strategy of UHVDC protection. Power System Technology, 2007, 31(10): 10-16.

[19] Zhang M, Shi Y, Sun Z. Influence of blocking and deblocking strategies of single 12-pulse converter group for UHVDC power transmission on reactive power impact to AC power grid. Power System Technology, 2007, 31(15): 1-7.

[20] Wang Q, Shi Y, Tao Y, et al. Simulation study on control strategy for balanced steady operation and block/deblock of dual 12-pulse converter groups in 800kV DC transmission project. Power System Technology, 2007, 31(17): 1-6.

Chapter 9 Advanced External Insulation Protection System

9.1 Transparent and Superhydrophobic Coating for the Solar Panels

After a rainy day, water droplets adopt a perfect round shape that roll on the surface of lotus leaf, and the rolling motion picks up and removes contaminants and dust. Inspired by this phenomenon, superhydrophobic surfaces with water contact angles greater than 150° and sliding angles below 10° have attracted tremendous attention because of their numerous applications in self-cleaning[1,2], anticorrosion[3,4], oil-water separation[5,6], and so forth. Extensive studies have revealed that the superhydrophobic surfaces result from the combination of rough surface with special micro-and nano-structure and low surface energy. A variety of artificial superhydrophobic surfaces have been developed by various approaches such as template-based methods, sol-gel process, lithography, layer-by-layer deposition, etc. However, these artificial surfaces tend to lose their self-cleaning properties when they are knife-scratched, abraded, or contaminated by oil[7].

Recently, there have been increasing interests in transparent superhydrophobic coatings which expand the range of potential applications such as windshields for automobiles, solar cell panels, safety goggles, and windows for electronic devices. Many research groups have made great efforts to enhance the mechanical stability and transparency of artificial superhydrophobic surfaces. For instance, Lai et al. fabricated transparent superhydrophobic TiO_2-based coatings by electrophoretic deposition technique[8]. Li et al. produced transparent superhydrophobic coating based on a nanoscale porous structure spontaneous assembly from branched silica nanoparticles. Xu et al. prepared transparent superhydrophobic coating by an electrochemical template and the following annealing at 500℃ for 2h[9]. Wang et al. fabricated superhydrophobic hybrid nanoporous coating by a simple solidification-induced phase-separation method. Nevertheless, to achieve transparent and abrasion-resistant superhydrophobic surfaces at the same time is still a challenge. The main challenge is that transparency and surface roughness are generally competitive properties. When surface roughness decreases, the transparency often increases because of less diffuse reflection of light, whereas the hydrophobicity often decreases. Moreover, the rough surface structures formed by nanoparticle assemblies and hybrid composites are prone to be destroyed because of the weak interaction between particles and substrates[10].

In this work, we present a simple method to form transparent superhydrophobic coating using PDMS interlayer and hydrophobic silica nanoparticle suspension. This transparent superhydrophobic coating exhibits self-cleaning properties in either air or oil environment. Furthermore, this transparent superhydrophobic coating retains their performances after many kinds of damage, including knife-scratch, sandpaper-abrasion and strong acid/base attack.

9.1.1 Materials

Hydrophobic silica nanoparticles (R972, ~16nm in diameter) were purchased from Degussa. 1H, 1H, 2H, 2H-perfluorooctyltriethoxysilane ($C_8F_{13}H_4Si(OCH_2CH_3)_3$, FAS) were purchased from Aladdin Reagent Co., Ltd., Shanghai, China. One component room temperature curing polydimethylsiloxane (PDMS) was provided by Ausbond Co., Ltd. (A6000-100mL, viscosity: 50000cps, relative density: $1.30g/cm^3$ at 25℃). Ethanol (≥99.7%), Hexane (≥99.5%), sulfuric acid (95.0% to 98.0%), and NaOH (≥96.0%) were purchased from Sinopharm Chemical Reagent Co., Ltd., Shanghai, China. All chemicals were analytical grade reagents and were used as received.

9.1.2 Preparation of Silica Nanoparticle Suspension

In a typical procedure, 0.2g of FAS was added into 9.8g of ethanol with stirring under ambient temperature. Then 0.4g of hydrophobic silica nanoparticle was added to form the final suspension. The introduction of FAS could further enhance the hydrophobicity of silica nanoparticle through immobilizing fluoroalkylsilane groups on their surfaces, as shown in Figure 9-1.

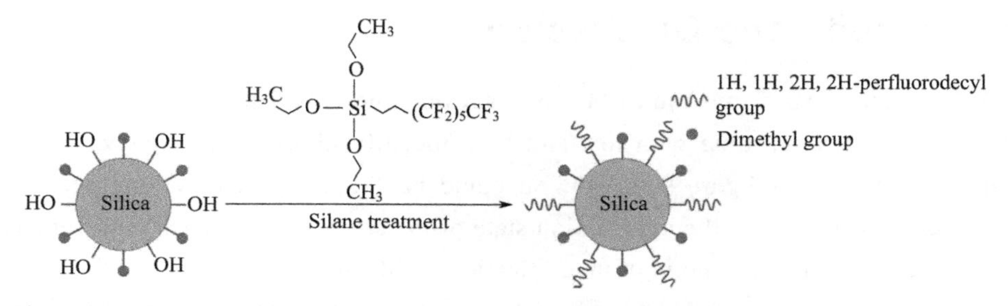

Figure 9-1 Schematic illustration of the synthesis of fluoroalkylsilane treated silica nanoparticle

9.1.3 Preparation of Superhydrophobic Coating

With the help of ZBQ four sides wet-film preparation device, PDMS interlayer was uniformly coated onto the glass substrate. After the PDMS interlayer was partially cured for 30min (25℃ and ~35% humidity), the silica nanoparticle suspension (SNS) was painted onto the PDMS interlayer. In order to further control the thickness of SNS, Japan OSP scraping ink bar was introduced. Because the PDMS interlayer was partially cured and soft, the scraping load was applied only by the weight of the scraping ink bar. The scraping velocity is 8m/min, and then a very thin suspension liquid layer could be deposited on the PDMS interlayer coated glass. After the solvent quickly evaporated, a layer of silica nanoparticle was deposited onto the partially cured PDMS interlayer. The function of PDMS interlayer was to promote the robustness of superhydrophobic coatings. The process is illustrated in Figure 9-2. Finally, the sample was set at room temperature for 24 hours before use in order to let the PDMS interlayer compeletely cured.

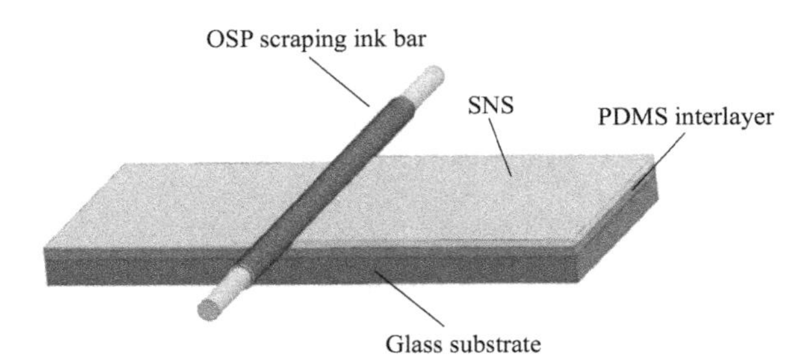

Figure 9-2 Schematic illustration of OSP scraping ink bar assisted SNS coating

9.1.4 Characterization

Scanning electron microscopy (SEM) was performed to measure surface morphology using a Hitachi S-3400N SEM at an accelerating voltage of 15kV. Prior to SEM measurements, a thin Au layer (ca. 5nm) was deposited on the specimens by sputtering. Atomic force microscopy (AFM) test was carried out using a Bruker Multimode 8 AFM in DFM mode. The water contact angles (CAs) and sliding angles (SAs) were measured with a SL200B apparatus at ambient temperature. 5μL water droplet was adopted in all measurements. The average water CA values were obtained by measuring the same sample at five different positions. Transmittance was measured using a UV-vis spectrophotometer (754, Shanghai Hengping Technology Co., Ltd., CA).

9.1.5 Results and Discussion

Figure 9-3(a) and (b) shows the SEM top-images of fluoroalkylsilane treated silica nanoparticles (FTSN) deposited substrate with low and high magnifications, respectively. From the low magnification SEM image (Figure 9-3(a)), it could be found that the coating surface was not smooth. It has been found that the aggregation state of the silica nanoparticles in the ethanol would lead to microscale roughness. The high magnification SEM image (Figure 9-3(b)) exhibits fractal-like nanoporous structures. Numerous void spaces among nanoparticles are observed, and nanoscale roughness is evident. The silica nano-particles ranged in size from dozens of nanometers to several hundreds of nanometers. The microscale roughness together with nanoscale roughness forms hierarchical structures, thus leading to superhydrophobicity. The surface uniformity of the sample was further assessed using AFM technique. It can be observed from the 2D AFM image that the surface has been coated homogeneously Figure 9-3(c). The RMS roughness value of FTSN deposited substrate can be calculated from the software, about 43.8nm, indicating the sub-100nm roughness of the coatng. Figure 9-3(d) and (e) shows the cross-sectional images of the sample. According to the relationship between the nanoparticles and PDMS interlayer, the nanoparticles were divided into three catalogs. The floating nanoparticles (blue nanoparticles in Figure 9-3(f)) did not embed into the PDMS interlayer at all, and the interparticle forces (van der Waals force and hydrogen bonds) contributed to their formation. The partially-embeded nanoparticles (yellow nanoparticles in Figure 9-3(f)) are the key components of the sample, and the support that the PDMS interlayer offered to the partially-embeded nanoparticles led to the robustness of the sample. The partially-embeded nanoparticles and the fully-embeded nanoparticles (gray nanoparticles in Figure 9-3(f)) showed that the embedment between nanoparticles and PDMS interlayer occurred during the formation of sample. We attributed this to the partially curing of PDMS when coating the silica nanoparticle suspension. As shown in Figure 9-4, the OSP scraping ink bar has two functions. One function is to achieve the uniform thin solution film with controlled thickness, which has been widely used in industry. Furthermore, as the PDMS interlayer was partially cured and soft, the scraping load offered by the weight of OSP scraping ink bar would embed the nanoparticles into the partially cured PDMS. Then, after evaporating the solvent (ethanol) and fully curing the PDMS, the robust superhydrophobic coating could be achieved because of partially-embeded nanoparticles.

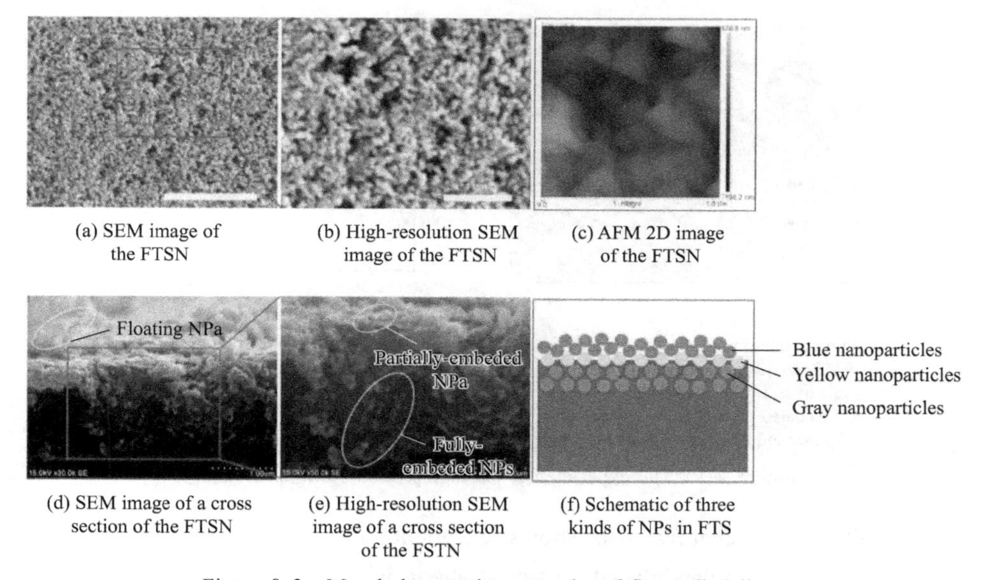

(a) SEM image of the FTSN
(b) High-resolution SEM image of the FTSN
(c) AFM 2D image of the FTSN
(d) SEM image of a cross section of the FTSN
(e) High-resolution SEM image of a cross section of the FSTN
(f) Schematic of three kinds of NPs in FTS

Figure 9-3 Morphology and topography of fluoroalkylsilane treated silica nanoparticles (FTSN) deposited substrate

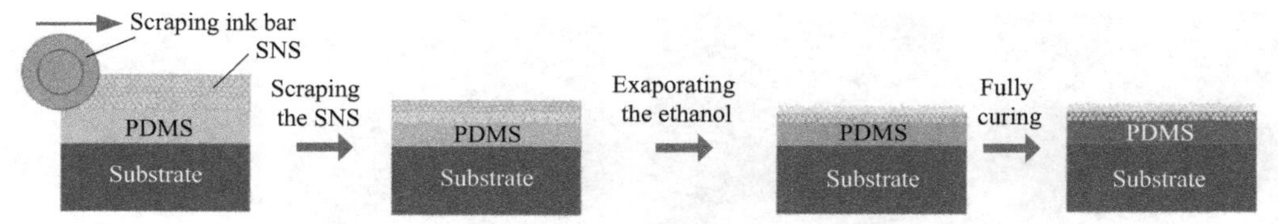

Figure 9-4 Schematic of formation mechanism of this superhydrophobic coating

The FTSN deposited substrate displayed transparency and superhydrophobic properties simutaneously (Figure 9-5(a)). The good transparency of the large-area superhydrophobic coating (about

$5 \times 2.5 cm^2$) could be directly proved by showing the clarity of letters beneath the FTSN deposited glass substrate (Figure 9-5 (a)). Many randomly distributed water droplets (~0.01ml) on the transparent coating showed nearly spherical shape, further indicating its good superhydrophobic properties. To quantitatively study the wetting state, we measured the water contact angles and sliding angles. The FTSN deposited glass substrate showed a high water contact angle of $163° \pm 1°$ (Figure 9-5 (a)). Meanwhile the water droplets don't come to rest on the surface when the sliding angle is 3°, suggesting low adhesion between water droplet and sample surface. The high contact angle and low sliding angle of the nano silica deposited paper indicates that the water droplets are suspended on not penetrate in the nanopores of the paper surface as the Cassie-Baxter state described. Superhydrophobicity together with transparency provides a coating with favorable optical applications. As evidenced by UV/V is transmittance spectra (Figure 9-5 (b)), the transmittance of the coating is reduced by less than 10% as compared to that of pristine glass for wavelengths above 450nm, which indicates that it is transparent to visible light. Furthermore, many researchers have found that the concentration of silica nanoparticles in the suspension affected. Therefore, the concentration of silica nanoparticles need to be further optimized for the specific application because transparency and surface roughness are competing properties.

(a) Photograph of water drops deposited on an FTSN coated glass slide with a contact angle of 163° and a sliding angle of 3° (The coated glass slide was placed on a printed paper,and the water was dyed by blue ink)

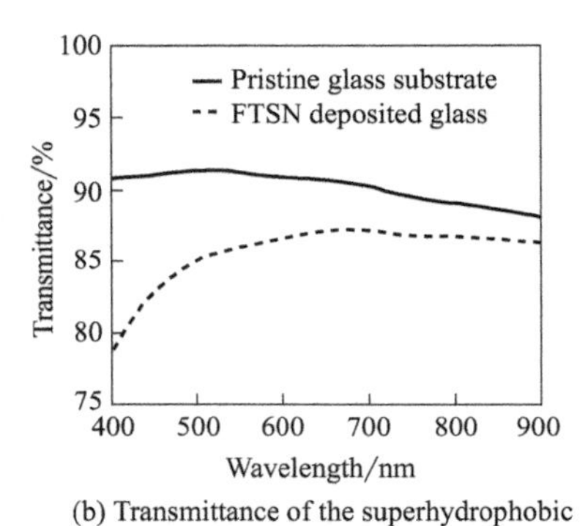

(b) Transmittance of the superhydrophobic coatings

Figure 9-5 The experimental photograph and the transmittance

This superhydrophobic coating also shows excellent self-cleaning property. As shown in Figure 9-6 (a), the substrate was placed on top at about 30° relative to horizontal line. The self-cleaning performance was investigated by applying soil as contaminant on the substrate. Water droplets were then poured onto the surface to test the self-cleaning property of the FTSN deposited

(a) The surface with graphite powder as a model of contaminant

(b) The contaminated surface with water drops on it

(c) The contaminated surface after water pouring process

Figure 9-6 Self-cleaning process on the FTSN deposited glass substrate

glass by removing the simulated contaminant (Figure 9-6(b)). When the water droplets contacted the soil powders, they just rolled off and carried the graphite powder away. After water pouring processes, water droplets brought away the contaminants completely and left a clean surface (Figure 9-6(d)).

Because most superhydrophobic surfaces are prone to lose their water repellency when even partially contaminated by oil, there were few researches focused on self-cleaning tests in oil. The reason is that the surface tension of the oil is lower than that of water, leading to the oil penetrating through the surfaces. Superamphiphobic surfaces may offer an effective route to solve this problem. However, superamphiphobic surfaces can not be applied in some instances which require both self-cleaning from water repellency and a smooth coating of oil, such as lubricating bearings and gears. Lu et al. prepared a kind of nanoparticle paint which retained self-cleaning properties in oil, but their paint was non-transparent. It was found that the superhydrophobic coating in this study exhibited excellent optical properties as well as self-cleaning properties in either air or oil. As shown in Figure 9-7(a), water droplets still maintained nearly spherical shape on the FTSN deposited surface when immersed in oil (hexane), which indicated that the surfaces would retain their superhydrophobic properties after being immersed in oil. As shown in Figure 9-7(a) and (b), a self-cleaning test was carried out the FTSN deposited glass substrate which was contaminated by oil (hexane), indicating self-cleaning property was retained even after being contaminated by oil. A dirt-removal test was further carried out on the FTSN deposited glass substrate to examine the self-cleaning property both in oil and air (Figure 9-7(c)). The treated glass substrate was partly inserted into oil; dirt (soil) was also sprinkled partly in oil and air onto the surface. Water droplets were poured in order to remove the dirt both in air and oil (Figure 9-7(c)). After the water pouring process, a clean surface was left (Figure 9-7(c)), indicating that self-cleaning property was retained both in air and oil environment.

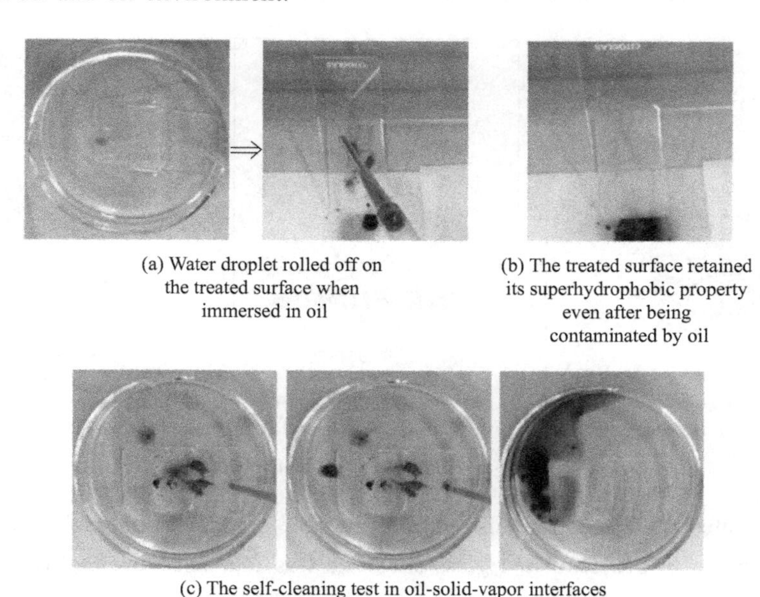

(a) Water droplet rolled off on the treated surface when immersed in oil

(b) The treated surface retained its superhydrophobic property even after being contaminated by oil

(c) The self-cleaning test in oil-solid-vapor interfaces

Figure 9-7 Self-cleaning tests after oil-contaminations

The surface robustness is a key factor which affecting the pratical application of the superhydrophobic surfaces. In this work, the knife-scratch test and sandpaper abrasion test were performed to evaluate the mechanical robustness of the as-prepared superhydrophobic surface. The methodology of knife-scratch test is illustrated in Figure 9-8(a). A knife was adopted to scratch the FTSN deposited glass substrate along the path of red dashed lines to examine the surface robustness. The FTSN deposited glass substrate retained the superhydrophobic and self-cleaning properties after the knife-scratch test. Moreover, sandpaper abrasion test was reported to be an effective route to evaluate the mechanical abrasion resistance of the superhydrophobic surfaces. In this research, a similar sandpaper abrasion test was carried out using 150 grit SiC sandpaper as an abrasion surface, as shown in Figure 9-8(b). The FTSN deposited glass with a weight of 50g above it was put face-down to sandpaper and moved for 10cm along the ruler (Figure 9-8(b) and Figure 9-9(a)); the sample was then contrarotated by 90° (face to the sandpaper) and moved for 10cm along the ruler (Figure 9-8(b) and Figure 9-9(a)). The three steps shown in Figure 9-8(b) is defined as one abrasion cycle, which guarantees the surface is abraded longitudinally and transversely in each cycle. Figure 9-9(b) shows water contact angles after each abrasion cycle. It can be found that the water contact angles were between 150° and 165°, which indicated that superhydrophobicity property was retained even after 9 sandpaper abrasion cycles. During the first cycle of abrasion test, some white powders might be found on the sandpapers. However, these white powders were hard to find from the second abrasion test cycle. Based on the cross-sectional SEM images of the sample, the white powders were assumed to be floating nanoparticles. Because the interparticles forces between nanoparticles were weak, they were easily abraded. Then, the partially-embeded nanoparticles tended to afford the abrasion from the second cycle. The PDMS interlayer would offer physical support to the partially-embeded nanoparticles, protecting them from being abraded. In this study, the maximum abrasion distance the superhydrophobic coating could endure was about 1.80m, and it is comparable with other reported mechanically robust superhydrophobic surfaces.

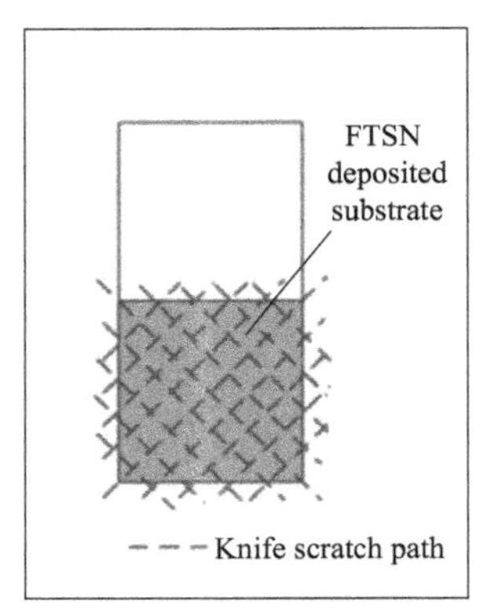

(a) Schematic of knife-scratch test

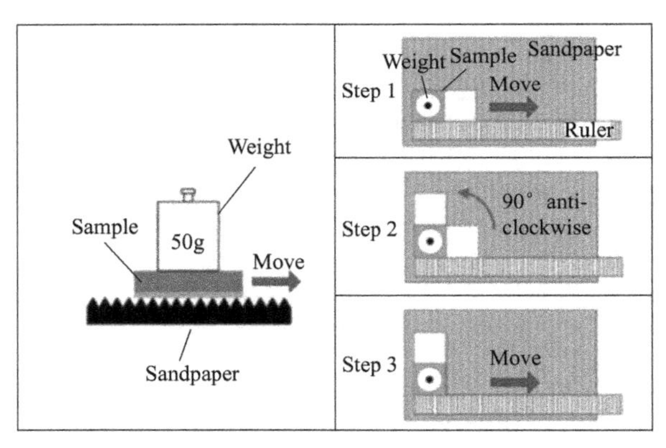

(b) Schematic of sandpaper abrasion test

Figure 9-8 Different experiment schematic

The FTSN deposited glass substrate also exhibited stable superhydrophobicity in strong acid and strong base. After the FTSN deposited glass substrate was immersed in 3mol/L H_2SO_4 or 3mol/L NaOH solution for 24 hours, it was rinsed with water and dried at room temperature for contact angle test. Figure 9-9(c) shows that the water droplets retained nearly spherical shape either after strong acid or strong alkali treatment. The contact angles change to $163^\circ \pm 3^\circ$ and $162^\circ \pm 2^\circ$ after 24 hours of immersing in 3mol/L H_2SO_4 or 3mol/L NaOH solutions, respectively.

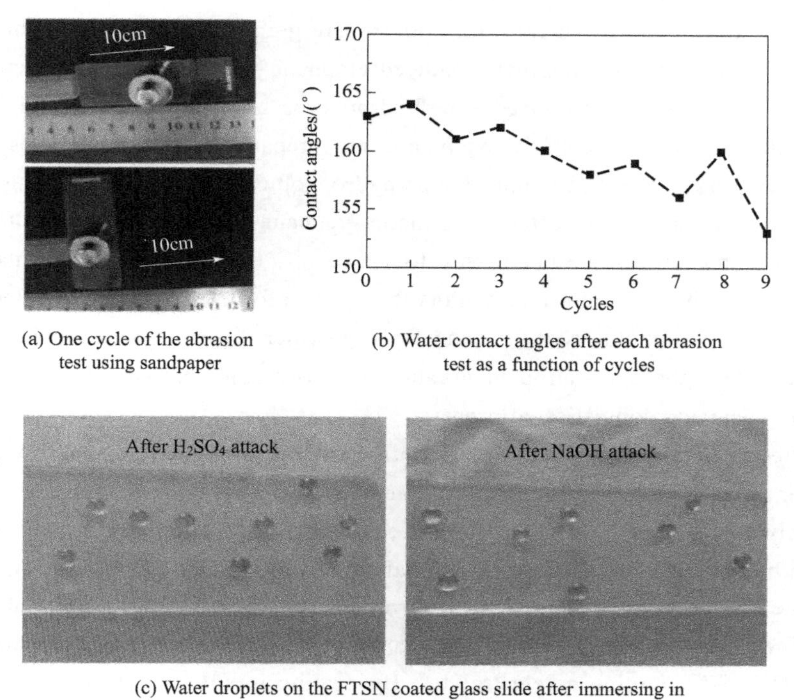

(a) One cycle of the abrasion test using sandpaper

(b) Water contact angles after each abrasion test as a function of cycles

(c) Water droplets on the FTSN coated glass slide after immersing in strong acid or strong base solution for 24h

Figure 9-9 Experimetal process and phenomenon

9.2 The Superhydrophobic Graphene Coating for Anti-Corrosion Application

Superhydrophobic surfaces, which display water contact angles greater than 150° and sliding angles below 10°, have attracted ever-growing attention because of their numerous applications in self-cleaning, anti-corrosion, anti-icing, oil-water separation, and so forth. Extensive studies have revealed that the superhydrophobic surfaces result from suitable surface roughness and low surface energy. A large amount of artificial superhydrophobic surfaces have been developed by various approaches such as layer-by-layer deposition, template-based methods, lithography, sol-gel process, etc. However, these artificial surfaces tend to lose their self-cleaning properties when they are subjected to impact of water and dust particles. Moreover, fluoropolymers such as perfluorooctanoic acid and perfluorooctyltrichlorosilane are conveniently utilized to prepare superhydrophobic surface. Recent researches revealed that long-chain polyfluorinated compounds have a documented ability to bioaccumulation and potential adverse effects on human offspring. Therefore, to prepare nonfluorinated superhydrophobic materials are more perferable for their nontoxic and low-cost benign.

Graphene, a single-atom-thick sheet with 2D sp^2 honeycomb structure, exhibits many exceptional physicochemical properties such as high thermal conductivity, good thermal/chemical stability and exceptional mechanical modulus. Besides their outstanding physical and chemical properties, graphene materials have recently been reported to possess intriguing hydrophobic properties. Many research groups have made great efforts to prepare graphene superhydrophobic surface. For instance, Rafiee et al. obtained thermally reduced graphene oxide with tunable wettability by solvent adjustment. Dong et al. produced superhydrophobic graphene films by microwave plasma CVD using a patterned Si substrate[11]. Nguyen et al. prepared graphene by ultrasonically exfoliating expanded graphite, and then obtained superhydrophobic graphene-based sponges by dip coating method[12]. Wang et al. fabricated biomimetic graphene surfaces with both superhydrophobicity and iridescence through a facile interference technique[13]. Choi et al. exhibited the fabrication of superhydrophobic graphene/Nafion hybrid films through conformational rearrangement. Asthana et al. fabricated superhydrophobic coating using fluoropolymer dispersions of graphene/carbon black composite. However, the application of graphene superhydrophobic surfaces was limited by the high cost of graphene and sophisticated process.

Recently, electrochemical exfoliation of graphite has been presented to be a low cost method to prepare graphene and attracted more and more attention[14]. Parvez et al. fabricated solution-processable, highly conductive electrodes using electrochemically exfoliated graphene (EEG)[15]. Wei et al. exhibited that the EEG has the potential application in energy storage. In this work, we present a simple and cost-effective method to form superhydrophobic coating based on EEG. The EEG was obtained from graphite in a one-step electrochemical approach using graphite rods as electrodes and $(NH_4)_2SO_4$ solution as electrolyte. By coating mixtures of EEG and polydimethylsiloxane (PDMS), the nonfluorinated superhydrophobic surface was fabricated in just one step. This superhydrophobic coating exhibited excellenct self-cleaning and anti-corrosive properties. Moreover, it maintains superhydrophobic property after water-impact and sand-impact test.

9.2.1 Materials

Graphite rods (~10cm in diameter) were purchased from Beijing Electric Carbon Co., Ltd., China. Polydimethylsiloxane (PDMS) which is soluble in ethanol was obtained from Hubei New Universal Co., Ltd., China. Ethanol (≥99.7%), 3-Aminopropyltriethoxysilane (APTES), and NaCl were purchased from Sinopharm Chemical Reagent Co., Ltd., Shanghai, China. All chemicals were analytical grade reagents and were used as received.

9.2.2 Preparation of EEG

The EEG was prepared based on the previous method with some modification. In brief, two graphite rods were utilized to form a conventional two-electrode system. The electrolyte usded was a 0.1mol/L $(NH_4)_2SO_4$ solution, and the interelectrode distance was kept at around 30mm. The graphene was then synthesized by electrochemically exfoliation under a constant voltage of 10V. After the graphite exfoliation was completed, the product was air-dried.

9.2.3 Preparation of Superhydrophobic Composite Coating

The superhydrophobic coating was prepared using the following method. Firstly, 0.2g of PDMS was added into 10g of ethanol and the solution was stirred for 6 hours under ambient temperature.

Then, 0.2g of EEG and 0.02g of APTES was added to form the final suspension. The suspension was stirred for another 6 hours before use. In the following experiments, the solution was painted onto the substrate (Al 6061 or stainless steel) using a disposable Pasteur pipette (1mL), and cured at 60℃ for 6h.

9.2.4 Characterization

The morphology and structure of the samples were investigated by SEM (Hitachi, S-3400N) and AFM (Bruker, Multimode 8). Prior to SEM measurements, a thin Au layer (ca. 5nm) was deposited on the specimens by sputtering. The water contact angles (CAs) and sliding angles (SAs) were measured with a SL200B apparatus at ambient temperature. 5μL water droplet was adopted in all measurements. The average water CA values were obtained by measuring the same sample at five different positions. The polarization curves were measured in 3.5% (wt) NaCl aqueous solution through a CHI600D electrochemical workstation (Shanghai CH instruments). A three-electrode system was adopted. The sample and a platinum electrode were employed as the working and counter electrodes, respectively. A saturated calomel electrode (SCE) was used as reference electrode. The scanning rate of polarization curves was 1mV/s.

9.2.5 Results and Discussion

The EEG was prepared by the electrochemical exfoliation of graphite. As shown in Figure 9-10 (a), the exfoliation process was conducted in a two-electrode system using two graphite rods as anode and cathode. After immersing the graphite rods into the 0.1mol/L $(NH_4)_2SO_4$ solution, SO_4^{2-} and H_2O tend to be intercalated into graphite layers[14]. When the voltage of 10V was applied, vigorous bubbles generated and anodic graphite began to dissociate into electrolyte. We attributed to the reduction of SO_4^{2-} anions and self-oxidation of water, which produced produce gaseous species such as SO_2, O_2, and others. These gaseous species can exert large forces on the graphite layers, which are sufficient to separate weakly bonded graphite layers from one another[15]. Finally, monolayer graphene sheets were formed exfoliating the weakly bonded graphited layers. In order to investigate the quality of the fabricated EEG, AFM and Raman spectroscopy tests were performed. Figure 9-10(b) shows a typical AFM image of an electrochemically exfoli-

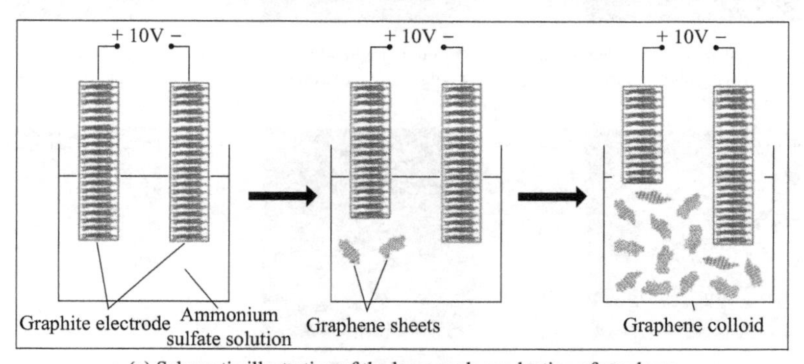

(a) Schematic illustration of the large-scale production of graphene sheets by electrochemical exfoliation using two graphite electrodes

Figure 9-10

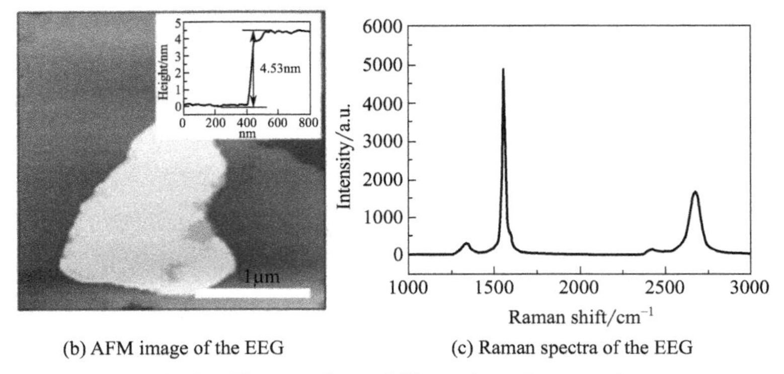

(b) AFM image of the EEG (c) Raman spectra of the EEG

Figure 9-10 The experimental illustration, image and spectra

ated graphene thin sheet (~4.53nm) drop-coated on a Si Substrate. Raman spectroscopy from EEG powders is shown in Figure 9-10(c). The D peak (~1336cm^{-1}) was caused by the breathing mode of the sp^2 carbon atoms and activated by the existence of defects such as structural disorders, functional groups, or edges. The ratio of I_D to I_G reflects the number of defects in graphene. In this study, the ratio of I_D to I_G is ~0.1, which is much smaller than chemically or thermally reduced graphene oxide (~1.1 - 1.5) and electrochemically exfoliated graphene (0.4) in acidic solution. The Raman spectrum and AFM test clearly exhibit the high quality of the synthesized graphene by electrochemical exfoliation methods in this research.

A paint was further created by mixing EEG in an ethanol solution containing PDMS and APTES. The introduction of PDMS has two functions. One is to offer low surface energy, and the other is to act as a binder to connect the EEGs with each other and with the substrate. According to the manufacturer of PDMS, APTES plays a role of improving the performance of PDMS. After extruding the paint onto the substrate, 6h at the temperature of 60℃ is needed for the evaporation of ethanol and the solidification of PDMS. Then, the superhydrophobic coating was obtained. As shown in Figure 9-11 (a), many randomly distributed water droplets (~10μL) on the coating

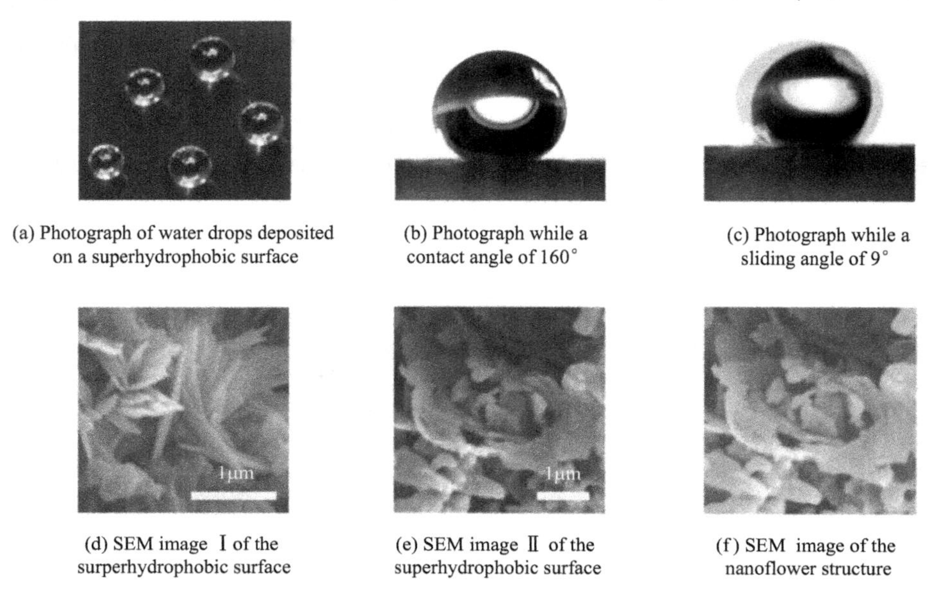

(a) Photograph of water drops deposited on a superhydrophobic surface (b) Photograph while a contact angle of 160° (c) Photograph while a sliding angle of 9°

(d) SEM image Ⅰ of the surperhydrophobic surface (e) SEM image Ⅱ of the superhydrophobic surface (f) SEM image of the nanoflower structure

Figure 9-11 The experimental photographs and images

showed nearly spherical shape, which indicates the good superhydrophobic property. To quantitatively study the wetting state, we measured the water contact angles and sliding angles. This kind of superhydrophobic coating shows a high water contact angle of 160°±2° (Figure 9-11(b)). Meanwhile the water droplets don't come to rest on the surface when the sliding angle is 9° (Figure 9-11(c)), suggesting low adhesion between water droplet and sample surface. The high contact angle and low sliding angle of this superhydrophobic coating indicates that the water droplets are suspended on not penetrate in the as-prepared surface as the Cassie-Baxter state described. SEM images were used to evaluate the structure of the superhydrophobic coating. Nanoplate structure can be found from Figure 9-11(d). What is more interesting, nanoflower structure can be found from Figure 9-11(e) and (f). Similar with ZnO, it was found that nanoplates and nanoflower structure could be form by EEG. These structures together with low surface energy offered by the PDMS lead to superhydrophobicity.

Most surfaces tend to get dirty owing to the accumulation of dust and dirt particles from the environment, and superhydrophobic coating is one potential solution. The self-cleaning performance of the graphene superhydrophobic coating was investigated using sand as contaminant dust particles (Figure 9-12). Firstly, a thin layer of sand was sprinkled onto the superhydrophobic coating (Figure 9-12(a)). Then, a water droplet was gently placed on this contaminated surface. The dust particles were immediately adsorbed on the surface of the water droplet as soon as they came into contact with the water droplet (Figure 9-12(b) and (c)). After several sliding and rolling processes, water droplets brought away the contaminants completely and left a clean surface (Figure 9-12(d)), which confirmed the excellent self-cleaning ability of the prepared graphene coating.

(a) The surface with sand powder as a model of contaminant

(b) The contaminated surface Ⅰ with water drops on it

(c) The contaminated surface Ⅱ with water drops on it

(d) The contaminated surface after water pouring process

Figure 9-12 Self-cleaning process on the graphene superhydrophobic surface

Not only excellent self-cleaning property, this superhydrophobic coating also exhibits intriguing anti-corrosive property. Recently, aluminum and its alloys have attained extensive industrial applications due to their cost efficiency, easy accessibility, high fatigue strength, high electrical conductivity, and excellent machinability. However, aluminum and its alloys corrode easily due to their high chemical and electrochemical activity. Here, the ability of the graphene superhydrophobic coating to protect Al6061 alloy from corrosion in a 3.5%(wt) NaCl aqueous solution was investigated by electrochemical experiment. Figure 9-13(a) shows the tafel polarization curves of bare Al6061 alloy substrate and superhydrophbic graphene composite coated Al alloy substrate. The corrosion potentials of the Al6061 substrate and graphene-coated surface were −0.731V and −1.141V, respectively. Moreover, the corresponding corrosion current density was 5.71×10^{-6} and 6.44×10^{-10} A/cm^2, respectively. It is worthwhile to notice that the corrosion current density of graphene coated Al alloy surface reduced 8866-fold form the bare Al alloy substrate. Such low current density indicates greatly improved corrosion resistance. This result is consistent with other researches that superhydrophobic surface can enhance corrosion resistance. When the graphene superhydrophobic surfaces are immersed in a corrosive solution, air is easily trapped in the pits and cavities of graphene composite, and the trapped air serves as a dielectric for a pure parallel plate capacitor and prevents the electron transfer between the electrolyte and the Al alloy substrate.

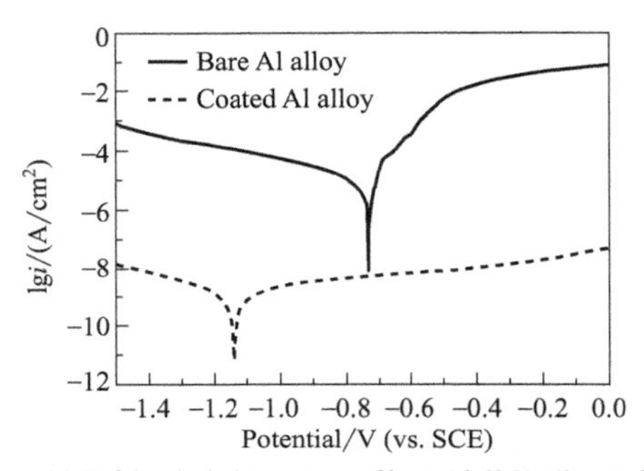

(a) Tafel polariztion curves of bare Al 6061 alloy and superhydrophobic graphene composite coated Al alloy

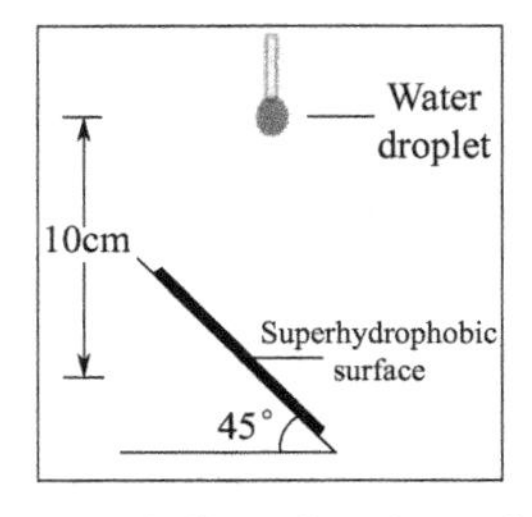

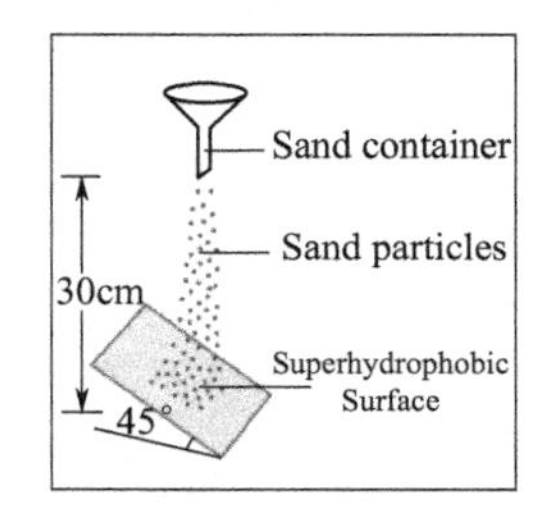

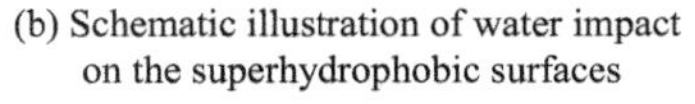

(b) Schematic illustration of water impact on the superhydrophobic surfaces

(c) Schematic illustration of sands impact on the superhydrophobic surfaces

Figure 9-13 The experimental curves and schematic illustration

For real applications in outdoor environments, superhydrophobic surfaces need to survive harsh conditions. To investigate the mechanical resistance of the graphene superhydrophobic coating, water-impact and sand-impact tests were performed. The methodology of long-term water-impact test is illustrated in Figure 9-13(b). The superhydrophobic coating was placed with a 45° tilted angle and water drops (~10μL) impacted the superhydrophobic surface from 10cm height. The graphene composite coating retained the superhydrophobic property after impinging 18000 water drops.

Moreover, the superhydrophobic surfaces tend to suffer from the impact of dust particles in daily life. To simulate the aforementioned situation, sand particles with a diameter of (150 ± 100) μm are impacted on the superhydrophobic surface at a height of 30cm in this research (Figure 9-13(c)). The graphene composite coating can maintain its superhydrophobicity even when the mass of the sand particles used were 50g.

9.3 The Superhydrophobic Steel for Anti-Corrosion Application

As well known, steel has achieved extensive industrial applications due to their easy accessibility, cost efficiency, high fatigue strength and excellent machinability. Nevertheless, steel is easily oxidized and highly susceptible to corrosion in damp environment. It has been founded that the corrosion of steel is partly due to the contact with air and moisture. One possible solution is to make the steel surface superhydrophobic, which not only provides corrosion resistance property but also adds self-cleaning function to the surface.

In nature, water droplets exhibit a nearly sphere shape on the surface of lotus leaves and immediately roll off because of low adhesion. The rolling motion carried away dust and contaminants, which shows self-cleaning property. Inspired with this phenomenon, superhydrophobic surfaces with large contact angle ($>150°$) and small sliding angle ($<10°$) have attracted a large amount of attention because of their potential applications in self-cleaning, oil/water separation, anti-icing, anti-corrosion, and so forth. Among these applications, to improve the anti-corrosion performance on metal surface is an important research direction. Many methods have been introduced to prepare superhydrophobic metal surface, such as electrodepositing film, nanocasting technique, coating SiO_2 or TiO_2 nanoparticle, and chemical etching. Because of the simplicity, cost efficiency, and suitable for industrial applications, chemical etching is thought to be an effective method for preparing large-area superhydrophobic steel surface. Wang et al. prepared a superhydrophobic surface on steel through the combined etching of H_2O_2 and HCl/HNO_3. Li et al. fabricated a superhydrophobic steel surface by HF/H_2O_2 etching. Nevertheless, preparing a superhydrophobic surface on steel using antiformin solution is still scarcely reported.

A crucial factor hindering the large scale application of superhydrophobic steel surfaces is their weak mechanical abrasion resistance. Although a large amount of superhydrophobic surfaces have been introduced, most of them are prone to be destroyed after a slight scratch, rubbing, and even finger contact. The reason may be that mechanical damage on the superhydrophobic surfaces tends to destroy the fragile micro-nano hierarchical structures. Recently, some researchers began to evaluate the mechanical durability. She et al. presented a robust superhydrophobic surface on magnesium alloy substrate, and the prepared surface showed a maximum abrasion distance of 0.70m under a 1.2kPa pressure. Wang et al. prepared a robust superhydrophobic steel surface by H_2O_2 and HCl/HNO_3 etching, which could not endure the abrasion distance over 1.10m under a 16kPa pressure. To the best of our knowledge, few studies about robust superhydrophobic steel surfaces under large pressure (24.5kPa) have been reported in the literature.

In this research, antiformin solution is introduced to prepare robust superhydrophobic steel surfaces. The fabrication process is composed from three steps: chemical immersion through the combination of antiformin and H_2O_2, ultrasonic treatment and surface modification. The present approach could be easily applied for large scale production, as the procedure is performed in an economical aqueous solution. In addition, the as-prepared surface exhibited superior anti-corrosion, anti-abrasion and self-cleaning properties, which get rid of the major neckbottle for its practical application.

9.3.1 Materials

1045 steel (composition: C, 0.42–0.50; Si, 0.17–0.0.37; Mn, 0.50–0.0.80) with the size of 20mm × 20mm × 1mm was used as the substrates. Ethanol (AR), acetone (AR), hexane (AR), H_2O_2 (30%(wt)) in water), antiformin (CP, free alkali 7.0%–8.0%, active chlorine ≥5.2%) were purchased from Sinopharm Chemical Reagent Co., Ltd., Shanghai, China. 1H, 1H, 2H, 2H-perfluorooctyltriethoxysilane ($C_8F_{13}H_4Si(OCH_2CH_3)_3$, FAS) were purchased from Aladdin Reagent Co., Ltd, Shanghai, China. Polydimethylsiloxane (PDMS) is a two part crosslinkable resin.

9.3.2 The Preparation of Micro Nano Roughness

Before using, steel substrates were ultrasonically cleaned in acetone, ethanol and deionized water, respectively, then dried in air for 30min. The etching solution was formed by adding 12.50mL antiformin solution into 10.00mL H_2O_2 solution. Then, the cleaned steel substrate was immersed gently into the etching solution for 4h. After reaction, the substrate was rinsed with deionized water, and then dried in air.

9.3.3 Ultrasonic Treatment

The steel substrate was immersed in deionized water, and then ultrasonically cleaned by a 240W ultrasonic cleaner for 30min. The ultrasonic cleaned steel was further dried in air.

9.3.4 Surface Modification

Firstly, the aforementioned sample was immersed in an ethanol solution which contains 2.0% (wt) FAS for 4h. Next, the sample was immersed in PDMS solution including 0.5g PDMS, 0.1g curing agent and 9.5g hexane. After 3min, the sample was taken out, and cured at 80℃ for 2h.

9.3.5 Characterization

The surface morphologies of as-prepared samples were observed with a scanning electron microscope (SEM, Hitachi S3400N). Prior to SEM measurements, a thin Au layer (ca. 5nm) was deposited on the specimens by sputtering. The water contact angles (CAs) and sliding angles (SAs) were measured with a 5L droplet of deionized water at ambient temperature on a contact angle measurement instrument (JC2000D, China). The average water CA and SA values were obtained by measuring the same sample at five different positions. The polarization curves were measured in 3.5%(wt) NaCl aqueous solution through a CHI600D electrochemical workstation

(Shanghai CH instruments). A three-electrode system was adopted. The sample and a platinum electrode were employed as the working and counter electrodes, respectively. A saturated calomel electrode (SCE) was used as reference electrode. The scanning rate of polarization curves was 0.5mV/s.

9.3.6 Results and Discussion

The surface morphology plays a key role in the preparation of superhydrophobic surfaces. Figure 9-14(a) and (b) show the SEM images of surface morphologies of untreated steel at low and high magnifications, respectively. Most areas of the surface were smooth, and a small amount of defects exist. After 4h of etching in the mixture of NaClO and H_2O_2 solution, the steel surface showed a porous structure. As shown in Figure 9-14(c) and (d), the pore size ranged from nanoscale to microscale. We attributed this to the existance of H_2O_2, which generated O_2 in the etching process. More interesting, ultrasonic treatment was adopted as a complementary texturing mechanism which can improve the surface's mechanical stability rather than degrade it. After ultrasonic treatment, micro sized cavities were presented on the surface as shown in Figure 9-14(e). Futhermore, nano sized features were created on the micro sized cavities as shown in Figure 9-14 (f). This hierarchical morphology extensively mimicks the lotus leaf surface, where the rough microscale stucture is covered by a nanoscale structure.

Chemical composition is another key factor to determine the superhydrophobicity of the surface. After ultrasonic treatment, the steel surface was further chemical modified to creat superhydrophobicity. The EDS spectrum of modified steel surface is displayed in Figure 9-15. From Figure 9-15(b),

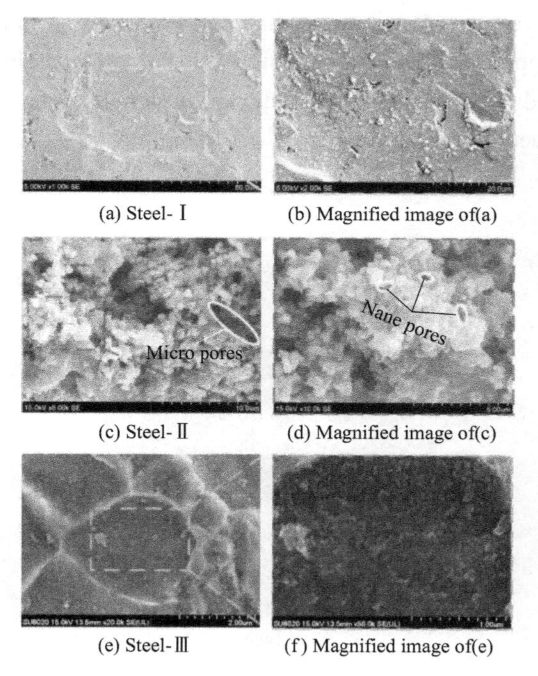

(a) Steel-Ⅰ (b) Magnified image of(a)

(c) Steel-Ⅱ (d) Magnified image of(c)

(e) Steel-Ⅲ (f) Magnified image of(e)

Figure 9-14 SEM images of bare steel (steel-I), chemical etched steel (steel-II), ultrasonic treated steel (steel-III)

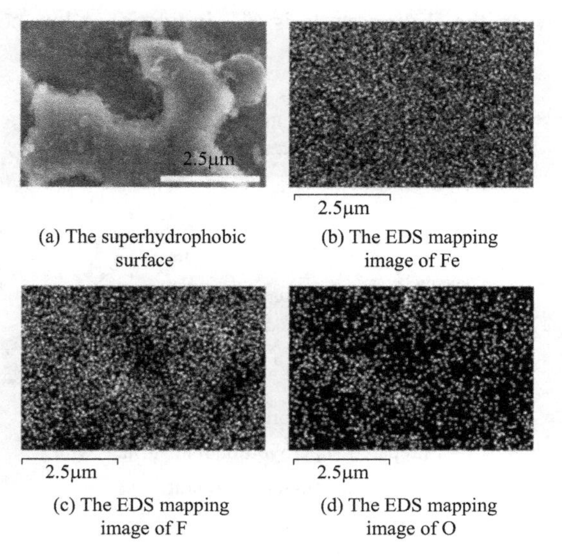

(a) The superhydrophobic surface (b) The EDS mapping image of Fe

(c) The EDS mapping image of F (d) The EDS mapping image of O

Figure 9-15 EDS mapping images of Fe, F, and O elements from the superhydrophobic surface

it is found that the Fe element exhibited a uniform distribution. The presence of fluorine element (Figure 9-15(c)) confirmed the successful grafting of the FAS molecules. Moreover, the F element showed a heterogeneous distribution, which is basically in coordination with the bright area shown in the SEM image. The O element also exhibited a heterogeneous distribution. Through the comparison of Figure 9-15(c) and (d), it is found that the distribution of O and F showed a consistency. Then it is hard to detect F element in the regions without O element. We attribute this to the—OH groups which existed on the oxidized surface, which is found to be the prerequisite for surface grafting of silane. Therefore, it is reasonable to deduce that the surface modification of FAS could only be performed at oxidized region, which is similar with other researches.

As known, both NaClO and H_2O_2 show strong oxidizing behaviour in the reacting solution, which means that Fe could be oxidized into Fe_2O_3. Due to the existence of defect and crystallinity in the steel surface, the oxidizing process is not uniform and micro-cavity structure tended to be formed. Moreover, a large amount of bubbles would generate in the etching process. We attribute this phenomenon to the existence of H_2O_2, which could generate O_2 gas in the oxidizing process. The generation of bubbles further led to the formation of porous structure in the steel surface. After ultrasonic treatment, the fragile structure was removed. Then the steel surface was changed to lotus-leaf-like hierarchical micro-nanostructures. After the surface treatment of FAS and PDMS, a layer with low surface energy covered the hierarchical micro-nanostructures. The formation mechanism of the as-prepared superhydrophobic surface with hierarchical structure is shown as Figure 9-16. In this research, the superhydrophobicity of the as-prepared surface derives from its rough surface with lotus-leaf-like hierarchical micro-nanostructures and the presence of low-surface-energy layer on it. Figure 9-17 shows the wetting state for water droplets with different diameters on the as-prepared superhydrophobic surface, while all droplets keep nearly spherical shapes. According to the contact angle measurement, this kind of superhydrophobic coating has a high water contact angle of $163° \pm 2°$ (Figure 9-18). In addition, the water droplets can easily roll down the sample surface when the sliding angle is 6°, indicating a low contact angle hysteresis.

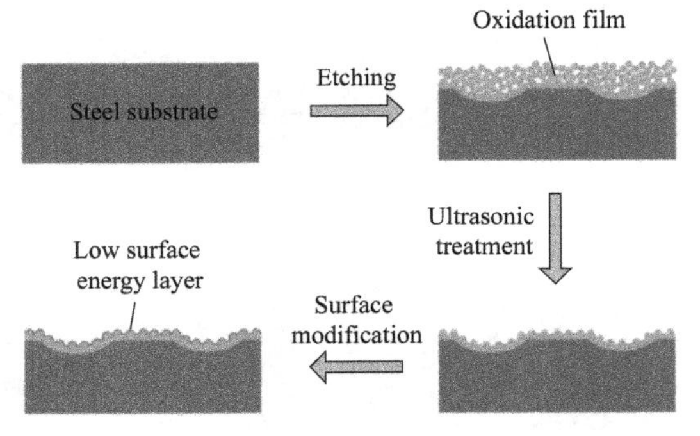

Figure 9-16 The formation mechanism of the as-prepared superhydrophobic surface with hierarchical structure

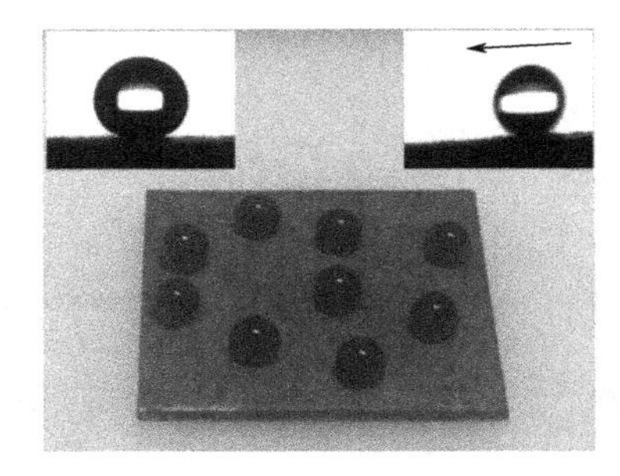

Figure 9-17 The wetting state for water droplets with different diameters on the as-prepared superhydrophobic surface

The mechanical durability of superhydrophobic surfaces is a crucial factor which limits the widespread applications. Nevertheless, superhydrophobic surfaces with special micro/nano structures of superhydrophobic surfaces are usually mechanically weak and easily destroyed. To solve this problem, lotus-leaf-like hierarchical micro-nanostructures were prepared on the steel surface. Using this structure, robust microscale bumps can provide protection to a more fragile nanoscale roughness. In this research, the sandpaper abrasion test were performed to systematically study the mechanically stability of the as-prepared superhydrophobic steel surface. As shown in Figure 9-18 (a), the superhydrophobic steel (2cm×2cm) was attached to a 1kg weight (a pressure of 24.5kPa) with the help of commercial adhesive. Then, the sample facing down sandpaper (grit No. 150) was moved for 28cm along the ruler by an external drawing force (Figure 9-18(b)). The water contact angles and sliding angles were measured after each abrasion test cycle. Figure 9-18 (c) exhibits the change in contact angles and sliding angles as a function of the number of abrasion cycles. It can be found that the water contact angles were between 150° and 165°, and sliding angles varied between 5° and 9° through the abrasion cycles. Therefore, the as-prepared steel surface retained superhydrophobicity after 9 sandpaper abrasion cycles (2.24m) under a pressure of 24.5kPa. It should be noted that the results displayed in this research showed better mechanical durability than other reported superhydrophobic metal surfaces, as shown in Table 9-1.

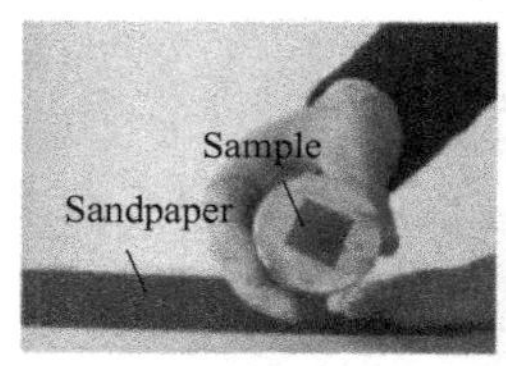

(a) Illustration Ⅰ of one cycle of the abrasion test for the as-prepared steel superhydrophobic surface under a 1kg loading weight on sandpaper

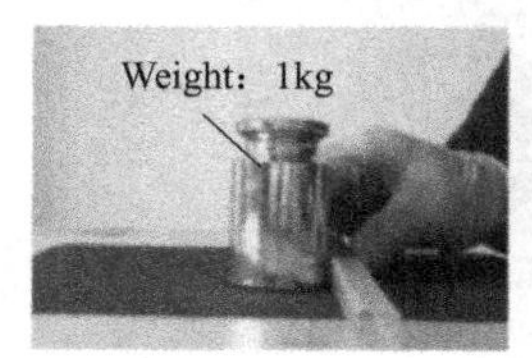

(b) Illustration Ⅱ of one cycle of the abrasion test for the as-prepared steel superhydrophobic surface under a 1kg loading weight on sandpaper

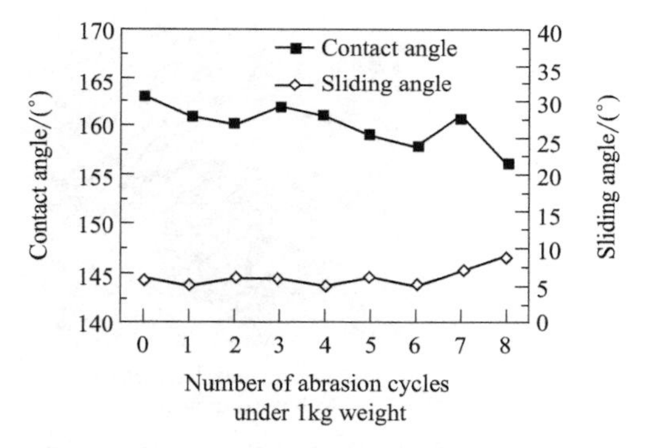

(c) The contact angles and sliding angles as a function of abrasion cycles for the as-prepared superhydrophobic surface

Figure 9-18 The experimental illustrations and plots

Table 9-1 Comparison of the mechanical durability of the prepared superhydrophobic coatings

Substrate	Method		Anti-abrasion test	
	Roughness	Surface modification	Load	Distance
Mg	electrodopositing Ni	stearic acid	1.2 kPa	0.70m
Cu	electrodepositing Ni	FAS	4.8 kPa	1.00m
Si	two step etching	FAS	3.45 kPa	2.00m
steel	HF and H_2O_2 etching	stearic acid	500g	1.00m
steel	HCl/HNO_3 and H_2O_2 etching	FAS	500g(16kPa)	1.10m
steel	NaClO and H_2O_2 etching	FAS	1kg(24.5kPa)	2.24m

The self-cleaning ability is a crucial character of superhydrophobic surfaces for practical application. A self-cleaning test for the as-prepared superhydrophobic surface was carried out (Figure 9-19). Here, the carbon black powder was utilized as characteristic dust particles. As shown in Figure 9-19(a), the dust particles were spreaded onto the superhydrophobic steel surface (Figure 9-19(b)). When a water droplet was dripped on the sample, they can smoothly roll down the surface and take away the dust at the same time (Figure 9-19(b) and (c)). After water pouring processes, water droplets carried away the dust particles completely and left a clean surface (Figure 9-19(d)). Therefore, the as-prepared steel superhydrophobic surface exhibited excellent self-cleaning property.

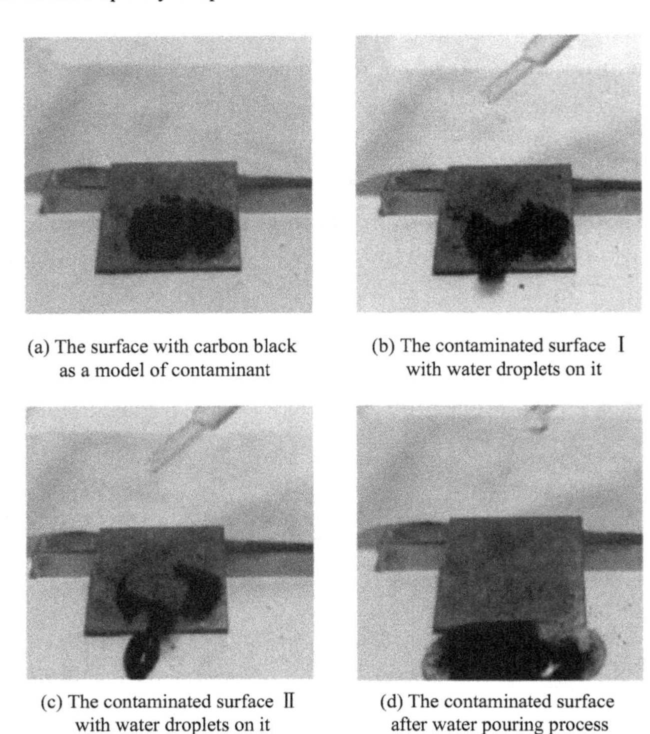

(a) The surface with carbon black as a model of contaminant

(b) The contaminated surface Ⅰ with water droplets on it

(c) The contaminated surface Ⅱ with water droplets on it

(d) The contaminated surface after water pouring process

Figure 9-19 Self-cleaning process on the as-prepared superhydrophobic surface

The polarization curves are useful methods to explore the impact of the superhydrophobic surfaces on the corrosion resistance of the steel substrate. Figure 9-20 shows the polarization curves of bare steel substrate and modified superhydrophobic steel surface. Parameters such as corrosion potentials (E_{corr}) corrosion current (I_{corr}) can be obtained using the Tafel extrapolation. As shown in Figure 9-20, the anti-corrosion ability was found to be improved on the superhydrophobic surface due to the lower I_{corr} (12.706m/cm^2) and higher E_{corr} (−0.511 V) as compared to those of bare substrate (I_{corr}=25.586m/cm^2, E_{corr}=−1.017V), suggesting a good corrosion protection for the steel substrate. This result is consistent with previously published reports that superhydrophobic surfaces exhibit excellent anti-corrosion ability. When the superhydrophobic surfaces are immersed in a corrosive solution, air tends to be trapped in the lotus-leaf-like hierarchical micro-nanostructure of the as-prepared superhydrophobic surface, and the trapped air behaves as a dielectric for a pure parallel plate capacitor and prevents the electron transfer between the electrolyte and the steel substrate.

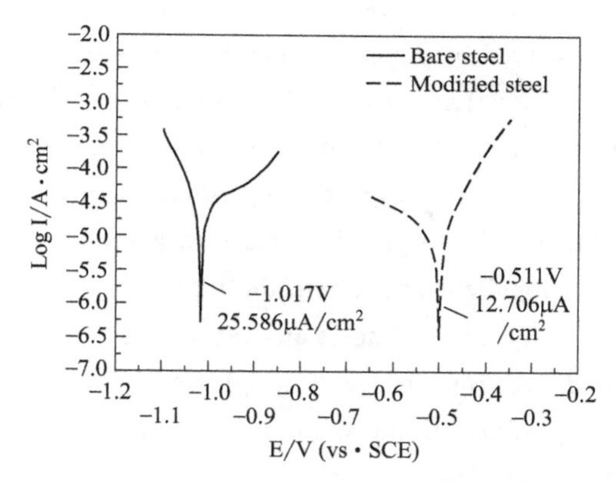

Figure 9-20 Polarization curves of bare steel and as-prepared superhydrophobic steel

9.4 The Graphene Semiconductor Superhydrophobic Coating for Anti-Icing Application

The accretion of ice on surfaces would pose a serious problem to many industrial applications, such as power transmission lines, wind turbines, and aircrafts. For instance, the longstanding freezing rain in 2008 damaged 37% of 500kV power transmission towers in Southern China. Many efforts have been done to eliminate the surface ice, such as ultrasound technique, adding anti-icing agent, slippery system, superhydorphobic surfaces and electrothermal technology. Due to the effectiveness, simplicity and cost efficiency, the electrothermal system is the traditional and most widely used method. Recently, the movable DC deicing devices have been commercialized to protect the power transmission lines. However, the high energy comsumption is still a challenge for DC deicing device. Many materials (i. e. graphite, carbon nanotube and graphene) have been introduced to improve the electrothermal efficiency. Particularly, graphene has received more and more attention due to its high surface area, exceptional electrical conductivity, superior thermal conductivity, and excellent mechanical strength. Dai et al. has found that the graphene exhibits much higher thermal conductivity (~5000W/(m·K)) than Carbon nanotube (~3000W/(m·K)) and amorphous carbon (~1W/(m·K)). Recently, Pan et al. modified the carbon fiber/PEEK composite with graphene, and found the graphene could improve the efficiency of converting Joule heating by up to 35%. Redondo et al. prepared a kind of epoxy coating doped with graphene nanoplatelets, which required a low electrical power of ~2.5W to melt the 3-4mm thick ice. Guadagno et al. fabricated a flexible graphene film which could melt a 1mm thick layer of ice in less than 7min. Zhang et al. constructed a conductive graphene-papers/glass-fiber reinforced epoxy composite, which could realize 10mm thick ice clean with energy consumption of ~0.03KW·h/(mm·m^2). Chen et al. found that the graphene coating could improve the heat-transfer efficiency by 70% compared with the existing anti/deicing method. Novoselov et al. further reported a scalable route for graphene-based deicing composites using Joule heating phenomenon.

Although considerable researches have been done on graphene deicing materials, one challenge is still under discussing: how to solve the re-freeze problem as the melted ice tend to stay on the surface. This problem can be solved by superhydrophobic surfaces, on which water droplets adopt a nearly round shape and roll off easily. Hence, the combination of Joule heating and superhydrophobic surfaces would be an effective route for deicing. Du et al. tried to add the carbon nanofillers into low surface energy polymers like high density polyethylene. Zhang et al. tried to put an electrothermal patch under the superhydrophobic surface to achieve synergistically deicing. Shiratori et al. combined the water-repellent and electrothermogenic properties by embedding PEDOT: PSS nanoparticles in polymer. Recently, Jiao et al. fabricated a film with 1H, 1H, 2H, 2H-perfluorodecyltrichlorosilane-modified SiO_2/reduced graphene oxide wrinkles[16]. This film could achieve superhydrophobicity by adjusting the wrinkle size, and exhibited excellent deicing and defrosting property.

It seems that the superhydrophobic material's Achilles heel is the mechanical durability. As low surface energy and micro/nano structure are two essential factors for constructing superhydrophobicity, most micro/nano structures are susceptible to mechanical abrasion. Kulinich et al. further pointed out that the ice-repellent properties of the superhydrophobic materials would reduce dramatically during icing/deicing cycles. The researchers have attempted many ways to solve this problem, such as hierarchical roughness surfaces and "Paint+Adhesive" method. However, there is still a long way to realize the practical application of superhydrophobic materials.

Recently, Lu et al. prepared a mechanically robust superhydrophobic surface by bonding dual-scale TiO_2 nanoparticles with adhesives, which showed a maximum abrasion distance of 8. 00m under a 100g load. Inspired by Lu's work, the hierarchical roughness was achieved by means of the tri-scale nature of graphene, carbon nanotubes and silica nanoparticles in this work. This hierarchical structure was further embedded into the substrate through a simple dissolution and resolidification method. The novelty of this research is realizing the combination of hierarchical and partially-embedded structure based on graphene. Benefited from the synergistic effect of hierarchical roughness, partially-embedded structure and graphene, the prepared surface could endure a maximum abrasion distance of 8. 00m under a 500g load. Moreover, the superhydrophobicity integrated with Joule heating could effectively avoid the formation of ice in glaze ice condition, and quickly remove the surface ice within 70s. Furthermore, this surface maintained icephobicity even after 30 icing/deicing cycles.

9.4.1 Materials

The graphene (xGnP, Grade M) were provided by XG Sciences, with an average size of 25μm and 6-8nm thickness. The multiwall carbon nanotubes (MWCNTs, length: $<10\mu$m, diameter: 10-20nm) were provided by Chengdu Organic Chemicals Co., Ltd., China. The hydrophobic silica nanoparticles (R812, diameter: 7-20nm) were purchased from Degussa. Polycarbonate (PC) plates (2mm thickness) were bought from Shenzhen Oudifu Material Co., Ltd., China. 1H, 1H, 2H, 2H-perfluorooctyltriethoxysilane ($C_8F_{13}H_4Si(OCH_2CH_3)_3$, FAS) was purchased from Aladdin Reagent Co., Ltd., Shanghai, China. Ethanol, acetone, tetrahydrofuran (THF) and methyl blue were purchased from Sinopharm Chemical Reagent Co., Ltd., Shanghai, China. All chemicals were used as received.

9.4.2 Preparation of Hydrophobic Powders

Firstly, 0.2g of FAS was added into 10g of ethanol, and the solution was stirred for 2h. Then 0.1g of graphene, 0.2g of MWCNT, and 0.2g of silica nanoparticles were sequentially added to form the turbid liquid. To ensure uniformity, the turbid liquid was ultrasonic treated for 30min and magnetically stirred for 4h. This turbid liquid was further coated onto the glass slide using a disposable Pasteur pipette (2mL). After natural air drying, the hydrophobic powders was obtained and collected. It should be noted that the parameters of the hydrophobic powders were determined after optimization. The optimization process can be found in the supporting information.

9.4.3 Dissolution and Resolidifcation Process to Construct Superhydrophobic Sample

Firstly, 5g of ethanol was added into 5g of THF. The mixture was further sealed in a 50mL glass bottle, and magnetically stirred for 1h. Then, 0.2g hydrophobic powders was added to the mixture and stirred for 2h to construct a uniform suspension. Next, the suspension was coated onto the polycarbonate sheet with the help of a disposable Pasteur pipette (1mL). After 8h of drying at room temperature, the superhydrophobic samples were obtained.

9.4.4 Characterization

The scanning electron microscope (SEM, TESCAN Vega3) was used to observe the micro/nano structures of samples. X-ray photoelectron spectroscopy (XPS, Thermo ESCALAB 250XI, USA) was utilized to determine the samples' chemical composition. The water contact angles (CAs) and sliding angles (SAs) were measured using JC2000D (Shanghai Zhongchen, China) apparatus. At least five positions were measured for each sample, and the average values were calculated. The dynamic bouncing test was measured in term of the high speed camera (Revealer 2F04).

9.4.5 Icing/Deicing Test

During all the icing/deicing process, a digital microscope (Andonstar-A1, China) was utilized to measure the freezing process and then determine the freezing time. Moreover, an infrared imager (FLIR-C2, FLIR Ltd., USA) was utilized to detect the surface temperature.

To measure the freezing process of a droplet, the samples were placed into the climate chamber (temperature: −15℃, relatively humidity: 30%). Then, a 5μL water droplet was carefully dropped onto the sample surface with a syringe.

For preparing glaze ice, the samples were placed into the climate chamber with the artificial cold rain environment (temperature: −5℃, relatively humidity: 30%). Then, a commercial moisturizer was placed into the climate chamber for generating artificial rain droplets. After putting the supercool water into the moisturizer, a thick layer of ice can be obtained on the surface of the samples. For the electrothermol deicing test, a DC voltage was applied to generate Joule heating. For the photothermal deicing test, a near-infrared laser (808nm, Model NO. LSR808H-FC-1W,

Lasever Inc., Ningbo, China) was utilized to irradiate the sample.

9.4.6 Results and Discussion

The suspension was prepared by mixing graphene (size: 5–30μm, thickness: 6–8nm), carbon naotubes (length: 5–10μm, diameter: 10–20nm) and silica nanoparticles (diameter: 7–12nm) in an ethanol/THF solution cotaining FAS. Due to tri-scale nature of inorganic fillers, the as-prepared samples exhibited hierarchical structure. As shown in Figure 9-21 (a), many microplates could be found, which could be ascribed to graphene. Furthermore, it could be found that some CNTs and silica nanoparticles adhered to the microplates (Figure 9-21 (b)). Compared with silica nanoparticles, the CNTs were irregularly distributed. We attributed this to the agglomeration of CNTs. It should be note that the addition of CNTs is essential, which not only favors the formation of hierarchical structure but also favors the construction of conductive nework. We further measured the surface roughness of the sample using true color confocal microscope (Figure 9-21 (c) and (d)). The roughness value was calculated to be 3.73μm, which indicates the microscale roughness and is consistent with the SEM measurement.

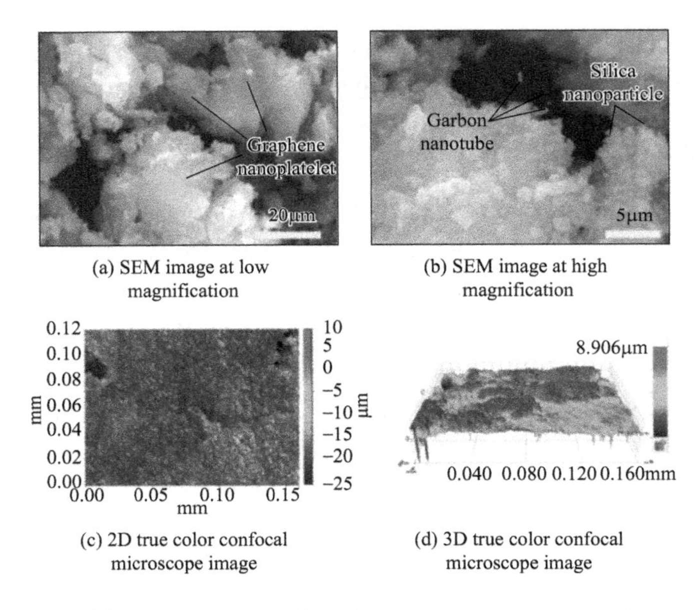

(a) SEM image at low magnification

(b) SEM image at high magnification

(c) 2D true color confocal microscope image

(d) 3D true color confocal microscope image

Figure 9-21 SEM images of the graphene composite at low and magnifications, and 2D and (d) 3D true color confocal microscope images of the graphene composite

The low surface energy is another key element for constructing superhydrophobicity. Here, the XPS tests were carried out to analyze the surface chemical compositon. As shown in Figure 9-22 (a), the survey spectrum is consisting of Si 2p, C 1s, O 1s and F 1s peaks. From the F 1s XPS spectra (Figure 9-22 (b)), the total peak is consisting of one main ingredient arising from CF_x (688.73eV) specie and one minor ingredients from SiF_x (689.80eV) specie. For Si 2p spectrum (Figure 9-22 (c)), the distinct peak locating at 103.08 eV can be attributed to—Si—OH/SiF_x, and the another peak at 104.10eV are assigned to SiO_2 based network. Furthermore, the C 1s (Figure 9-22 (d)) spectrum can be fitted to six peaks at 284.32eV, 284.99eV, 285.87eV, 286.65eV, 291.50eV, and 293.67eV, which can be ascribed to the C—Si, C—C, C—O, C—CF,

CF_2 and CF_3, respectively. It should be noted that the C—CF, CF_2 and CF_3 are low surface energy groups, which play a key part to construct superhydrophobic materials. Therefore, during the water bouncing test, the water droplets could leave the surface completely without any contaminants.

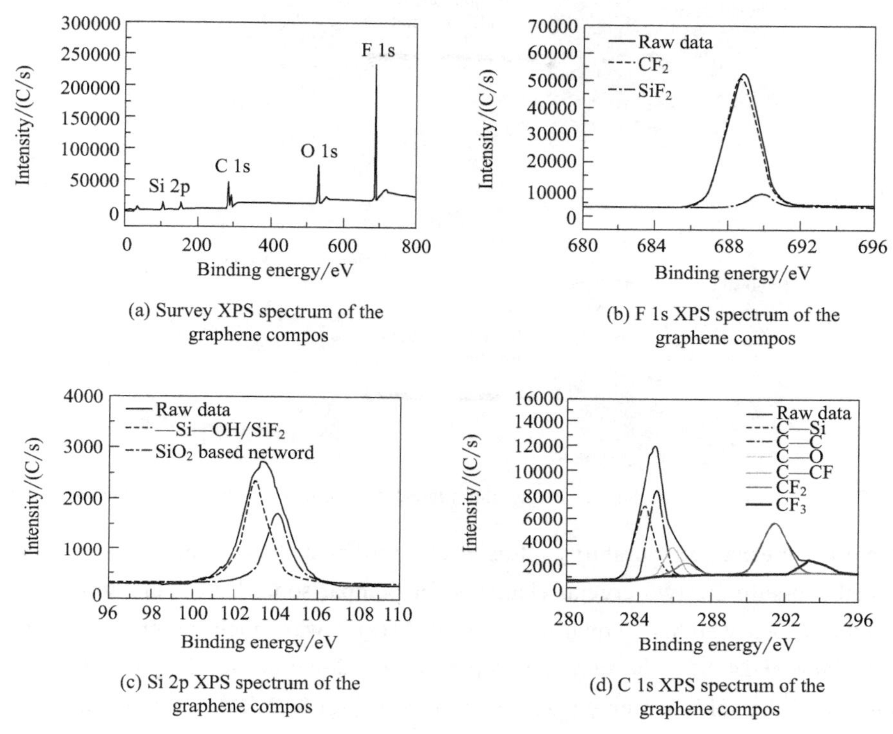

(a) Survey XPS spectrum of the graphene compos

(b) F 1s XPS spectrum of the graphene compos

(c) Si 2p XPS spectrum of the graphene compos

(d) C 1s XPS spectrum of the graphene compos

Figure 9-22 Different XPS Spectrum

Moreover, this graphene superhydrophobic sample exhibits superior self-cleaning properties. The water droplets stayed on the surface of bare glass and RTV (room temperature vulcanized silicone rubber) coated glass, whereas the graphene composite remained clean and dry after water pouring process.

In this research, an essential step is to partially embed the fillers into a substrate to enhance mechanical durability, which would be realized by the dissolution and resolidifcation process (Figure 9-23). As is known to all, THF is a good solvent to polycarbonate. Thus, the THF based suspension of micro/nano filler was adapted in our initial experiment. After painting the suspension onto the polycarbonate, the interface between the polycarbonate and solvent became viscous during the dissolution process. Then, the fillers could get into the interface. With the evaporation of THF, the interface would be gradually resolidificated again. It was found that the THF dissolve the polycarbonate so good that the fillers would be fully embedded into the substrate. Then, we even could not obtain the superhydrophobicity. Hence, ethanol was added into THF to reduce the solubility. After trial and error, THF/ethanol solution (5∶5, $V:V$) was chosen due to the best mechanical results from the anti-abrasion test.

The durability is an essential factor limiting the real-life application of superhydrophobic surfaces. Here, some physical and chemical methods were employed to determine the durability of the samples. Firstly, sandpaper-abrasion test was applied because it is the most widely used meth-

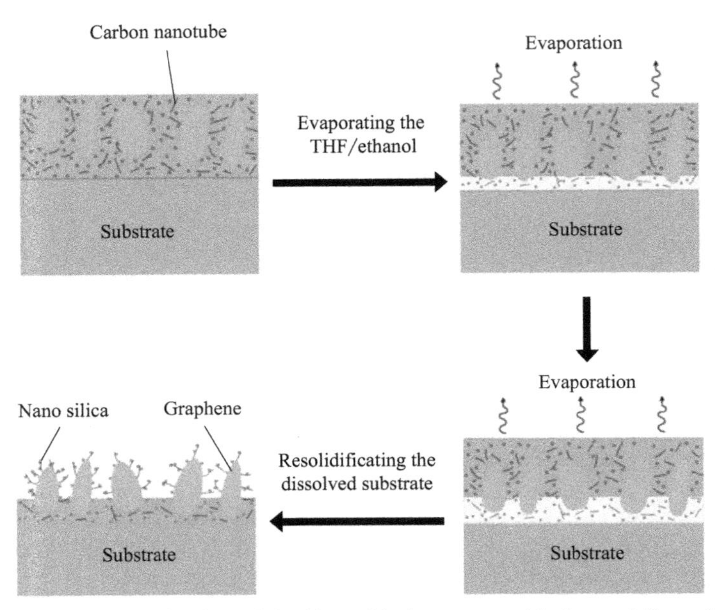

Figure 9-23 The forming mechanism of the hierarchical structure with the partially-embeded graphene

od to evaluate the mechanical durability. Ras et al. further deduced that abrasion distance and the applied normal pressure are two crucial elements for comparison. Here, the applied normal pressure of 12.5kPa (500g weight, shown in Figure 9-24(a)) was much larger that other researches. As shown in Figure 9-24(b), the sample was placed face down to the 150 grid SiC sandpaper, and moved 20cm along the ruler as one cycle. As shown in Figure 9-24(c), the water CAs gradually decrease and SAs gradually increase as the increase of abrasion cycles. However, the CAs were

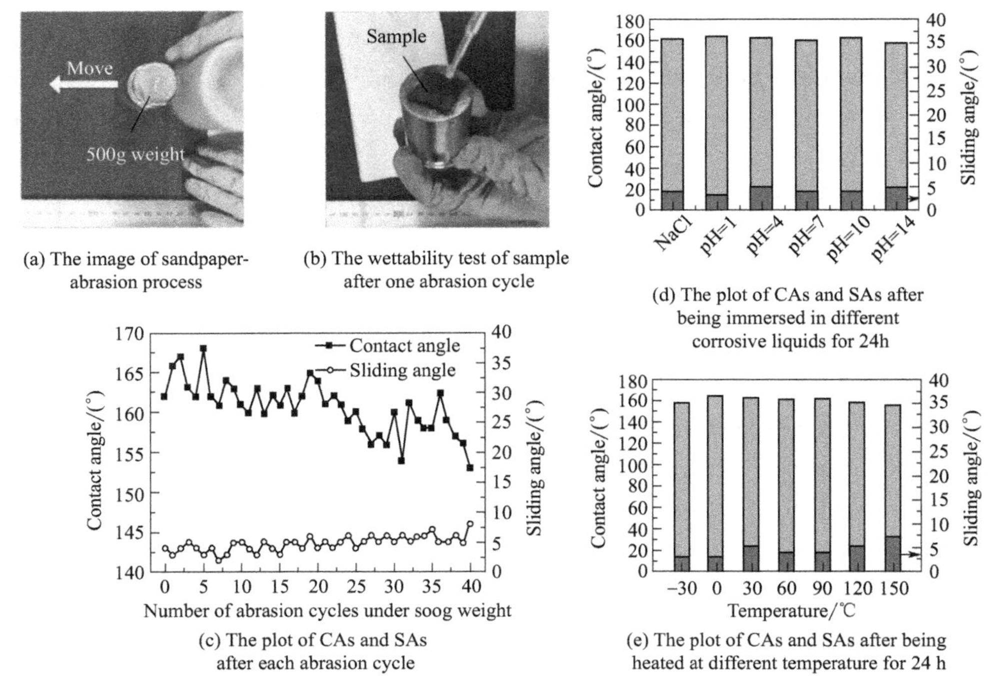

(a) The image of sandpaper-abrasion process

(b) The wettability test of sample after one abrasion cycle

(c) The plot of CAs and SAs after each abrasion cycle

(d) The plot of CAs and SAs after being immersed in different corrosive liquids for 24h

(e) The plot of CAs and SAs after being heated at different temperature for 24 h

Figure 9-24 The experimental process and plots

still larger than 150° and the SAs were still smaller than 10° after 40 abrasion cycles, indicating that the superhydrophobicity was retained. For a more comprehensive comparison, we also abraded the superhydropobic sample under a relatively low load of 2.5kPa (100g weight), and the sample maintains superhydrophobic for 400 abrasion cycles (80.0m).

Besides harsh mechanical damage, superhydrophobic materials may meet the invasion of corrosive liquids in real-life applications. Here, the superhydrophobic samples were separately immersed into six kinds of liquids (3.5%(wt) NaCl, aqueous solution with the pH adjusted to 1, 4, 7, 10 and 14) to test the chemical durability. After 24h immersion, the samples were rinsed with water and the surface wettability was measured. As shown in Figure 9-24(d), the water CAs were between 155° and 165°, and SAs varied between 3°and 7°. Thus, the graphene superhydrophobic materials demonstrated superior chemical durability. Thermostability is also an essential element to evaluate superhydrophobic materials' environmental adaptability. As shown in Figure 9-24(e), the CAs and SAs of the graphene superhydrophobic sample were measured from −15℃ to 150℃ for 24h. Many researchers have reported that the surface tended to lose superhydrophobicity when the surface temperature was below −10℃. Suprisingly, when the surface temperature varied from −15℃ to 150℃, the change range of CA and SA for graphene composite were very small, indicating high thermostability. Here, we further investigated the low temperature wettability by setting the control group (Ⅰ. the standard sample with electrical heating; Ⅱ. the standard sample without electrical heating; Ⅲ. the superhydrophobic cement according to the reference; Ⅳ. the sample without adding graphene and CNT; Ⅴ. the sample without adding silica nanoparticles). With the help of electrical heating, the surface temperature of sample I will be above 0℃. The sample Ⅰ and Ⅱ maintained superhydrophocity even when the environmental temperature was −15℃. However, the sample Ⅲ, Ⅳ and Ⅴ lose superhydrophobicity. Thus, we attribute the low temperature superhydrophobicity of the standard sample in this research to the synergistical effects of graphene, CNT and silica nanoparticles, which led to the construction of hierarchical structures.

Due to the superhydrophobicity at subzero temperature, this graphene composite could delay the freezing time of supercooled water, which is a crucial indicator for anti-icing. Figure 9-25(a) exhibits the photographs of individual water droplets on bare glasses and graphene superhydrophobic samples in the freezing process. The water droplet on the bare glass demonstrated a hemispherical shape, and it was completely freezed after 57s. On the other hand, the surface of the gra-

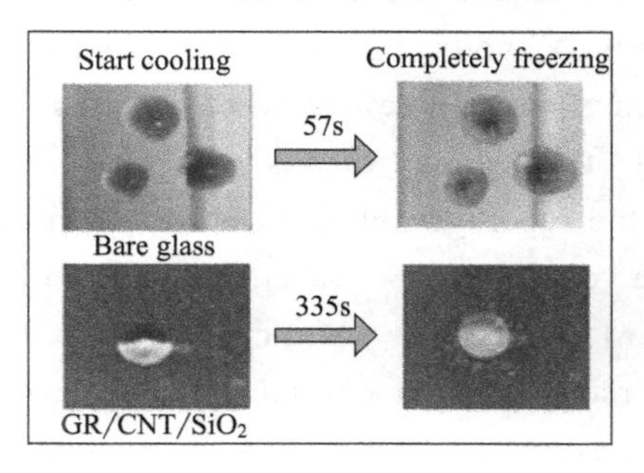

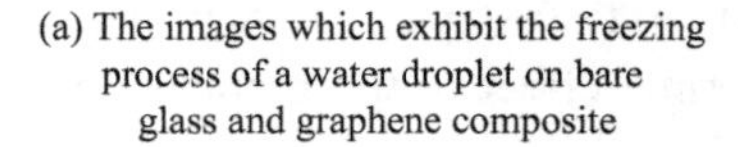

(a) The images which exhibit the freezing process of a water droplet on bare glass and graphene composite

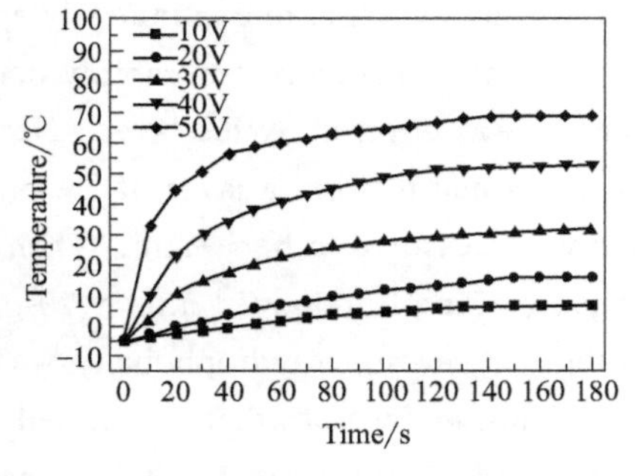

(b) The time dependent temperature profile of the graphene superhydrophobic composite under different voltages

Figure 9-25 The experimental images and profile

phene superhydrophobic sample had a nearly spherical water droplet on it, and exhibited a delay freezing time of 335s. Even after completely freezed, the droplet maintained the spherical shape (Figure 9-25(a)).

Due to the superiorthermal conductivity, this superhydrophobic material would exhibit comparable electrothermal performance to other graphene film. Here, the time dependent temperature profile of this superhydrophobic material was measured by applying different DC voltage. As shown in Figure 9-25(b), the temperature could increase from subzero temperature (−5℃) to 16.2℃ under 10V voltage. Furthermore, the surface temperature could reach the saturation temperature of 69.1℃ from −5℃ rapidly after applying 50V.

As Jiang et al. has proved that the glaze ice is the most dangeous type among the surfce icing accidents, the glaze ice test was performed in this research. The glaze ice was simulated in the artificial climate chamber according to reference with some minor modifications. The bare glass slide, RTV (room temperature vulcanized silicone rubber) coated glass slide (sample RTV) and graphene superhydrophobic sample (sample SH) were placed in the artificial climate chamber at the tiling angle of 30°. Figure 9-26 demonstrated the ice accretion on the three aforementioned samples within 2h duration. Although the graphene composite exhibits freezing-delay property, it can't probibit the ice accretion after long time duration of 30min. Thus, an obvious layer of ice could be found from the bare glass, RTV, and SH sample.

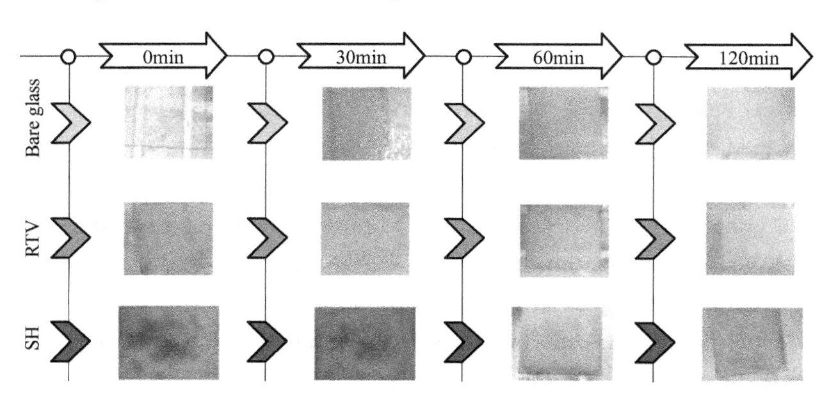

Figure 9-26 The images of ice accretion on bare glass, RTV and superhydrophobic (SH) within 2h

Although the synergistical effect of superhydrophobic/electrothermal can effectively prohibit the ice accretion, it is unfavorable to continuously apply DC voltage from the viewpoint of energy conservation. One possible solution is to rapidly deicing by Joule heating when the graphene superhydrophobic surface was covered by ice layer. Here, the deicing test was performed. The glaze ice test was first performed to form a layer of ice on the superhydrophobic surface, and the thickness of the icelayer was measured to be ~3mm. Then a voltage of 50V was applied. From Figure 9-27, the ice layer can be rapidly removed after ~70s. Moreover, no water droplets stayed on the surface, indicating that the superhydrophobcity was remained after one icing/deicing cycle. The repetitive icing/deicing test was further performed to evaluate the durability of graphene composite. As shown in Figure 9-28(a), the heating temperature was between 65° and 70°, and the deicing time was between 65℃ and 75℃. Thus the eletrothermal effect of the graphene composite was retained. Moreover, the CA gradually decreased from 167° to 153° and the SA gradually increased from 2° to 7°, indicating that superhydrophobicity was not lost (Figure 9-28 (b)). Therefore, our graphene superhydrophobic surface has potential application for long-term outdoor deicing with high efficiency.

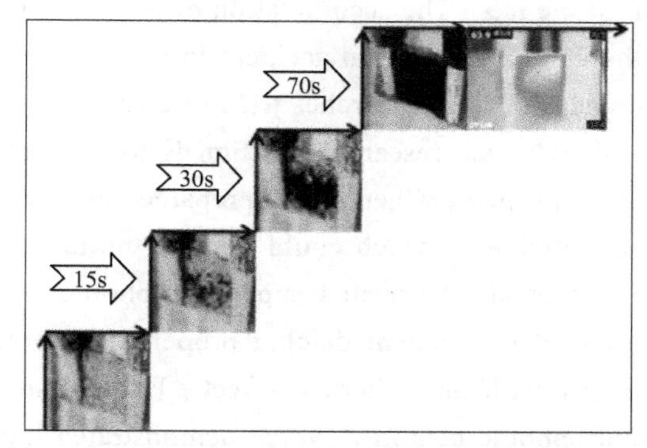

Figure 9-27 The rapidly electrothermal deicing process of graphene superhydrophobic composite.

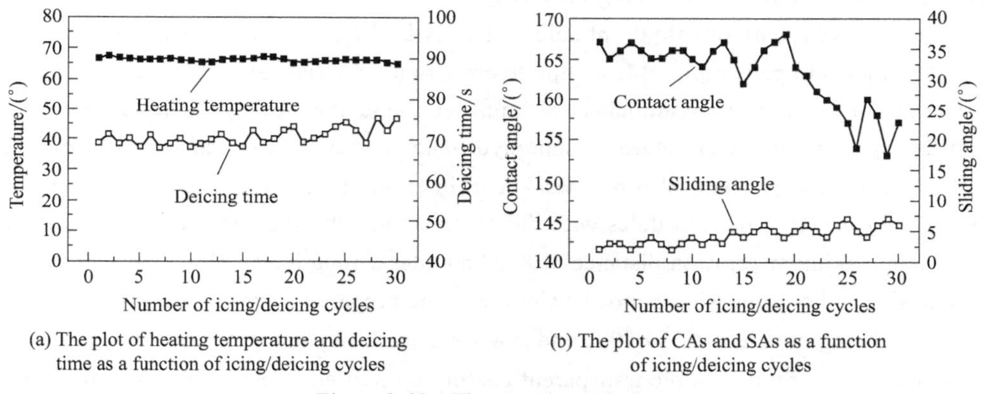

(a) The plot of heating temperature and deicing time as a function of icing/deicing cycles

(b) The plot of CAs and SAs as a function of icing/deicing cycles

Figure 9-28 The experimental plots

9.5 The Self-Healable Graphene Coating for Anti-Icing Application

The accumulation of ice tended to create life-threatening hazards in our daily life. For instance, the ice accretion will generate extra weight which increases the burden of aircraft; the ice formation between umbrella skirts of the insulator in power line will generate new conductive path which tends to cause flashover. Many methods have been introduced for anti-icing/deicing, such as heating, releasing anti-icing agent, liquid-infused materials, and superhydrophobic surfaces. Among these methods, the passive anti-icing method inspired by superhydrophobic surfaces has garnered considerable attention due to their advantages of energy-saving and high efficiency.

A typical superhydrophobicity can be found in the lotus leaves in nature. The water droplets exhibit a nearly round shape on the surface of lotus leaves, and roll off easily under a tiny sliding angle. The further researches reveal that the combination of low surface energy and suitable micro/

nano structure can capture air pockets between the interfaces, which leads to superhydrophobicity. Furthermore, it is found that the air pockets underneath the water droplets or ice could effectively delay freezing time and decrease the adhesion force. Nevertheless, the superhydrophobic surface can only delay the formation of the ice. The accumulation of ice in the long-lasting terrible weather is still inevitable, such as the 2008 freezing rain accident in China.

In order to overcome this obstacle, the researches tried to combine the superhydrophobic materials with active deicing technology. One research direction is to integrate the superhydrophobicity with joule heating method. For instance, Chen et al. prepared superhydrophobic conductive paper based on ketjen black and polyethylene, which could remove the ice using the electrothermal and non-adhesive performance. Jiao et al. fabricated superhydrophobic film using reduced graphene oxide wrinkles, which demonstrated excellent deicing property. Another research direction is to integrate the superhydrophobicity with photothermal effect. For example, Jiang et al. fabricated SiC/carbon nanotube superhydrophobic coating, which demonstrated high light-to-heat conversion efficiency. Therefore, A superhydrophobic/electrothermal/photothermal synergistically anti-icing strategy will further improve the deicing efficiency.

The biggest drawback of superhydrophobic materials is the poor mechanical robustness, which prohibits the practical application. Most superhydrophobic materials are fragile to sandpaper abrasion, or even finger contact. Kulinich et al. further found that the cyclic icing/deicing process tended to damage the surface structure of superhydrophobic materials which led to lose of icephobicity. To improve the mechanical robustness, Parkin et al. invented a "Paint+Adhesive" method. By bonding the TiO_2 nanoparticles with the commercial adhesive, the superhydrophobic coating exhibited a maximum abrasion distance of 8.00m under a 100g load. Guo et al. utilized the inorganic adhesive to bond the SiO_2 nanoparticles, and the prepared surfaces maintained superhydrophobicity after being abraded for 10.00m under a 200g load. In our previous work, we prepared abrasion-resistant superhydrophobic transparent coating by partially-embedding the silica nanoparticles into the PDMS matrix. Meanwhile, Ras et al. introduced hierarchical structure which attempted to use the relatively robust microstructure to guard the fragile nanostructure. Nevertheless, the superhydrophobic materials prepared by the aforementioned methods could not withstand severe mechanical damage, such as wide and deep scratches.

When we accidentally broke our skin, the wound could be spontaneously repaired in our daily life. In fact, the capability to repair damaged organisms is a fundamental function for the living organisms. Inspired by this phenomenon, many kinds of self-healable materials have been prepared by the introduction of reversible or exchangeable bonds. Recently, Sun et al. fabricated self-healable superhydrophobic film by drop-casting Ag nanowires onto the healable film. In this research, we tried to partially embed micro-scale carbon powders and nanoscale carbon nanotubes into healable elastomer. The base elastomer endows this composite self-healable function, and the fluoroalkylsilane modified dual scale carbon materials endow the composite superhydrophobic/electrothermal/photothermal properties simultaneously. The further anti-icing test demonstrated the superiority of this superhydrophobic/electrothermal/photothermal synergistically anti-icing strategy. Furthermore, the as-prepared materials combined the partially-embedded and hierarchical structure, which maintained superhydrophocity even after abraded for 16.00m under a load of 500g.

9.5.1 Materials

TrifunctionalPoly (propylene glycol) (PPG, Mn=6000) was bought from Bayer Materials Science. The PPG with low molecular weight (330N) was purchased from Jining Huakai Resin Co., Ltd., China. 1H, 1H, 2H, 2H-perfluorooctyltriethoxysilane ($C_8F_{13}H_4Si(OCH_2CH_3)_3$, FAS), 4-Aminophenyl disulfide (APS, 98%), isophorone diisocyanate (IPDI, 99%), tetrahydrofuran (THF, 99%), and carbon powder (≥200 mesh) were purchased from Aladdin Reagent Co., Ltd., Shanghai, China. The dibutyltin dilaurate (DBTDL, 95%) was purchased from Sinopharm Chemical Reagent Co., Ltd., Shanghai, China. The multiwall carbon nanotubes (MWCNTs) with an average diameter of 30 -50nm (TNM7, ≥98% purity) were bought by Chengdu Organic Chemicals Co., Ltd., China. All chemicals were used as received.

9.5.2 The Synthesis of Prepolymer A

The prepolymer A was synthesized based on reference with some modification. Firstly, the PPG 6000 (39g, 65mmol) was poured into a 500mL four-necked flask equipped with a mechanical stirrer and a vacuum inlet. Then, The PPG 6000 was stirred under vacuum at 120℃ for 1h to remove residual moisture. After cooling the temperature to 70℃, IPDI (4.545g, 204.5mmol) was added, and the mixture was stirred under vacuum at 70℃ for 10min. Finally, DBTDL (2mg) was added, and the mixture was further heated at 70℃ for 45min under vacuum and stirring to obtain prepolymer A.

9.5.3 The Synthesis of Prepolymer B

Firstly, PPG330N (25g, 125mmol) was poured into a 500mL four-necked flask, and heated at 120℃ under stirring and vacuum for 1h to remove internal moisture. After the temperature was cooled to 60℃, IPDI (5.55g, 250mmol) was added. Then, the mixture was stirred under vacuum at 60℃ for 10min. In the next step, 1.5mg of DBTDL was added. Finally, the mixture was further stirred under vacuum at 60℃ for 70min to obtain prepolymer B.

9.5.4 Preparation of Hydrophobic CNT Powders

Firstly, 0.3g of FAS was dissolved in 25g of THF, and the solution was magnetically stirred for 2h. Then 1g of MWCNTs were added to the solution. In order to ensure the dispersion of MWCNTs, the mixture was ultrasonic treated for 1h and then magnetically stirred for 6h. Finally, the mixture was placed in a vacuum drying box. After vacuum drying, the hydrophobic MWCNT powders were collected.

9.5.5 Preparation of Hydrophobic CNT Solution

Firstly, 0.4g of FAS was dissolved in 30g of THF, and the solution was magnetically stirred for 2h. In the next step, 1g of carbon powders was added to the solution. Then, the mixture was magnetically stirred at room temperature for 6h to obtain uniform hydrophobic CNT solution.

9.5.6 Preparation of Self-Healing Superhydrophobic Coating

Firstly, the prepolymer (A) and the prepolymer (B) were mixed in a 100mL glass container. Then, a solution of 4-Aminophenyl disulfide (20.6mmol) in THF (3mL) was added, and the mixture was mechanically stirred to obtain self-healing polymer C. In the next step, the mixture was poured into a tetrafluoroethylene mold and degassed for 20 minutes in a vacuum environment to remove bubbles. Then, hydrophobic carbon powders were uniformly sprinkled on the surface of polymer C with the help of a copper mesh (200-mesh). After further heating the sample at 70℃ for 3h, the polymer C was semi-cured. Then, the hydrophobic CNT solution was sprayed onto the sample to improve the conductivity. Finally, the self-healing superhydrophobic sample was fabricated by removing the sample from the mold after 24h.

9.5.7 Characterization

The scanning electron microscope (SEM, TESCAN Vega3) was utilized to measure the surface morphology. X-ray photoelectron spectroscopy (XPS, Thermo ESCALAB 250XI, USA) was adapted to detect the samples' chemical composition. We further measured the water contact angles and sliding angles in term of JC2000D (Shanghai Zhongchen, China) apparatus. At least five positions were measured for each sample, and the average values were calculated. The dynamic bouncing test was investigated by the means of high-speed camera (Revealer 2F04).

9.5.8 Icing/Deicing Test

The icing/deicing tests are based on references with some minor modi fications. During all the icing/deicing process, a digital microscope (Andonstar-A1, China) was adopted to observe the freezing process and then determine the freezing time. Moreover, an infrared imager (FLIR-C2, FLIR Ltd., USA) was used to measure the surface temperature.

For the freezing test of a droplet, the samples were put into the climate chamber (temperature: −25℃, relative humidity: 30%). Then, a 5μL water droplet was carefully put onto the sample surface using a syringe.

For the glaze ice test, the samples were put into the climate chamber at the temperature of −25℃ and the relative humidity of 80%. A commercial moisturizer was also put onto the climate chamber and served as the generator of water droplets. Then, 1L of supercool water (~0℃) was put into a commercial moisturizer. With the help of moisturizer, a large amount of microscale water droplets could be generated and fell onto the surface of samples.

For the electrothermal (ET) deicing test, the moisturizer stopped work and a direct current (DC) voltage of 15V was applied to generate Joule heating. For the photothermal (PT) deicing test, the near-infrared laser (808nm, Model NO. LSR808H-FC-1W, Lasever Inc., Ningbo, China) was utilized to irradiate the surface of sample. For the ET/PT synergistically deicing test, both DC resource (15V) and near-infrared laser worked to accelerate the melt of the ice.

9.5.9 Results and Disscussion

9.5.9.1 The Morphology and Chemical Composition

Here, the keypoint is to achieve self-healing function and mechanical robustness simultaneously, which was achieved by partially embedding hierarchical structure into a self-healing substrate (Figure 9-29 (a)). Both microscale filler (carbon powder, 10-100μm size) and nanoscale filler (MWCNT, 30-50nm diameters) were utilized in this research. Because of the dual-scale nature of the fillers, the hierarchical structure would be constructed by sparsing the fillers onto the substrate. By further sparsing the filler onto the semi-cured substrate, the partially-embedded structure could be obtained which has been confirmed by our previous research. Here, the poly (urea-urethane) elastomer was chosen due to its excellent self-healing function. Therefore, the hierarchical structures together with partially-embedded structure lead to mechanical robustness, and the self-healing substrate endows the sample healable function.

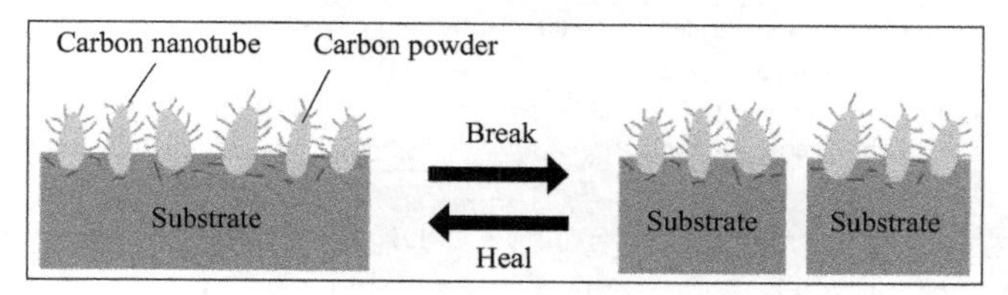

(a) The schematic of the self-healing superhydrophobic sample

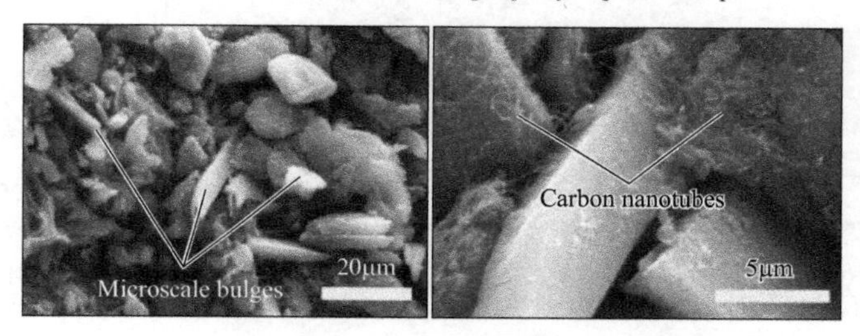

(b) SEM image of superhydrophobic sample at low magnifications

(c) SEM image of superhydrophobic sample at high magnifications

Figure 9-29 The schematic and images of sample

The hierarchical structure can be confirmed by the SEM observation. Many microscale bulges could be found in Figure 9-29(b) and (c). We attributed these to the carbon powders. Moreover, it was found that some CNTs were on top and between the microscale bulges (Figure 9-29(c)). The CNTs play a vital role in the sample, which not only favors the construction of the hierarchical structure, but also favors the formation of conductive network.

The low surface energy is another indispensable factor for preparing superhydrophobic materials. Here, we analyzed the surface chemical composition using the XPS test. From Figure 9-30(a), Si 2p, C 1s, O 1s and F 1s peaks were detected. From the F 1s core-level spectra (Figure 9-30 (b)), a dominant peak at 689.2eV was measured. We ascribed this to the fluorine bond as CF_x, which indicates that F is present in same bonding environment as that of FAS. In the case of Si 2p spectrum (Figure 9-30(c)), the raw data could be further fitted to two peaks, which are ascribed to —Si—OH or SiF_x species at 104.5eV and SiO_2-based network at 103.4eV. Moreover, the C 1s (Figure 9-30 (d)) spectrum can be deconvoluted into six peaks at 284.80, 285.90, 286.86, 292.11 and 294.16eV, which can be assigned to the —CH, C—O, C—CF, $—CF_2$ and $—CF_3$, respectively. In this research, the C—CF, $—CF_2$ and $—CF_3$ groups offer low surface energy, which is crucial for the formation of superhydrophobicity. Then, this superhydrophobic sample demonstrates outstanding self-cleaning function. The surface of the superhydrophobic sample was clean and dry because the water droplets could leave the surface completely and carry the contaminants away. On the other hand, the water droplets could not leave the surface of bare glass and RTV (room temperature vulcanized silicone rubber) coated glass, which led to dirty and wet surfaces.

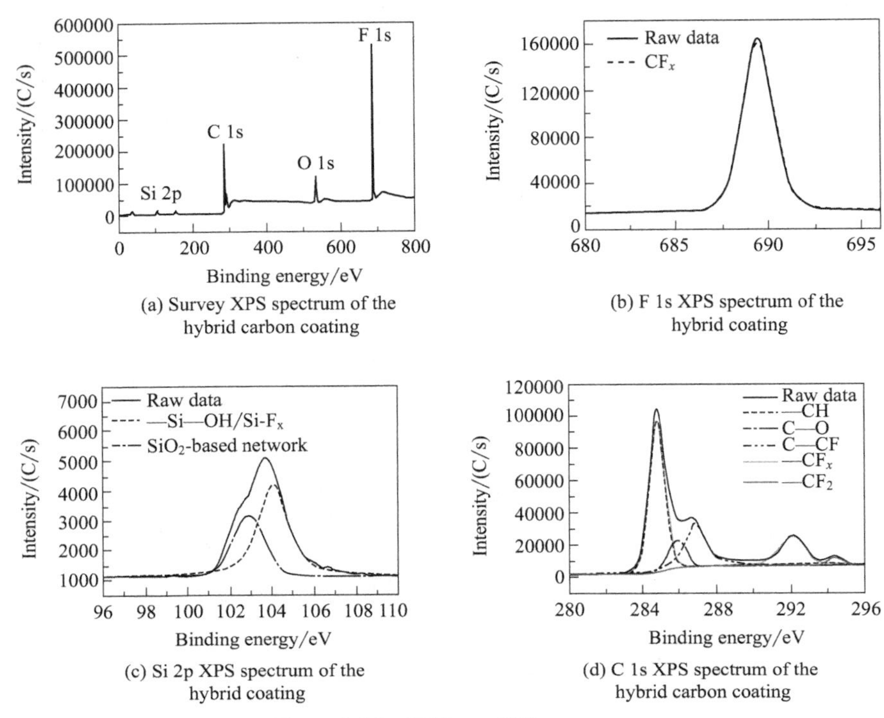

(a) Survey XPS spectrum of the hybrid carbon coating

(b) F 1s XPS spectrum of the hybrid coating

(c) Si 2p XPS spectrum of the hybrid coating

(d) C 1s XPS spectrum of the hybrid carbon coating

Figure 9-30 Different XPS spectrum

9.5.9.2 The Self-Healing Ability

This superhydrophobic sample has advantages in restoring the loss of superhydrophobicity caused by cutting. Here, the original sample was cut in half using scissors (Figure 9-31(b)). Then, we mended the sample by simply contacting the cutted parts at room temperature (Figure 9-31(d)). After 1h of contact, the mended sample retained superhydrophobicity. Moreover, the mended sample demonstrates excellent mechanical properties, which could withstand the weight of 500g. The further SEM observation confirmed that the cutted polymers contact with each other again. Therefore, the self-healing mechanism can be ascribed to the covalent bonds caused by the aromatic disulfide metathesis.

9.5.9.3 The Anti-Icing Ability

In this research, both microscale carbon powders and nanoscale CNTs were utilized as the fillers. All of them have been widely studied as thermal heater for efficiently transforming electrical energy into Joule heating energy. To investigate the electrothermal property, a hybrid carbon coating (25mm×16mm size) was connected to a source of direct current (DC). Then, we detected the time dependent temperature profile under different DC voltage using an infrared imager (Figure 9-32(a)). It could be found that the hybrid carbon coating could attain a saturation temperature of 14.9℃ from subzero temperature (−5.0℃) after applying 9V voltage. Moreover, the surface temperature could increase rapidly from −5.0℃ to 74.6℃ under a voltage of 15V (Figure 9-32(b)).

(a) The original sample (b) Cutting the sample

(c) The cutted sample (d) The healed sample

Figure 9-31 The self-healing process of the superhydrophobic sample

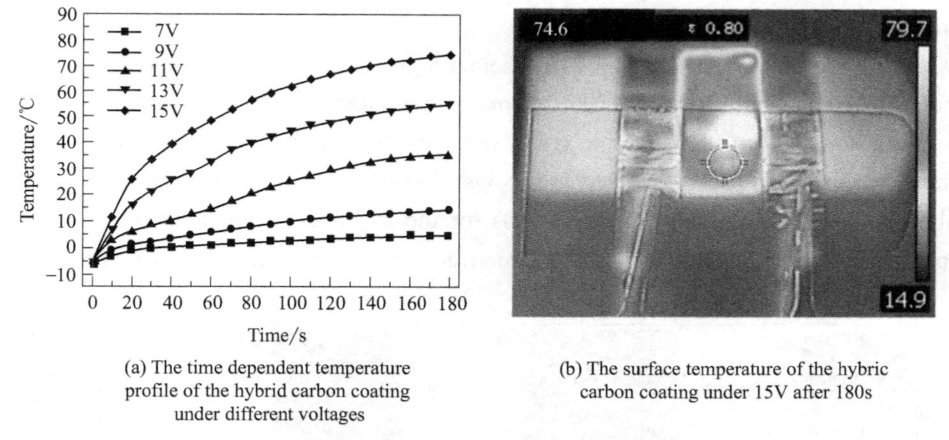

(a) The time dependent temperature profile of the hybrid carbon coating under different voltages

(b) The surface temperature of the hybric carbon coating under 15V after 180s

Figure 9-32 The temperature profile and the surface temperature

The superior electrothermal (ET) property together with superhydrophobicity endows a unique anti-icing ability to the hybrid carbon coating. We first investigated the removing ability of the frozen droplet. A droplet (~5μL) was carefully placed on the sample, and gradually frozen under −25℃ temperature. Then a voltage of 15V was applied. From Figure 9-33(a), the frozen water droplet can be melted after ~148s, and then rolled off completely without any residue. It is worth noting that the environment temperature of deicing test was set to be −25℃. Thus, the time required to melt the ice is relatively long. Besides the electrothermal effect, the photothermal (PT) performance of the carbon-based materials were also attracting more and more attention. Recently, He et al. prepared superhydrophobic surfaces based on candle soot, which can rapidly melt the accumulated frost and ice in 300s under 1 sun. Here, we tried to investigate the photothermal effect of the hybrid carbon coating using the near-infrared laser (1W). Within 265s of illumination, the frozen droplet on the hybrid carbon coating melted (Figure 9-33(b)).

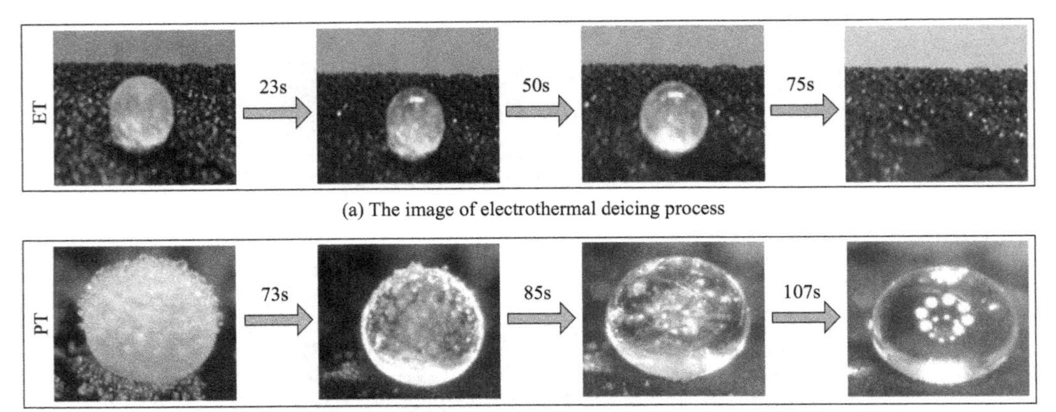

(a) The image of electrothermal deicing process

(b) The image of photothermal deicing process of a fozen droplet

Figure 9-33 The images of different deicing process

Nevertheless, the deicing process of frozen droplet can not completely reflect the actual freezing weather. Recently, Jiang et al. systematically analyzed the surface icing accidents and found that the glaze ice is the most dangerous situation. In this research, we prepared the glaze ice in the artificial climate chamber. An ice layer (~3mm thick) was frozen on the surface of the hybrid carbon coating.

It is worth noting that CNTs are a kind of fascinating material which demonstrates excellent electrothermal (ET) and photothermal (PT) properties simultaneously. Here, the ET, PT and ET/PT deicing test were all investigated and compared. We first applied a voltage of 15V onto the hybrid carbon coatings. The ET effect led to the fast rise of the surface temperature, which further melted the ice (Figure 9-34(a)). It takes 530s for the ice layer to be completely melted. In the next step, a near-infrared laser was utilized to irradiate the surface. The carbon-based materials

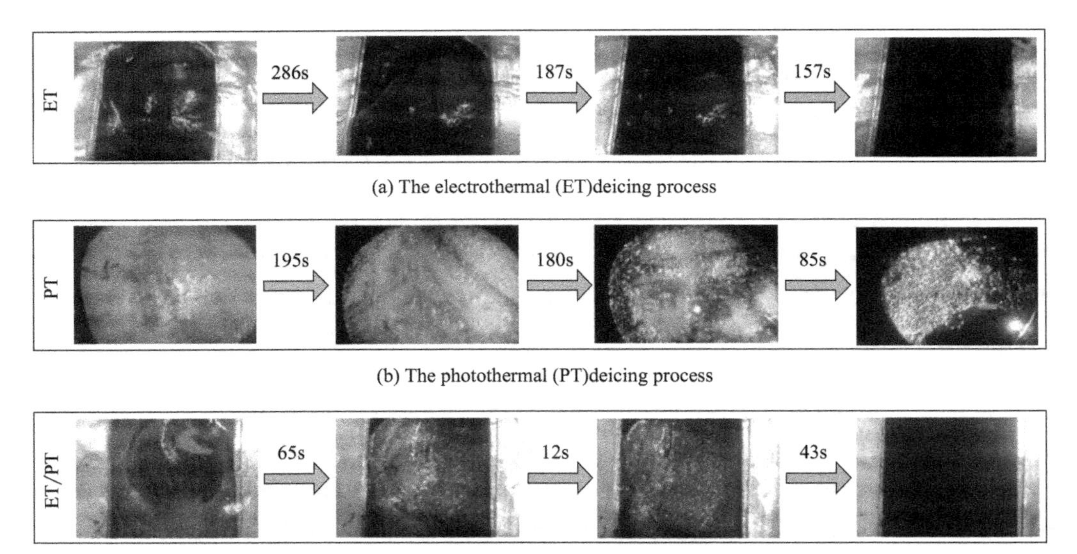

(a) The electrothermal (ET)deicing process

(b) The photothermal (PT)deicing process

(c) The ET/PT synergistically deicing process of an thick ice layer

Figure 9-34 The different deicing process

would absorb the near-infrared light, and the heat was generated. All of the ice on the hybrid carbon coating melted, and the water droplets rolled off, leaving a dry and clean surface after 460s (Figure 9-34(b)). Finally, the ET/PT synergistically deicing strategy was investigated. We let the DC current resource and the near-infrared laser worked at the same time. The heat generated from ET and PT could be accumulated, which would effectively speed up the deicing time (120s), as shown in Figure 9-34(c).

9.5.9.4 The Mechanical Robustness

Although superhydrophobic materials demonstrated exciting performance in self-cleaning, anti-icing, oil-water separation and so on, most of them can be easily destroyed by a slight scratch, rubbing, and even finger contact. Thus, the poor mechanical robustness has been regarded as Achilles' heel which prohibited the practical applications of the superhydrophobic materials. With the development of mechanical robust superhydrophobic materials, the sandpaper abrasion test has been regarded as the most widely used method. Ras et al. further pointed out that the abrasion distance and applied normal pressure were two crucial indicators for the convenience of comparison. Furthermore, Kulinich et al. found that repetitive icing/deicing would gradually destroy the micro/nano structure of the superhydrophobic surfaces, and then led to the loss of ice-repellence. Therefore, sandpaper abrasion test and cyclic icing/deicing test of the hybrid carbon coating were investigated in the research.

Firstly, commercial sandpaper (Starcke 200#, Germany) was utilized in the sandpaper abrasion test. The hybrid carbon coating was coated onto an Al substrate as the sample. Then, the sample was placed on the sandpaper under a weight of 500g, and moved for 20cm as one cycle (Figure 9-35 (a) and (b)). A gradual decrease in the contact angles and increase in sliding angles could be observed with the increase of abrasion cycles (Figure 9-35(c)). We attributed this to two factors: some micro/nano fillers on the surface were abraded, which might change the surface morphology; the destruction of surface chemistry during the abrasion process. Nevertheless, the contact angles and sliding angles were still in the range of superhydrophobicity after 80 abrasion cycles, indicating the relatively strong mechanical robustness (abrasion distance: 16.00m, ap-

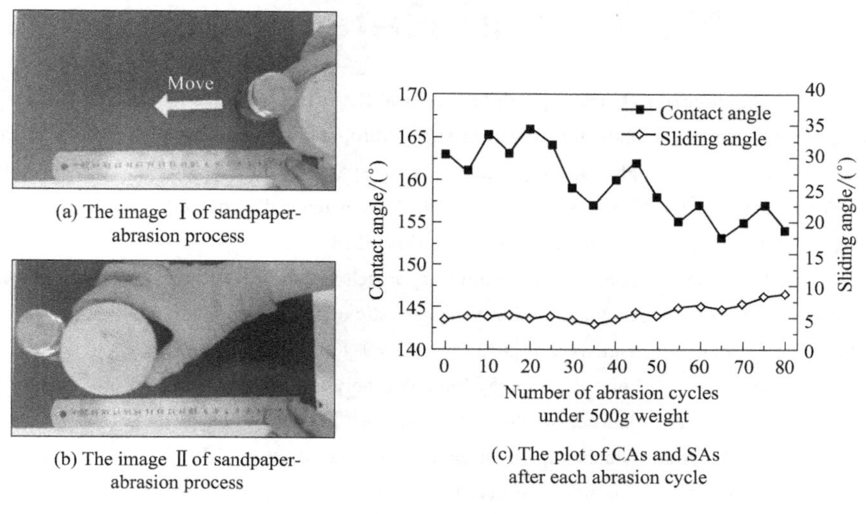

(a) The image Ⅰ of sandpaper-abrasion process

(b) The image Ⅱ of sandpaper-abrasion process

(c) The plot of CAs and SAs after each abrasion cycle

Figure 9-35 The experimental images and plots

plied pressure: 12.50kPa). We attribute the aforementioned abrasion-resistant superhydrophobicity to our collaborative strategy of hierarchical structure and partially-embedded structure. On the one hand, the microscale bulges in the research were relatively strong which could protect the nanofillers from being abraded. On the other hand, the micro/nano fillers were partially embedded into the substrate, and then the substrate could offer support to the fillers.

As a second mechanical robustness test, the repetitive icing/deicing test was further investigated. Here, a layer of glaze ice (~3mm thickness) was formed on the surface of the hybrid carbon coating. We further applied a voltage of 15V for electrothermal deicing. Then, the timerequired to completely melt the ice was recorded (Figure 9-36(a)). After fully removing the ice, the surface temperature, contact angles and sliding angles were measured Figure 9-36(a) and (b). As shown in Figure 9-36(a), the deicing time gradually increased from 530s to 554s, and the surface temperature gradually decreased from 69.0℃ to 65.9℃. Furthermore, 30 icing/deicing cycles led to a gradual drop in contact angle from 163° to 152°, and a tiny increase of sliding angle from 4° to 7°. However, the hybrid carbon maintained superhydrophobicity even after 30 icing/deicing cycles, indicating excellent mechanical robustness.

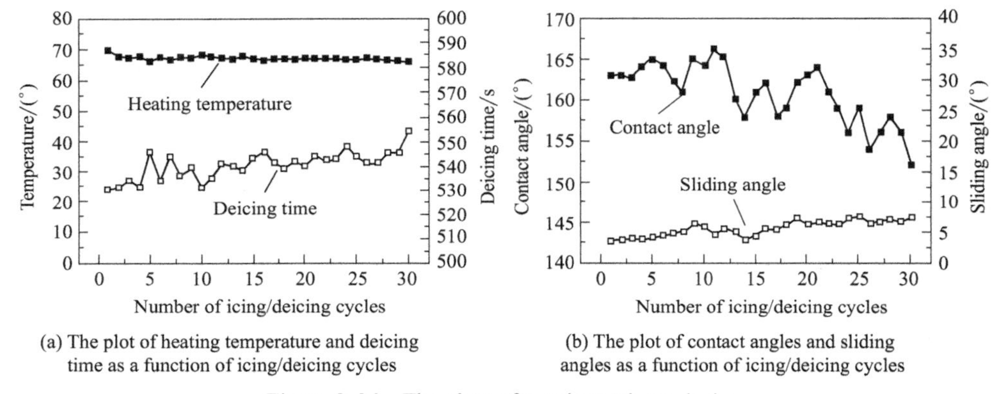

(a) The plot of heating temperature and deicing time as a function of icing/deicing cycles

(b) The plot of contact angles and sliding angles as a function of icing/deicing cycles

Figure 9-36 The plots of exprimental conclusion

Notes and References

[1] Liu K, Jiang L. Bio-inspired self-cleaning surfaces. Annual Review of Materials Research, 2012, 42: 231-263.

[2] Ogihara H, et al. Simple method for preparing superhydrophobic paper: spray-deposited hydrophobic silica nanoparticle coatings exhibit high water-repellency and transparencty, Langmuir, 2012, 28: 4605-4608.

[3] Xiao F, Yuan S, Liang B, et al. Superhydrophobic CuO nanoneedle-covered copper surfaces for anticorrosion. Journal of Materials Chemistry A, 2015, 3: 4374-4388.

[4] Chang C, Hsu M, Weng C, et al. 3D-bioprinting approach to fabricate superhydrophobic epoxy/organophilic clay as an advanced anticorrosive coating with the synergistic effect of superhydrophobicity and gas barrier properties. Journal of Materials Chemistry A, 2013, 1: 13869-13877.

[5] Crick R C, Gibbins J A, Parkin I P. Superhydrophobic polymer-coated copper-mesh; membranes for highly efficient oil-water separation. Journal of Materials Chemistry A, 2013, 1: 5943-5948.

[6] Li J, Shi L, Chen Y, et al. Stable superhydrophobic coatings from thiol-ligand nanocrystals and their application in oil/water separation. Journal of Materials Chemistry A, 2012, 22: 9774-9781.

[7] Deng X, Mammen L, Butt H, et al. Candle soot as a template for a transparent robust superamphiphobic

coating. Science, 2012, 335: 67-70.

[8] Lai Y, Tang Y, Gong J, et al. Transparent superhydrophobic/superhydrophilic TiO-based coatings for self-cleaning and anti-fogging. Journal of Materials Chemistry A, 2012, 22: 7420-7426.

[9] Xu L, Zhu D, Lu X, et al. Transparent, thermally and mechanically stablesuperhydrophobic coating prepared by anelectrochemical template strategy. Journal of Materials Chemistry A, 2015, 3: 3801-3807.

[10] Park K, Choi H J, Chang C, et al. Nanotextured silica surfaces with robust superhydrophobicity and omnidirectional broadband supertransmissivity. ACS Nano, 2012, 6: 3789-3799.

[11] Dong J, Yao Z, Yang T, et al. Control of superhydrophilic and superhydrophobic graphene interface, Scientific Reports, 2013, 3: 1733.

[12] Nguyen D D, Tai N H, Leea S B, et al. Superhydrophobic and superoleophilic properties of graphene-based sponges fabricated using a facile dip coating method. Energy Environ. Sci. , 2012, 5: 7908-7912.

[13] Wang J N, Shao R Q, Zhang Y L, et al. Biomimetic graphene surfaces with superhydrophobicity and iridescence. Chemistry-An. Asian Journal, 2012, 7: 301-304.

[14] Chen K, Xue D. Preparation of colloidal graphene in quantity by electrochemical exfoliation. Journal of Colloid & Interface Science, 2014, 436: 41-46.

[15] Parvez K, Wu Z S, Li R, et al. Exfoliation of graphite into graphene in aqueous solutions of inorganic salts. Journal of the American Chemical Society. , 2014, 136: 6083-6091.

[16] Chu Z M, Jiao W, Huang Y, et al. FDTS-modified SiO_2/rGO wrinkled films with a micro-nanoscale hierarchical structure and anti-icing/deicing properties under condensation condition, Advanced Materials Interfaces, 2019, 7.

Chapter 10
Thermal Ageing Mechanism of Advanced Electrical Insulation Materials—Taking Oiled Paper Insulation as an Example

10.1 Introduction of Oiled Paper Insulation

In this chapter, the oil-paper insulation surface discharge model and the tip discharge model are designed to simulate the case of transformer oil-paper insulation with a strong tangential electric field and a strong normal electric field. Power transformer is one of the key equipment of power system. It undertakes many functions such as power transmission and transformation. However, large power transformers are generally in the operation state of high voltage and high current, and are vulnerable to wind, frost, rain and snow. With the increase of operation time, transformer faults are inevitable, and insulation deterioration is the main cause of transformer faults. According to the statistics of transformer faults of State Grid Corporation of China from 2002 to 2003, the results show that the faults caused by insulation deterioration account for 83.3% of the total transformer faults, and the faults mainly occur at the longitudinal insulation and main insulation of the transformer[1]. Therefore, judging the insulation condition of transformer and accurately warning the insulation fault can effectively improve the safety and stability of transformer operation, which is of great significance to improve the reliability and economy of power system.

Oil paper insulation is mainly composed of insulating paper or insulating paperboard and transformer oil. It is a typical composite insulation form. Oil paper insulation not only has good insulation performance, but also has excellent heat dissipation and arc extinguishing performance[2,3]. The main form of insulation of large power transformer is oil paper insulation [4]. During the design and manufacture of transformer oil paper insulation, its insulation performance and margin are fully considered, so it is difficult for the general oil paper insulation to directly cause insulation failure. However, due to the manufacturing process, operating environment and related faults, the insulating paperboard will inevitably have defects such as moisture, bubbles and burrs, which is easy to produce local high field strength areas. Under the action of this local high field strength region, oil paper insulation is prone to partial discharge (referred to as partial discharge)[5]. At the same time, with the increase of transformer operation time, the oil paper insulation will gradually age, resulting in the decline of its performance and affecting the insulation level. At present, it is easy to detect and judge the performance of transformer oil. When the transformer oil is aged and deteriorated, it is also easy to deal with or even replace it. However, it is difficult to treat and replace the insulating paperboard, which is one of the important components of oil paper insulation. When the oil paper insulation is aged, the probability of partial insulation deterioration under the action of impulse overvoltage and other faults will increase significantly, and the possibility of partial discharge will also increase greatly.

When partial discharge occurs, it mainly has the following destructive effects on insulation[5]:

(1) Partial discharge makes charged particles continuously impact the surface of insulating medium and cut off its molecular structure.

(2) Under the action of partial discharge, it is easy to increase the local temperature of insulating materials, cause local overheating and accelerate the deterioration of insulating materials.

(3) Under the action of partial discharge, it is easy to produce oxidizing substances, which makes the insulating medium oxidized and deteriorated.

Of course, the destructive effect of partial discharge on insulating materials develops slowly, and generally will not cause insulation failure in a short time. However, with the continuous discharge time and the development of the discharge process, the degree of insulation degradation will be deeper and deeper, which will seriously shorten the insulation life and even cause insulation accidents such as breakdown.

The research on the law of partial discharge degradation of transformer oil paper insulation can improve the effectiveness of transformer oil paper insulation performance evaluation based on partial discharge, and lay a foundation for the evaluation and judgment of the development degree of oil paper insulation degradation based on partial discharge.

It must be pointed out that during the operation of transformer, with the increase of operation time and the influence of various related faults, the oil paper insulation will inevitably age. There are many reasons for ageing, mainly including thermal ageing, mechanical ageing, environmental ageing, etc[6]. Thermal ageing is the main cause of transformer oil paper insulation ageing. In actual operation, the performance of transformer oil is often tested. The performance of transformer oil is easy to monitor, and it is easy to handle or replace when the transformer oil deteriorates. For the insulating paperboard, one of the important components of oil paper insulation, it is difficult to directly monitor, process and replace when it is aged. Therefore, in order to be consistent with the actual operation of the transformer, the comprehensive influence of insulation paperboard ageing should be considered when analyzing the partial discharge deterioration law of transformer oil paper insulation. This chapter mainly analyzes the partial discharge deterioration law of transformer oil paper insulation under the ageing state of insulating paperboard.

10.2 Experimental Plan Design

10.2.1 Typical Defect Model and Experimental Wiring of Oil-Paper Insulation

In this chapter, the oil-paper insulation surface discharge model and the tip discharge model are designed to simulate the case of transformer oil-paper insulation with a strong tangential electric field and a strong normal electric field.

The surface discharge model adopts the column plane model, as shown in Figure 10-1. Column and plate electrodes are made of aluminum, and the surface is polished. The diameter of column electrode is 25mm, the height is 20mm, and the diameter of plate electrode is 90mm. The square insulating paperboard is added between the column plate electrodes, with a side length of 80mm and a thickness of 2mm.

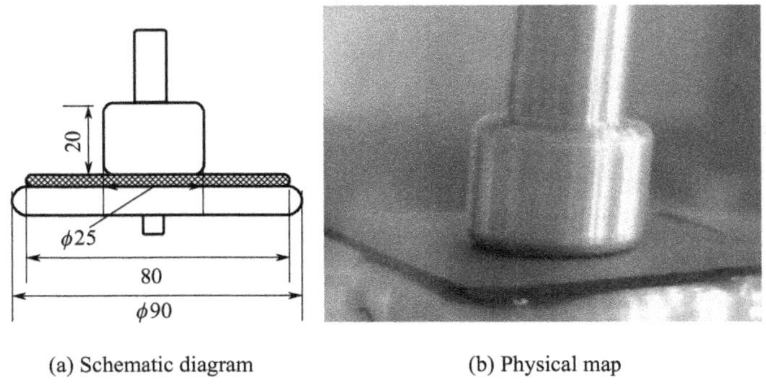

(a) Schematic diagram (b) Physical map

Figure 10-1 Oil-pressboard surface discharge model

The tip discharge model adopts the needle-plate model, as shown in Figure 10-2. The size and material of the plate electrode are the same as those of the plate electrode in the column plate model. The inside of the needle electrode is copper, and the surface is gold-plated. The diameter of the needle electrode is 1mm, and the needle tip angle is 30 degrees. Between the needle board models is still a square insulating paperboard with a side length of 80mm, with a thickness of 2mm, and the needle tip lightly rests on the upper surface of the insulating paperboard.

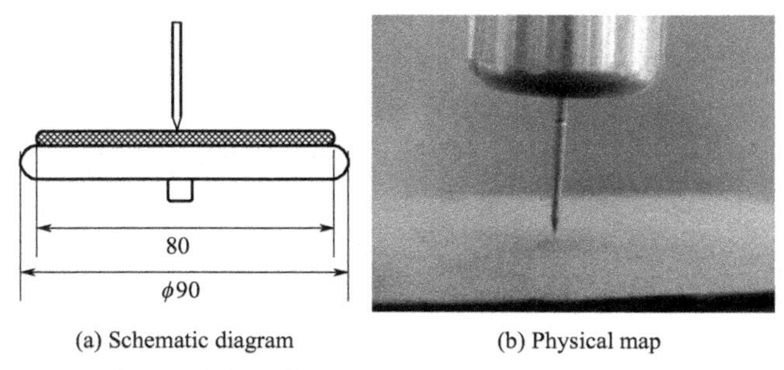

(a) Schematic diagram (b) Physical map

Figure 10-2 Oil-pressboard point discharge model

This section is based on the pulse current method to measure the partial discharge signal under the surface discharge and tip discharge defect of oil-paper insulation, and uses a digital partial discharge instrument to save and analyze the partial discharge signal. Figure 10-3 and Figure 10-4 are the experimental wiring diagram and experimental device diagram respectively. The test transformer adopts the TDTW 50/250 oil-immersed experimental transformer, with a rated voltage of 250kV, a rated capacity of 50kVA. Under the rated voltage, the partial discharge of the test transformer is less than 5pC. The protection resistance is 10.5kΩ. The capacitive divider is used to extract the voltage signal, and the divider ratio is 1500∶1. The HCPD-2022 portable PD (partial discharge) tester produced by Baoding Huachuang electric company limited is used to collect the PD signal. The PD tester is operated by full touch screen, with the function of automatic selection of range and automatic selection of signal amplification ratio. In the experiment, shielding layer is added to the wiring to shield external interference. During the experiment, the sampling frequency was set to 25MHz, and a full power frequency cycle discharge signal was collected and saved every 12s. At the same time, real-time shooting of the discharge is performed through a video camera.

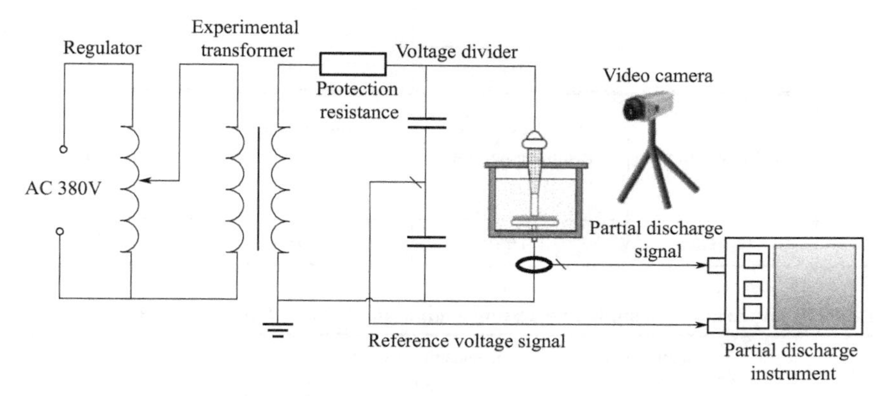

Figure 10-3 Experimental wiring diagram

In this section, the high frequency current transformer (HFCT) based on the principle of Rogowski coil is used to collect partial discharge pulse signals. The sensitivity is less than 1pC and the measurement frequency band is 100kHz~100MHz, as shown in Figure 10-5.

Figure 10-4 Experimental device diagram

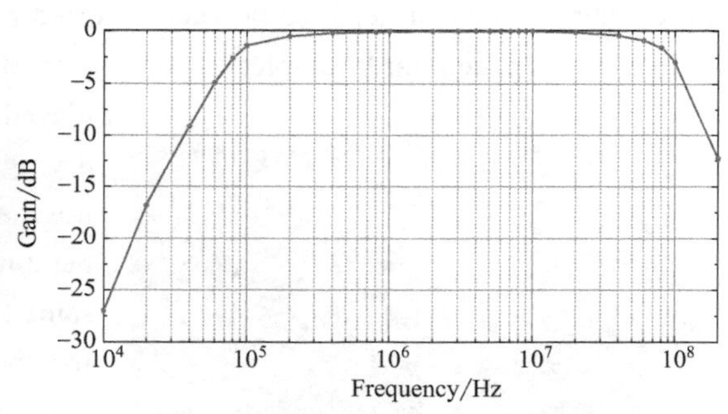

Figure 10-5 Amplitude-frequency characteristic of HFCT

10.2.2 Selection and Treatment of Oil-Paper Insulation Samples

Oil-paper insulation consists of insulating paper or insulating board dipped in transformer oil. In this section, the insulating paperboard sample selected is the insulating paperboard for UHV (ultra-high voltage) transformers produced by Weidmann Company of Switzerland, which has a large market share, and its thickness is selected as 2mm. The transformer oil selected in this article is No. 25 transformer oil produced by Xinjiang Karamay Petrochemical Company. The technical indicators of the selected insulating paperboard and transformer oil are shown in Table 10-1 and Table 10-2.

Table 10-1 Training sample and testing sample for PD development process recognition

Thickness	Thickness	Longitudinal tensile strength ≥	Transverse tensile strength ≥	Longitudinal elongation ≥	Transverse elongation ≥
+5%	$1.20g/cm^3$	112.24MPa	97.52MPa	7.44%	7.68%
Longitudinal shrinkage ≤	Lateral shrinkage ≤	Thickness shrinkage ≤	Moisture ≤	Ash ≤	Conductivity ≤
0.5%	0.7%	5.1%	7.1%	0.29%	3.04mS/m
pH value	Oil absorption	Compressibility (Compression ratio) ≤	Compressibility (Recoverable part) ≥	Electric strength in oil ≥	Electric strength in the air ≥
6.45	12.12%	4.48%	57.58%	35.59kV/mm	13.47kV/mm

Table 10-2 Training sample and testing sample for PD development process recognition

Density(20℃)	Kinematic viscosity (40℃)	Kinematic viscosity (-30℃)	Pour point	Flash point
$884.1kg/m^3$	$10.05m^2/s$	—	-42℃	142℃
Dissipation factor(90℃)	Aniline point	Acid value	precipitation	Breakdown voltage
<0.001	78℃	0.07mgKOH/g	0.03	68kV

Before using the newly-made insulating paperboard, in order to remove impurities such as moisture contained in it, it needs to be vacuum dried pretreatment. The specific method: After cutting the insulating paperboard samples according to the size requirements specified above, they are placed in a vacuum drying box, and the vacuum drying box is evacuated by an ion pump, so that the internal air pressure is lower than 50Pa, and the temperature of the vacuum drying box is adjusted at the same time vacuum drying the cardboard sample at 90℃ for about 48h. After the vacuum drying is completed, put the insulating paperboard sample in a vacuumoiling machine, and inject it into the transformer oil sample that has been filtered and processed, while maintaining the temperature at 80℃ and vacuum immersion for 24 hours. Figure 10-6 shows the vacuum drying oven.

Figure 10-6 Vacuum drying box

10.2.3 Accelerated Ageing of Insulating Paperboard

Thermal ageing is the main reason for the ageing of insulation materials, and it mainly depends on the operating environment temperature of the insulation materials. Studies have shown that every time the operating temperature increases by 6 to 8℃, the life of the insulating material will be reduced by about half. When thermal ageing occurs in insulating materials, the Monsinger thermal ageing rule shown in formula (10-1) can be used to estimate the remaining life.

$$T = T_0 e^{-\alpha(\theta-\theta_0)} \tag{10-1}$$

θ_0 is the base working temperature of the insulation material; T_0 is the insulation life at the reference working temperature; θ is the actual working temperature of the insulation material; T is the insulation life at the actual working temperature; α is the thermal ageing coefficient, it is generally determined by the structure and properties of the insulating material. For oil-immersed transformers in China, the thermal ageing coefficient α is generally equal to 0.1155.

It can be seen from formula (10-1) that increasing the actual working temperature θ can accelerate the thermal ageing of insulating materials. Therefore, by increasing θ, that is, heating ageing can be used to obtain oil-paper insulation samples with different degrees of ageing in a short time. At this time, θ is also called the heating ageing temperature. Two factors should be considered when determining heating ageing temperature θ: (1) If the selection of θ is too low, the speed of heat ageing will decrease, and the time required to reach the specified degree of ageing will be too long. From the perspective of improving the efficiency of the experiment, the temperature of heat ageing should be higher. (2) If the selection of θ is too high, when the temperature is close to or exceeds the flash point temperature of the transformer oil, it is easy to cause the transformer oil to catch fire and explode. It is stipulated in GB/T 7595—2008 that the flash point of transformer oil should be ≥135℃, and it can be seen from Table 10-2 that the flash point temperature of the transformer oil sample selected in this article is 142. Considering the two aspects of experimental efficiency and experimental safety comprehensively, the heating ageing temperature is selected as 130℃ in this section. Under normal conditions, the reference operating temperature θ_0 of transformer oil-paper insulation is about 80℃. It can be seen from formula (10-1) that 1 day, 6 days, 16 days, 32 days and 48 days of operation at 130℃ can be equivalent to 0.88, 5.3, 14.1, 28.2 and 42.4 years of operation under normal conditions of transformer.

The insulating paperboard samples treated by vacuum drying and soaking and the transformer oil samples treated by filtration and purification were mixed according to the mass ratio of 1∶10, and the oil-paper insulation samples were obtained and sealed in the explosion-proof beaker. The oil-paper insulation test sample was heated and aged by vacuum oven at 130℃. When the ageing time reaches 1 day, 6 days, 16 days, 32 days and 48 days respectively, take out an appropriate amount of insulating paperboard and soak it in the filtered and purified new transformer oil for 2 days. After the soaking is completed, replace with the new transformer oil and continue soaking 2 days, then repeated vacuuming three times to fully extract the aged oil sample from the cardboard. Through the above method, the insulation paperboard samples with different ageing degrees were obtained. Degree of polymerization (DP) can be used to directly characterize the ageing degree of insulation paperboard. Generally speaking, the DP value of the new insulation paperboard after vacuum drying and oil immersion treatment is about 900. When DP value drops to about 500, its insulation life drops to about half. When the DP value decreases to about 200, the insulation life of the insulation paperboard reaches the late stage. This section measures the degree of polymerization according to the method specified in IEC 60450—2007, and the basic steps are as follows: (1) Use a rope soxhlet to perform liposuction on the insulating paperboard sample to remove the transformer oil from the insulating paperboard; (2) Weigh a certain mass of three samples of insulating paperboard obtained after liposuction operation, each with a mass of about 50mg, and dissolve them with copper ethylene diamine aqueous solution. (3) Using an Ubbelohde viscometer, measure the viscosity of the solution in step (2), and then calculate the degree of polymerization of the insulating paperboard from this. Figure 10-7 shows the Soxhlet extractor and Ubbelohde viscometer.

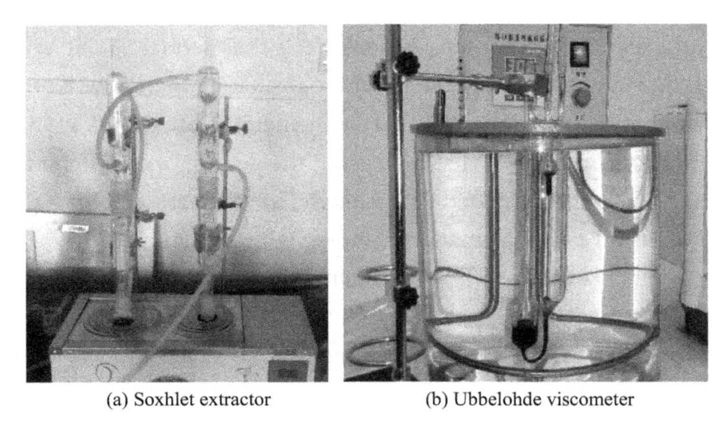

(a) Soxhlet extractor (b) Ubbelohde viscometer

Figure 10-7 Soxhlet extractor and Ubbelohde viscometer

Figure10-8 shows the measurement results of the degree of polymerization of the insulating paperboard for each ageing degree. From the results in Figure 10-8, when the ageing time is 16 days, the DP value is close to 500, and the insulation life is only about half left. When the ageing time reaches 48 days, the DP value is 231, which is close to the end of the insulation life.

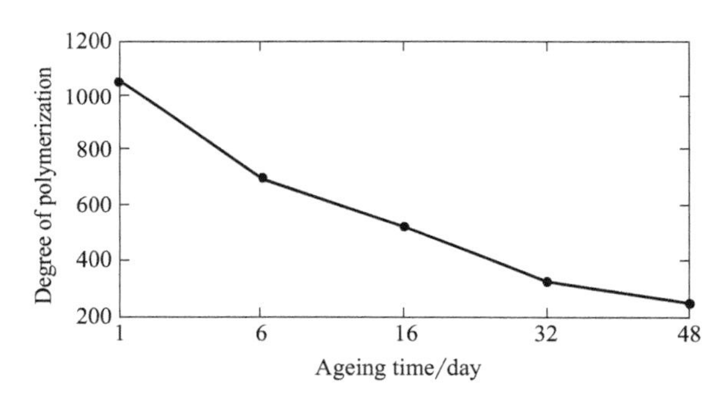

Figure 10-8 Measurement results of the degree of polymerization

10.3 Surface Discharge Deterioration Law of Transformer Oil-Paper Insulation

10.3.1 Initial Discharge Voltage and Discharge Endurance Time of Transformer Oil-Paper Insulation

Put the aged insulating paperboard sample into the new transformer oil to form the transformer oil paper insulation sample. In order to ensure sufficient oil immersion, the insulating cardboard should be placed in the transformer oil for one day before the test. The initial discharge voltage of surface discharge of transformer oil-paper insulation was measured. The specific method was as follows: the voltage was slowly increased, and the voltage remained unchanged after the discharge signal appeared, and maintained for 10min. If there was still a stable discharge signal at this time, this voltage was the initial discharge voltage of surface discharge of the test sample. Insulation paperboards with different ageing degrees all contain three samples. The initial discharge voltage along the surface of each sample is recorded and the average value is calculated, as shown in Table 10-3. From the results in Table 10-3, it can be seen that with the deepening of the ageing of the insulating paperboard, the initial discharge voltage of the transformer oil-paper insulation decreases slightly, but the amplitude is smaller.

Table 10-3 Surface discharge inception voltage

Ageing time	1 day	6 days	16 days	32 days	48 days
Initial discharge voltage/ kV	18.6	18.3	18.2	17.9	17.5

The constant pressure method was used to carry out a long-term pressurization experiment on each sample. The voltage applied during the experiment was 1.4 times the initial voltage of the insulating paperboard after ageing for one day, that is, 26kV. Table 10-4 is the surface discharge tolerance time of oil-paper insulation under different ageing degree insulation board. From the results in Table 10-4, it can be seen that with the increase of the ageing degree of the insulating paperboard, the surface discharge endurance time of the transformer oil-paper insulation is gradually shortened.

Table10-4 Surface discharge tolerance time

Ageing time	1 day	6 days	16 days	32 days	48 days
Tolerance time	24h 37min	24h 14min	23h 31min	22h 12min	20h 15min

During the experiment, it was found that the surface discharge development law of the transformer oil-paper insulation samples with the same degree of ageing of the insulating paperboard was similar. Therefore, when the subsequent research on the deterioration phenomenon and law of the transformer oil-paper insulation surface discharge, the transformer oil-paper insulation with the same degree of ageing of the insulating paperboard, only select a sample and describe its surface discharge phenomenon and discharge law.

10.3.2 Description of Surface Discharge Deterioration Phenomenon of Oil-Paper Insulation of Transformer and Division of Discharge Development Degree

This article uses video and audio long-term recording and real-time observation by experimenters to describe the development and deterioration of surface discharge on transformer oil-paper insulation with different ageing degrees.

During the experiment, the surface discharge deterioration phenomenon of oil-paper insulation samples with different ageing degrees of insulating paperboard is similar, so only the surface discharge phenomenon of oil-paper insulation with insulation paperboard ageing degree of 1 day is described here.

At the beginning of the discharge, there was no obvious discharge spark and discharge sound; with the increase of the discharge time, when the pressurization time reached about 2h 40min, the discharge spark was occasionally generated at the edge of the column electrode, and it was accompanied by the occasional discharge spark. Weak and short discharge sound, as shown in Figure 10-9(a) ; with the further progress of the discharge, the frequency of spark discharge increases, and the discharge sound is further increased. At the same time, there are occasional precipitates at the insulating paperboard at the edge of the column electrode; When the pressurization time is about 8h, as shown in Figure10-9(b) , the edge of the column electrode produces a dendritic discharge along the surface of the insulating paperboard, accompanied by a large sharp and short discharge sound, but the dendritic discharge creeps up at this time. The electrical distance and duration are short, and the energy is weak (shown as low brightness) , and the interval between two adjacent dendritic discharges is longer, and spark discharge is still mixed; as the experiment time continues to increase, as shown in the figure 10-9(c) , the creepage distance of the dendritic discharge gradually increases, and the time interval between two adjacent dendritic discharges is gradually shortened, the brightness gradually increases, and the frequency of the discharge sound gradually increases; when the pressure time reaches at about 22h, as shown in Figure 10-9(d) , the number of dendritic discharges at the edge of the column electrode is increasing, and the appearing areas are more and more scattered. At the same time, it is accompanied by bright cluster-shaped discharge areas. Accompanied by a loud squeaking discharge sound, accompanied by a high-frequency sharp sound, and more and more rapid. When the discharge reaches 24 hours, continue the experiment. At this time, the cardboard discharge sound is already loud, and there are more and more precipitates near the cardboard, and there is obvious oil layer disturbance; as the discharge progresses, the surface of the insulating cardboard suddenly produces a through discharge channel. The cardboard flashed along the surface, accompanied by strong luminescence and sound. At the same time, thick black smoke emerged from the electrodes and spread rapidly.

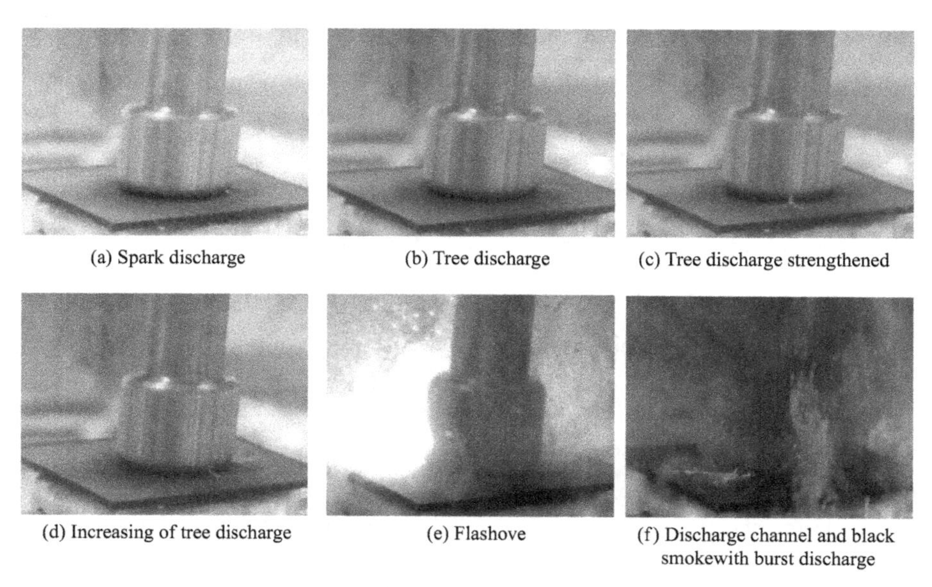
(a) Spark discharge (b) Tree discharge (c) Tree discharge strengthened

(d) Increasing of tree discharge (e) Flashove (f) Discharge channel and black smokewith burst discharge

Figure 10-9 The process of discharge

For the oil-paper insulation samples aged 6 days, 16 days, 32 days, and 48 days of the insulation board, the surface discharge deterioration phenomenon is similar to the surface discharge deterioration of the oil-paper insulation with the insulation board ageing degree of 1 day, but each phenomenon appears. The time is slightly different, as shown in Table10-5.

Table 10-5 Discharge phenomena of oil-pressboard surface discharge

Discharge phenomenon	Approximate time of occurrence or time				
	Ageing 1 day	Ageing 6 days	Ageing 16 days	Ageing 32 days	Ageing 48 days
During the initial discharge stage, there is no obvious discharge spark or discharge sound	0-2h	0-2h	0-3h	0-3h	0-2h
The color of the black carbonized area gradually becomes darker and the area gradually becomes larger; the number of precipitated bubbles increases and floats up in an inverted cone shape; the intensity of the discharge spot becomes stronger and the size becomes larger; a weak discharge sound begins to form and becomes larger	2-8h	2-8h	3-7h	3-7h	2-7h
Intermittent dendritic discharge, weak energy, and low frequency, there are small precipitates near the cardboard. With the discharge progress, the creeping distance of dendritic discharge increases gradually, the brightness increases, the discharge sound frequency increases, and the precipitates increase	8-22h	8-22h	7-21h	7-21h	7-19h
The occurrence frequency of dendritic discharge is higher, the brightness is getting bigger and bigger, and accompanied by cluster discharge, there is continuous discharge sound accompanied by rapid and sharp discharge sound, and the precipitation is more obvious	22-24h	22-24h	21-23h	21-22h	19-20h
A penetrating discharge channel is produced, flashover occurs along the surface, accompanied by strong luminescence and sound, and thick black smoke is produced	24h 19min	24h 7min	23h 45min	22h 21min	20h 22min

According to the phenomenon differences in the development process of surface discharge of oil-paper insulation of various ageing degrees, this paper will divide the surface discharge development stage into three stages: the initial stage of discharge, the development stage of discharge and the dangerous stage of discharge.

In the initial stage of discharge, the discharge is relatively weak, and spark-like discharge is the mainstay. In the development stage of the discharge, the carbonization traces of the discharge increase rapidly, the discharge sound becomes larger and larger, and the dendritic discharge appears. In the discharge risk stage, the occurrence frequency of dendritic discharge is high, and there are obvious precipitates on the surface of insulating paperboard. Table 10-6 shows the development stages of surface discharge of oil-paper insulation with different ageing degrees, and the division of the pressure time period.

Table10-6 Division of surface discharge process for oil-pressboard of various ageing degree

Discharge development process	Pressurized time period				
	Ageing 1 day	Ageing 6 days	Ageing 16 days	Ageing 32 days	Ageing 48 days
Initial stage of discharge	0–8h	0–8h	0–7h	0–7h	0–7h
Discharge development stage	8-22h	8-22h	7–21h	7–21h	7–19h
Dangerous stage of discharge	22–24h	22–24h	21–23h	21–22h	19–20h

10. 3. 3 Variation Law of Characteristic Quantity of Surface Discharge of Transformer Oil-Paper Insulation

10. 3. 3. 1 The Change Rule of the Total Discharge Amount of the Surface Discharge

Taking 1h as the statistical period, the total discharge along the surface of oil-paper insulation in each statistical period is calculated, and the variation curve of the total discharge along the surface with the pressure time is obtained, as shown in Figure10-10.

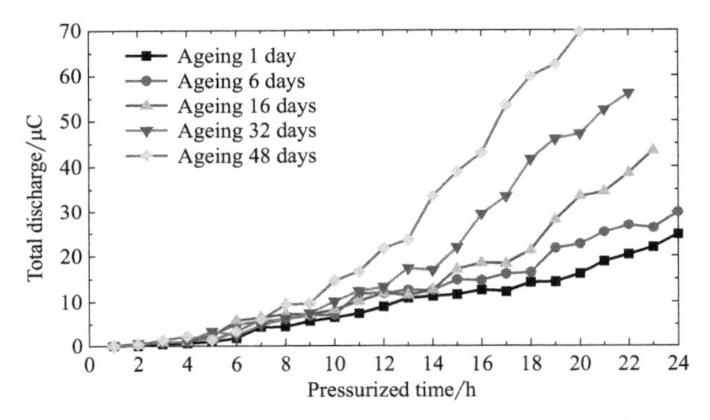

Figure 10-10 Total discharge quantity development law of surface discharge

From the results in Figure 10-10, it can be seen that in the early stage of discharge, that is, the initial stage of discharge, the total discharge volume of the surface discharge of the sample with each ageing degree increases slowly. With the increase of the experiment time, the surface discharge gradually developed, and the total discharge amount of the surface discharge of the samples with various ageing degrees showed an increasing trend. For the oil-paper insulation sample made of insulating paperboard aged 1 day and 6 days, the total discharge amount basically increased in a straight line, and there is no obvious mutation point. For the oil-paper insulation sample made of insulating paperboard aged 16 days, press for 18 hours the rate of increase of total discharge increased rapidly. For the oil-paper insulation test sample made of insulating paperboard aged for 32 days, the increase rate of total discharge increased significantly after pressure for about 14 hours. For the oil-paper insulation test made of insulating paperboard aged for 48 days, when the product is pressurized for about 13 hours, the curve of the total discharge volume has experienced a turning point, and the rising speed has increased significantly. A comprehensive comparison of the curves in Figure 10-10 shows that after 8 hours of pressurization, the surface discharge of the oil-paper insulation of the transformer enters the discharge development stage and the dangerous discharge stage. The ageing degree of the insulating paperboard has a significant impact on the total discharge of the oil-paper insulation: insulation the more serious the ageing of the paperboard, the faster the total discharge of the oil-paper insulation surface discharge increases; the same pressure time, the total discharge of the oil-paper insulation creeping discharge increases significantly with the increase of the ageing degree of the insulating paperboard. In order to more intuitively see the relationship between the total discharge volume and the degree of ageing at the initial stage of discharge, Figure 10-11 shows the change rule of the total discharge volume for 1 to 8 hours of pressure (the ordinate is processed by taking the logarithm). From the results in Figure 10-11, it can be clearly seen that at the initial stage of discharge, the total discharge of oil-paper insulation surface discharge has no obvious relationship with the degree of ageing of the insulation board.

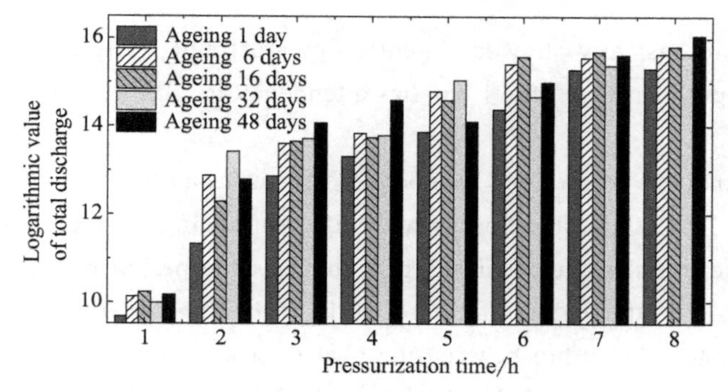

Figure 10-11 Total discharge quantity development law of surface discharge (1～8h)

10.3.3.2 Variation Law of Total Discharge Time of Surface Discharge

Taking 1h as the statistical period, calculate the total discharge times of oil-paper insulation surface discharge in each statistical period, and obtain the curve of the total discharge times of surface discharge with the pressure time, as shown in Figure 10-12. Figure 10-13 shows the change rule curve of the total discharge times of surface discharge when the pressure is 1-8h.

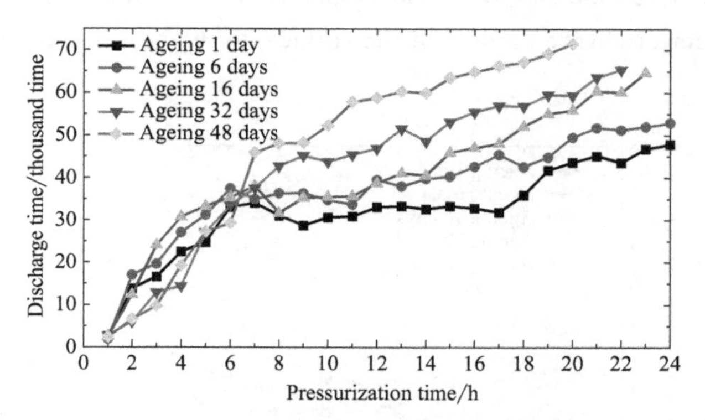

Figure 10-12 Total discharge number development law of surface discharge

From the results in Figure 10-12, it can be seen that in the initial stage of discharge, the total number of discharges along the surface of the samples with different ageing degrees all increase rapidly. With the increase of the experiment time, the surface discharge of oil-paper insulation gradually developed, and the rising speed of the total number of surface discharges of each sample de-

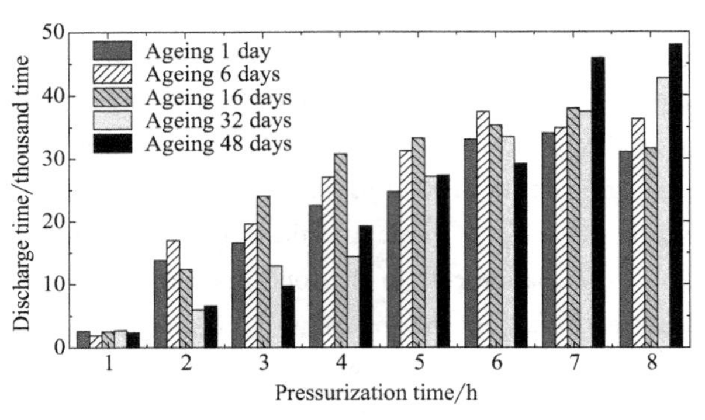

Figure 10-13 Total discharge number development law of surface discharge (1-8h)

creased obviously, and basically showed a gentle upward trend. The oil-paper insulation sample made of insulating paperboard aged for 1 day has a tendency to significantly reduce the total number of surface discharges.

A comprehensive analysis of the total number of discharges along the surface of the oil-paper insulation composed of insulating paperboards with different ageing degrees in Figure 10-12 and Figure 10-13 shows: the initial stage of discharge, that is, the period of rapid increase in the total number of discharges, and the degree of ageing of the insulating paperboard is related to the oil-paper. There is no obvious relationship between the total discharge times of insulation surface discharge; the discharge development stage and the discharge dangerous stage, that is, the total discharge-time increase and the gentle period, the ageing degree of the insulating paperboard has a greater influence on the total discharge times. The total discharge times of oil-paper insulation surface discharge increases with the insulation, and the degree of ageing of the cardboard increases.

10. 3. 3. 3 Variation Law of Maximum Discharge of Surface Discharge

Taking 1h as the statistical period, calculate the maximum discharge of oil-paper insulation surface discharge in each statistical period, and obtain the curve of the maximum discharge of surface discharge with the pressure time, as shown in Figure 10-14. Figure 10-15 shows the curve of the change of the maximum discharge volume of the surface discharge when the pressure is applied for 1-8h.

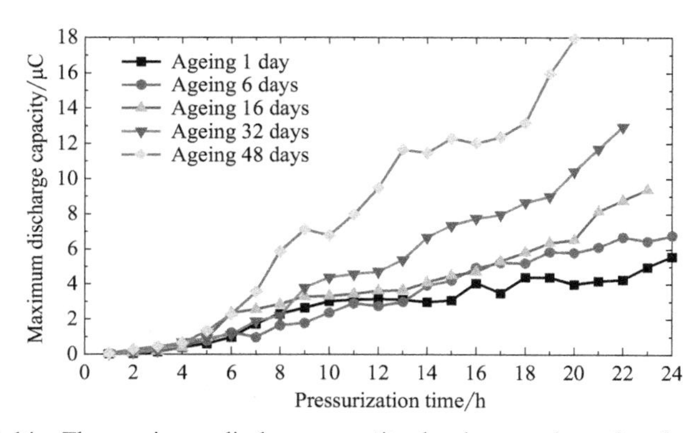

Figure 10-14 The maximum discharge quantity development law of surface discharge

It can be seen from the results in Figure 10-14 that the maximum discharge volume curves of the

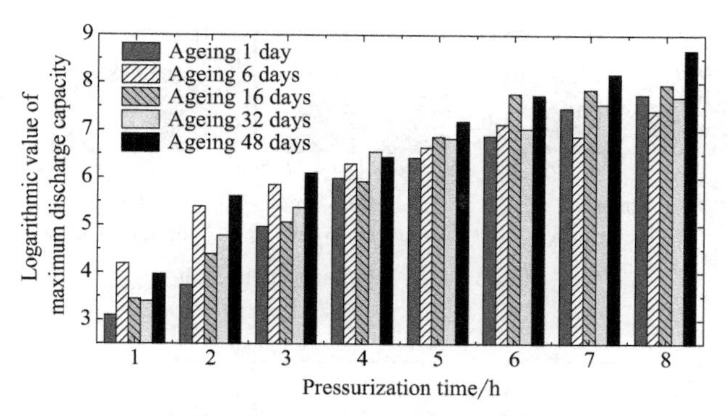

Figure 10-15 The maximum discharge quantity development law of surface discharge (1-8h)

samples with different ageing degrees are similar. In the initial stage of the experiment and the initial stage of discharge, the maximum discharge capacity of the surface discharge of each sample slowly increased. With the progress of the experiment, the surface discharge has entered a stage of discharge development and discharge danger, and the maximum discharge capacity of each sample surface discharge basically increases linearly.

Comprehensive analysis of the results of Figure10-14 and Figure 10-15 shows that: when the pressure is 1-8h, that is, the initial stage of discharge, the maximum discharge capacity of oil-paper insulation surface discharge has no obvious relationship with the degree of ageing of the insulation paperboard. After 8 hours of pressurization, which is the discharge development and discharge dangerous stage, the ageing degree of the insulating paperboard has a greater influence on the maximum discharge capacity of the oil-paper insulation surface discharge, and the more serious the ageing degree of the insulating paperboard, the faster the maximum discharge capacity of the oil-paper insulation surface discharge increases. Pressing time, the maximum discharge capacity increases with the increase of the ageing degree of the insulating paperboard.

10.3.3.4 Discharge Phase Change Law of Surface Discharge

This paper summarizes the changes in the phase change of the surface discharge of the oil-paper insulation of the transformer when the insulation board is aged Figure 10-16 to Figure 10-20 show the scatter plot of surface discharge q-φ of oil-paper insulation under different ageing degrees (the amplitude is normalized). It can be seen from Figure 10-16 to Figure 10-20 that the initial discharge stage of surface discharge, the discharge is mainly distributed in 30°-90°, 210°-270°. As time increases, the discharge phase gradually expands to the 2, 4 quadrants, and finally extends to the entire phase cycle. For insulating paperboard aged for 1 day and ageing for 6 days, the discharge is basically extended to the entire phase cycle when pressurized for about 13 hours. For insulating paperboard aged for 16 days and 32 days, the discharge is basically extended to the entire phase cycle when

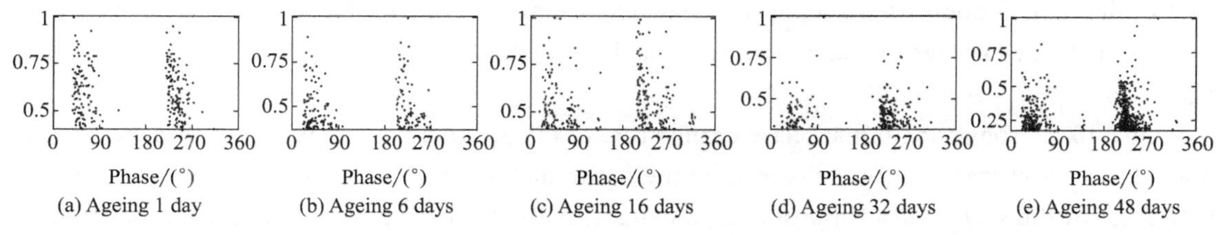

(a) Ageing 1 day (b) Ageing 6 days (c) Ageing 16 days (d) Ageing 32 days (e) Ageing 48 days

Figure 10-16 Voltage applied 1h

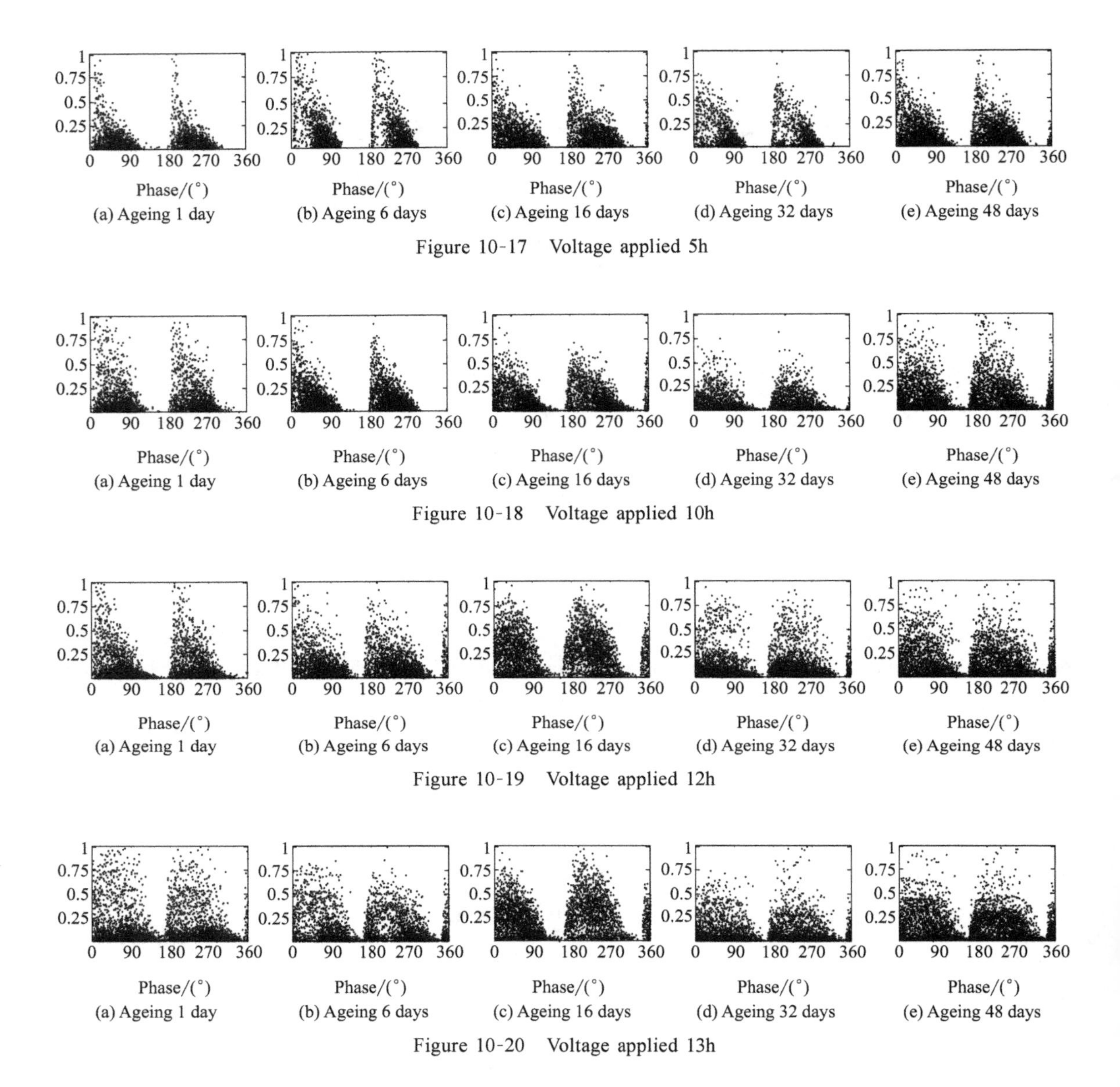

Figure 10-17 Voltage applied 5h

Figure 10-18 Voltage applied 10h

Figure 10-19 Voltage applied 12h

Figure 10-20 Voltage applied 13h

pressurized for about 12 hours. For the 48-day ageing insulation paperboard, the discharge will extend to the entire phase cycle after 10 hours of pressure.

After the discharge is extended to the entire period, it will be difficult to determine the start and end phases of the discharge. Therefore, when analyzing the phase characteristics of the surface discharge of the oil-paper insulation of the transformer, the insulation paperboard only analyzes the data of the first 13 hours after ageing for 1 day and ageing for 6 days. Only the first 12h data is analyzed for the 32-day insulation paperboard, and the first 10h data is only analyzed for the 48-day ageing insulation paperboard.

Figure 10-21 shows the initial phase change rule of the positive half-cycle of the oil-paper insulation surface discharge. It can be seen from the figure that the initial discharge phase of the positive half cycle of the surface discharge shows an exponentially decreasing trend with the development of the discharge. At the same time, comparing the change law of the curves in Figure 10-21, it can be seen that the more severe the ageing degree of the insulating paperboard is, the smaller the initial phase of the positive half cycle of the oil-paper insulation surface discharge is.

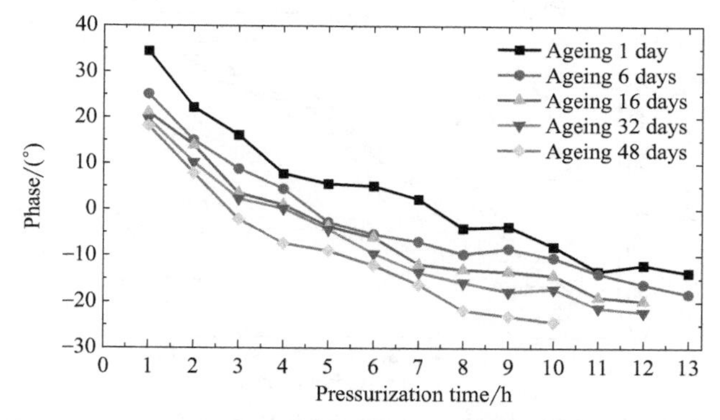

Figure 10-21 Starting phases of positive half cycle for surface discharge

Figure 10-22 shows the initial phase change rule of the negative half cycle of the oil-paper insulation surface discharge. Similar to the change rule of the initial phase of the positive half cycle of the surface discharge, the initial phase of the negative half cycle of the surface discharge of each sample also showed an exponential decrease trend with the development of the discharge. At the same time, comparing the change rules of the curve gauges in the figure, it can be seen that the ageing degree of the insulating paperboard has a certain influence on the initial phase of the negative half cycle of the oil-paper insulation surface discharge, and the more serious the ageing degree of the insulating paperboard, the smaller the initial phase of the negative half cycle of the oil-paper insulation surface discharge.

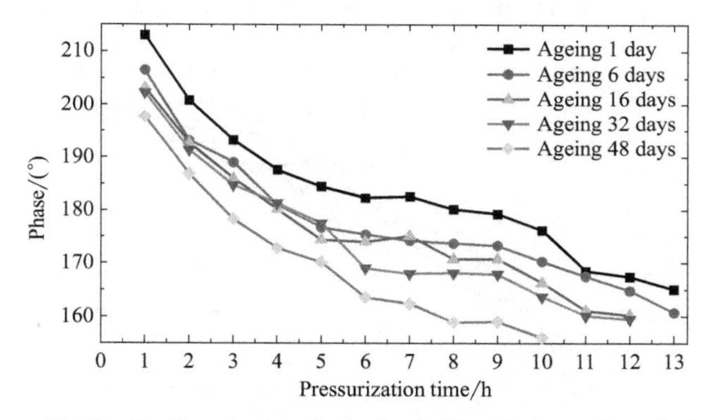

Figure 10-22 Starting phases of negative half cycle for surface discharge

Figure 10-23 shows the change rule curve of the termination phase of the positive half cycle of the oil-paper insulation surface discharge with the discharge time. It can be seen from the figure that with the increase of the discharge time, the phase of the positive half cycle of each test sample shows an increasing trend. At the beginning of the experiment, its phase is between 90° and 100°, when the pressurizing time is 10 to 13 hours, its phase is between 155° and 165°. With the increase of the discharge time, the increasing rate of the termination phase of the positive half cycle of the surface discharge of each sample gradually decreased. However, comparing the changing laws of the curves, it can be seen that there is no obvious relationship between the termination phase of the positive half cycle of the oil-paper insulation surface discharge and the ageing degree of the insulation paperboard.

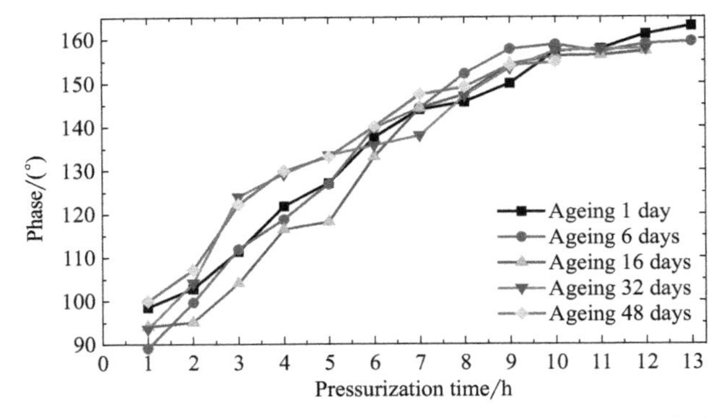

Figure 10-23 Ending phases of positive half cycle for surface discharge

Figure 10-24 shows the variation curves of the negative half cycle termination phase of the surface discharge of oil-paper insulation with the discharge time. It can be seen from the figure that with the increase of discharge time, the negative half-cycle termination phase of each sample showed an increasing trend. At the beginning of the experiment, the phase was between 265° and 285°. When the pressure was 10–13h, the phase was between 330° and 345°. Similar to the regular pattern of positive half-cycle termination phase, with the increase of discharge time, the increase rate of negative half-cycle termination phase along the surface of insulation paperboard with different ageing degrees decreases gradually. By comparing the variation law of each curve, it can be seen that the negative half-cycle termination phase of surface discharge of oil-paper insulation has no obvious relationship with the ageing degree of insulation paper board.

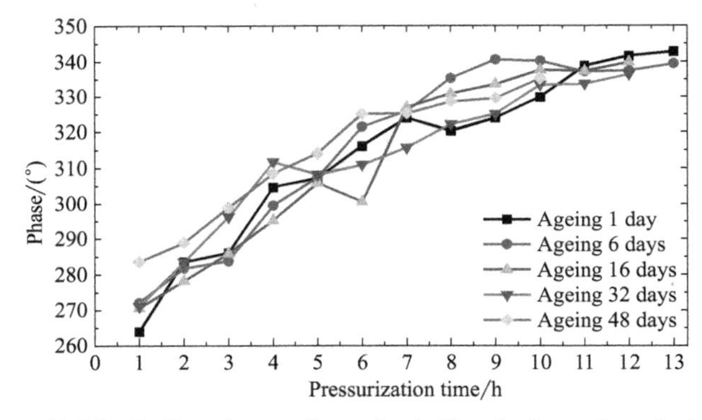

Figure 10-24 Ending phases of negative half cycle for surface discharge

10.3.3.5 The Change Rule of Carbon Marks in Surface Discharge Insulation Board

With the increase of the experiment time, the discharge carbon marks on the surface of the insulating paperboard of various ageing degrees will gradually increase, as shown in Figure 10-25. The development of the discharge carbon mark on the surface of the insulating paperboard can reflect the destructive effect of the partial discharge on the surface of the insulating paperboard and the development degree and speed of the partial discharge to a certain extent. In order to study the change characteristics of the discharge carbon marks during the development of the surface discharge of the insulating paperboard with various ageing degrees, the paperboard was taken out every two hours and the carbon marks on the paperboard were measured.

(a) Voltage applied 5h (b) Voltage applied 13h (c) Voltage applied 21h

Figure 10-25 Surface discharge carbon tracking

The simplest and most direct method to measure the discharge carbon trace is to measure its inner and outer diameters. However, during the measurement, it was found that the discharge carbon mark is not a standard concentric circle, and the width of the carbon mark is not consistent. The different measurement paths will cause large errors. It is difficult to accurately determine the boundary of the discharge carbon mark by vision, and the carbon mark different boundary determination methods will also produce large errors. At the same time, experiments have found that the discharge carbon traces increase slowly at the beginning of the discharge, and the error caused by the above-mentioned reasons even exceeds the increase in the actual size of the discharge carbon traces. Therefore, it is more difficult to directly measure the size of the discharge carbon traces to characterize the change law of the discharge carbon trace. Since the surface of the insulating cardboard is mainly brown, dark brown or brownish yellow, and the discharge carbon traces are mainly black or light black, the two colors are relatively strong, so the discharge carbon traces can be extracted by image processing, and then the discharge traces the law of development is described. Based on the above thoughts, this paper designs a method for measuring the area of the discharge carbon traces, and based on the measurement results, realizes the characterization of the variation of the discharge carbon traces.

Firstly, use the carbon mark image acquisition system shown in Figure 10-26 to collect the discharge carbon mark image. The system is mainly composed of an adjustable bracket and an adjustable focal length wide-angle high-definition camera. During the experiment, the insulating paperboard to be tested was placed on the platform, and the camera lens was perpendicular to the insulating paperboard. By adjusting the position of the bracket and the focal length of the camera, a clear picture of the overall discharge carbon trace of the insulating paperboard was achieved. Figure 10-27 shows the color image of the carbon traces of the insulated cardboard after ageing for one day.

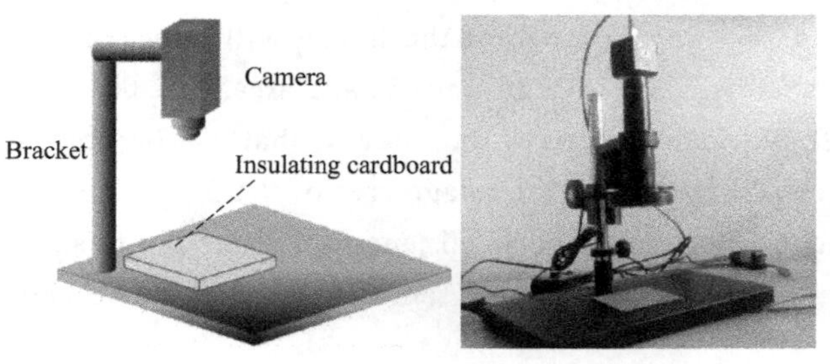

Figure 10-26 Carbon tracking image acquisition system

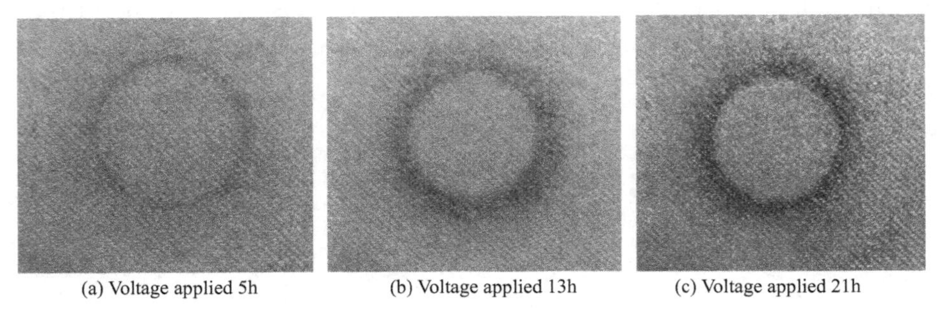

(a) Voltage applied 5h (b) Voltage applied 13h (c) Voltage applied 21h

Figure 10-27 Carbon tracking colour image for the pressboard ageing 1 day

Gray-scale transformation was carried out on the above discharge carbon mark color image to obtain the discharge carbon mark gray image. The gray image is binarized. By setting the appropriate threshold, the gray value of the dark pixels (the discharge carbon mark) is 0, and the gray value of the rest positions (the discharge carbon mark is not covered) is 1. Then the binarized image is obtained, as shown in Figure 10-28. Through the above method, discharge carbon mark can be clearly visible.

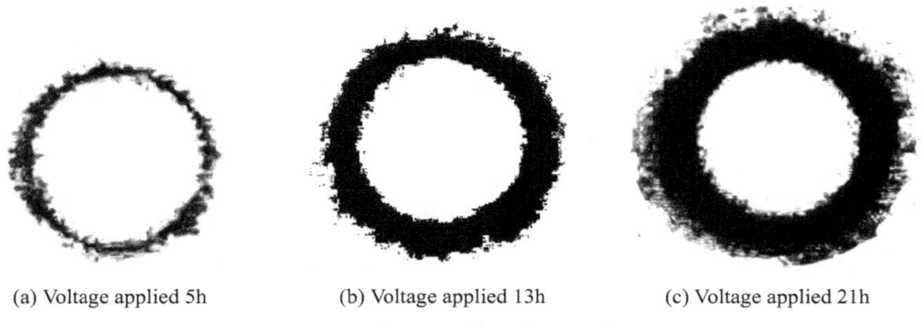

(a) Voltage applied 5h (b) Voltage applied 13h (c) Voltage applied 21h

Figure 10-28 Carbon tracking image after extracting

Since the gray value of the pixel in the carbon mark area in the discharge carbon mark binarization image is 0, the area of the discharge carbon mark is characterized by counting the number of pixels with a gray value of 0. In this section, the total number of pixels of the discharge carbon mark (pixel) is called the image area of the discharge carbon mark, as shown in formula (10-2):

$$S = MN - \sum_{m=1}^{M} \sum_{n=1}^{N} h(m, n) \tag{10-2}$$

M is the total pixel of the image width; N is the total pixel of the image length; $h(m,n)$ is the gray value of the pixel (m, n) (its value is 0 or 1).

The focal length and shooting distance of the image will affect the magnification of the image, and then affect the discharge carbon trace image area. At the same time, it can be seen from the formula (10-2) that the size of the image, that is, the total width of pixels and the total length of pixels, will also affect the image area of the discharge carbon trace. Therefore, after adjusting the shooting distance and focal length to make the image clear, keep the shooting distance and focal length fixed. At the same time, this article chooses the image size as 1000×900 pixels.

During the experiment, samples of insulating paperboard were taken every 2h and the discharge carbon trace image area was measured by the above method. Figure 10-29 shows the change rule of the image area of the discharge carbon trace of the insulating paperboard at various ageing degrees. It can be seen from the results in the figure that the pattern of changes in the image area of the surface discharge carbon traces of each sample is similar: at the initial stage of discharge, the area of the discharge carbon trace image of each sample increases relatively slowly; as the experiment time increases, the area of the discharge carbon trace image is also rapid increase; with the further increase of the experiment time, the increasing trend of the discharge carbon trace image area has slowed down significantly. At the same time, comparing the curves, it can be seen that the discharge development and discharge danger stage (after 7 hours of pressure in the figure), after the same length of partial discharge degradation, the higher the ageing degree of the insulating paperboard, the larger the discharge carbon trace image area. This shows that the discharge development and discharge dangerous stage, the higher the ageing degree of the insulating paperboard, the faster the development speed of the oil-paper insulation creeping discharge, and the more obvious the destructive effect of the partial discharge.

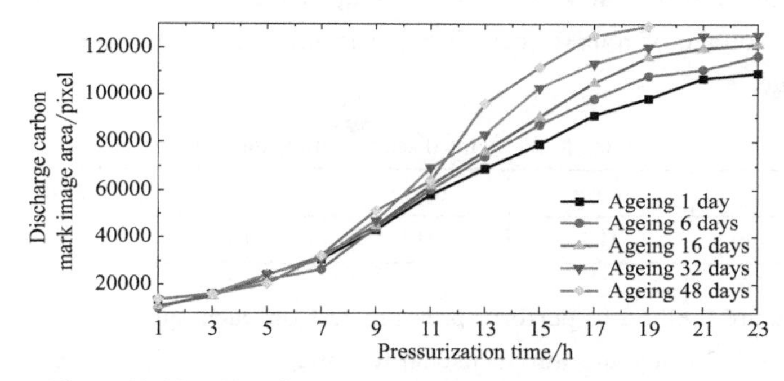

Figure 10-29 Changing rules for the area of the carbon tracking

Figure 10-30 is a column chart of the area of the discharge carbon mark image at the initial stage of the discharge. The results show that there is no obvious monotonic correlation between the ageing degree of the insulating paperboard and the surface discharge carbon mark image area of oil-paper insulation at the initial stage of the discharge. In the initial stage of discharge, the ageing of the insulating paperboard has little effect on the development of surface discharge deterioration of the oil-paper insulation of the transformer.

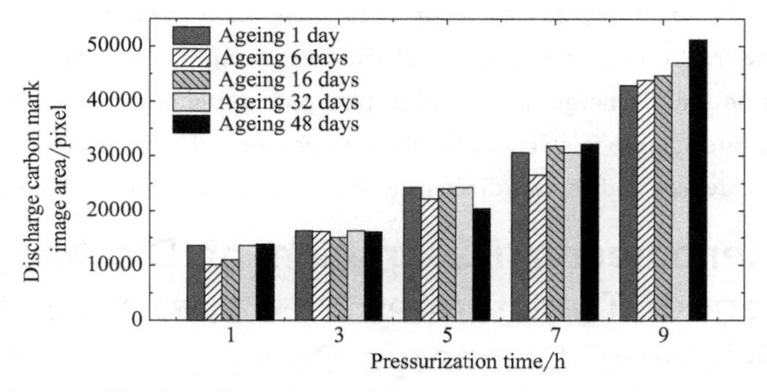

Figure 10-30 Changing rules for the area of the carbon tracking (1-9h)

10.4 Development Law of Tip Discharge Deterioration of Transformer Oil-Paper Insulation

10.4.1 Initial Discharge Voltage and Discharge Endurance Time of the Transformer Oil-Paper Insulation Tip Discharge

Before the pressure test of the oil-paper insulation tip discharge, the test method is the same as that of the surface discharge test. Put the pre-treated insulating paperboard sample in the new transformer oil and let it stand for 1 day until the insulating paperboard sample is soaked fully. Firstly, measure the initial discharge voltage at the tip of the insulating paperboard oil-paper insulation sample with different ageing degrees, and the method is still consistent with the measurement method of the initial discharge voltage of the oil-paper insulation surface discharge. Similarly, insulation paperboards with different ageing degrees contains three samples. Record the initial discharge voltage of the tip of each sample and find the average value, as shown in Table 10-7. From the results of Table 10-7, it can be seen that with the deepening of the ageing of the insulating paperboard, the initial discharge voltage of transformer oil-paper insulation tip decreases gradually, but the decline is still not large.

Table 10-7 Point discharge inception voltage

Ageing time	1 day	6 days	16 days	32 days	48 days
Initial discharge voltage / kV	11.7	11.3	10.8	10.1	9.7

Use constant pressure method to perform pressure test on each sample. During the experiment, the applied voltage is 1.2 times the initial discharge voltage of the tip of the insulating paperboard after ageing for 1 day, that is, 14kV. Table 10-8 shows the tip discharge endurance time of each sample. It can be seen from the results in Table 10-8 that with the increase of the ageing degree of the insulating paperboard, the discharge endurance time of the oil-paper insulating tip gradually shortens.

Table10-8 Point discharge tolerance time

Ageing time	1 day	6 days	16 days	32 days	48 days
Tolerance time	16h 3min	14h 27min	13h 15min	12h 16min	10h 31min

Similarly, due to the same degree of ageing of the insulating paperboard, the tip discharge development law of the transformer oil paper insulation test sample is similar. Therefore, when the subsequent research on the discharge deterioration phenomenon and law of the transformer oil-paper insulation tip, only one sample is selected for the transformer oil-paper insulation with the same insulation board ageing degree and the tip discharge phenomenon and the discharge law are described.

10.4.2 Phenomenon Description of Discharge Development of Transformer Oil-Paper Insulation Tip and Division of Discharge Development Process

Similar to the surface discharge of oil-paper insulation, the tip discharge deterioration of oil-paper insulation samples with different degrees of ageing of insulation board age is similar. Therefore, this paper only describes the tip discharge phenomenon of oil-paper insulation samples with ageing degree of 1 day.

At the beginning of the experiment, there was no obvious discharge spark or discharge sound, as shown in Figure 10-31 (a). After 40 minutes, an intermittent discharge light spot with lower brightness and smaller size appeared at the tip of the needle electrode, and a large number of bubbles with a smaller diameter were precipitated at the junction of the needle electrode and the cardboard, and they spread rapidly. The surface of the insulating cardboard at the tip of the needle electrode produces a black carbonized area, but the area was small and the color was lighter, as shown in Figure 10-31

(a) Initial discharge
(b) Generating bubbles and black
(c) Increasing of bubble carbonization area
(d) Bubbles floating
(e) Increasing of black carbonization
(f) Generating bubbles outside inverted conearea and bubble diameterthe carbon area
(g) Bubbles shift to one side
(h) Discharge stope and bubble
(i) Break down floating as hyperbolic
(j) Pale white marks in pressboard surface and black smoke

Figure 10-31 Monitoring image of point discharge

(b). With the increase of the experiment time, a weak discharge sound appeared, the number of precipitated bubbles began to increase and basically floated up to the surface of the oil layer in an inverted cone shape. The area of the black carbonized area on the surface of the insulating cardboard at the tip of the needle electrode gradually increased, and the color gradually became darker, as shown in Figure 10-31 (d). When the discharge experiment was carried out for 7 hours, a bright discharge spot area was generated near the tip of the needle electrode. At this time, the bubbles basically diffuse and rise rapidly with the needle electrode as the axis in a hyperbolic law. The area of discharge and carbonization at the interface of the needle electrode and the cardboard was large and the color was dark, as shown in Figure 10-31 (e). At this time, the discharge spot near the tip of the needle electrode was already very bright and had a large area. With the increase of time, the precipitated bubbles clearly shifted to the source of the bubbles outside the carbonization area, as shown in Figure 10-31 (g), and basically floated up in a semi-cylindrical shape, and the crackling discharge sound became louder and faster. At the same time, the discharge would stop intermittently, as shown in Figure 10-31 (h). When the discharge was stopped, the precipitated bubbles were significantly weakened and rise in a hyperbolic shape with the needle electrode as the axis. The experiment continued. At this time, the discharge sound was more intense, the bubbles near the electrode rose faster and faster, and the oil layer disturbance became more and more intense. More fine particles would precipitate in the black carbonized area, and after the oil layer was disturbed, they would float up to a certain position and fall on the surface of the cardboard far away from the needle electrode. When the experiment progressed to about 16h 6min, there was a sudden and crisp breakdown sound, a strong bright light was generated near the electrode, the cardboard broke down, and thick black smoke was precipitated on the surface of the cardboard, which quickly diffused into the oil layer. For oil-paper insulation test samples aged for 6 days, 16 days, 32 days and 48 days of insulation board, the tip discharge deterioration phenomenon was similar to that of oil-paper insulation samples aged for 1 day of insulation board, but the occurrence time of each phenomenon was slightly different.

According to the different phenomena in the discharge development process of oil-paper insulation tip according to the ageing degree, this paper divides the discharge development stage of oil-paper insulation tip into discharge initial stage, discharge development stage and discharge dangerous stage. In the initial stage of discharge, the discharge was relatively weak, the size of the bubbles on the surface of the cardboard was small, the number was large, and they diffuse rapidly, and the black carbonized area on the surface of the cardboard gradually increased. In the discharge development stage, a bright discharge spot appeared near the tip of the needle electrode, the discharge sound was sharper, the number of precipitated bubbles was reduced but the size was larger, the discharge sound increased, and the frequency increased; In the dangerous stage of discharge, bubbles began to stably precipitate somewhere outside the black carbonized area, and there were intermittent crackling and louder discharge sounds. The precipitated bubbles began to shift to the direction of the precipitation source outside the black carbonized area of the insulating cardboard. The bubbles were arranged in a semi-cylindrical shape and float up. Disturbance of the oil layer near the electrode. Particles were precipitated in the black carbonized area and mostly fell on the surface of the insulating cardboard. Table 10-9 shows the development stages of surface discharge of oil-paper insulation with different ageing degrees, and the division range of the pressure period.

Table10-9 Division of point discharge process for oil-pressboard of various ageing degree

Discharge development process	Pressurized time period				
	Ageing 1 day	Ageing 6 days	Ageing 16 days	Ageing 32 days	Ageing 48days
Initial stage of discharge	0–7h	0–6h	0–6h	0–6h	0–5h
Discharge development stage	7–14h	6–12h	6–11h	6–11h	5–9h
Dangerous stage of discharge	14–16h	12–14h	11–13h	11–13h	9–10h

10.4.3 Variation Law of Characteristic Quantities of Discharge at the Tip of Transformer Oil-Paper Insulation

10.4.3.1 The Changing Law of the Total Discharge Volume of The Tip Discharge

Taking 1h as the statistical period, calculate the total discharge volume of the oil-paper insulated tip discharge in each statistical period, and obtain the curve of the total discharge volume of the tip discharge with the pressurized time, as shown in Figure 10-32. Figure 10-33 shows the change rule of the total discharge volume of the tip discharge of each sample under pressure for 1-7h (for clarity of comparison, the ordinate has been processed by taking the logarithm).

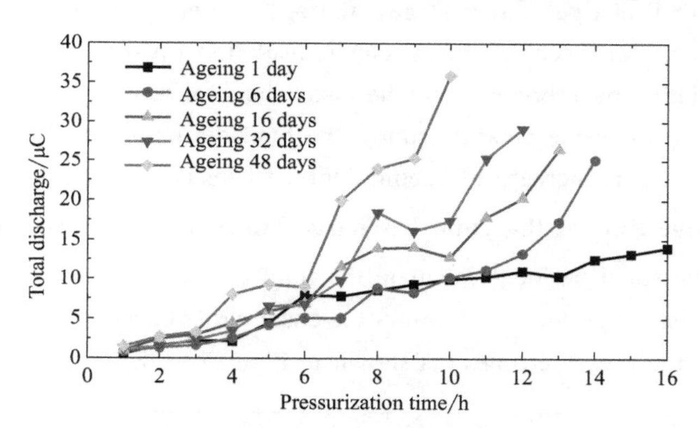

Figure 10-32 Total discharge quantity development law of point discharge

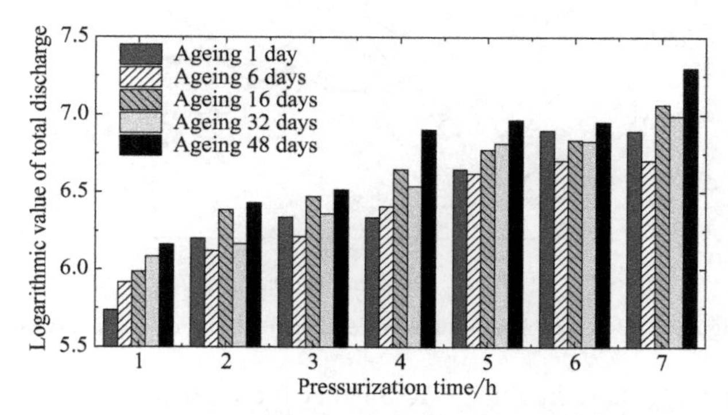

Figure 10-33 Total discharge quantity development law of point discharge (1-8h)

It can be seen from Figure 10-32 that the total discharge of each sample tip discharge is on the rise, but slightly different. During the period of 1-3h of pressurization, the total discharge at the tip of each test sample increased slowly. With the increase of the experimental time, when the insulating paperboard was aged for 1 day, the total discharge rate of the test product had increased from the 4th hour, and the later period basically showed a linear increase pattern, with occasional decrease. When the insulating paperboard was aged for 6 days, the total discharge rate of the sample increased slightly in 4-11h, and the total discharge rate increased further in 12-13h. When the insulation board aged for 16 days, the total discharge amount of the sample increased linearly in 1-7h, decreased in 8-10h, and increased rapidly in 10-13h. When the insulating paperboard was aged for 32 days, the total discharge volume of the sample also basically rose linearly in the first to 7h, and the rising rate was relatively slow. From 8h to 10h, the total discharge volume oscillated and rose, and the rising rate was accelerated. The total discharge capacity increased at a faster rate. When the insulating paperboard was aged for 48 days, the total discharge volume of the test product oscillates and rises within 1-6h, and the rising speed of the total discharge volume increased rapidly in the 6th to 9th hours, and the rising speed would gradually slow down with the increase of time. From the 9th to 10th hours, the increase rate of total discharge increased further.

At the same time, a comprehensive comparison of the results of Figure 10-32 and Figure 10-33 shows that at the initial stage of discharge, the ageing degree of the insulating paperboard has no obvious influence on the total discharge capacity of the oil-paper insulating tip. In the discharge development stage and discharge danger stage, the ageing degree of insulating paperboard had a significant effect on the total discharge of oil-paper insulation tip discharge: the more serious the ageing degree of insulating paperboard was, the faster the total discharge of oil-paper insulation tip discharge increased. At the same pressure time, the total discharge of oil-paper insulation tip increased significantly with the increase of ageing degree of insulation board.

10. 4. 3. 2 The Change Rule of the Total Discharge Times of the Tip Discharge

Taking 1h as the statistical period, calculate the total discharge times of oil-paper insulated tip discharges in each statistical period, and obtain the change rule curve of the total discharge times of the tip discharge with the pressure time, as shown in Figure10-34.

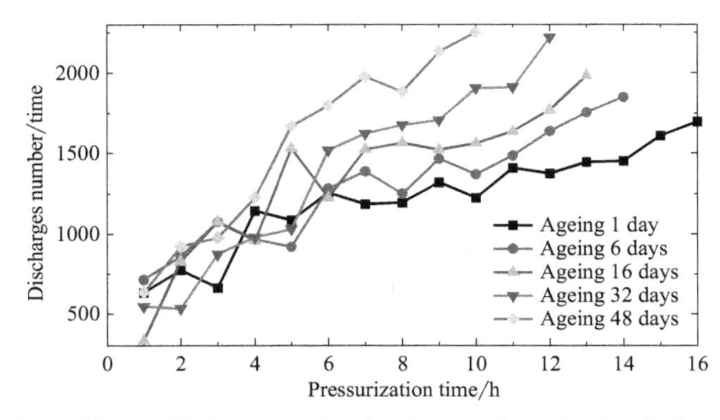

Figure 10-34 Discharge number development law of point discharge

It can be seen from the results in Figure 10-34 that the total number of discharges at the tip of each sample shows an oscillating upward trend as the discharge time increases. The change rule of the total number of discharges at the tip of each sample is similar, with slight differences: when the insulating paperboard was aged for 1 day, the total number of discharges fluctuates rapidly in the sample within 1-6h; when the pressure time was 7-13h, the total number of discharges increased slowly, and the number of discharges decreased in certain periods. It was more obvious. When the pressure time was 14-16h, the rate of increase of the total number of discharges increased to a certain extent. When the insulating paperboard was aged for 6 days, the total discharge times of the test product fluctuates rapidly within 1-6h, and the decrease was obvious in 3-5h; when the pressurized time was 7-11h, the total discharge times change trend curve was relatively flat, and itwas basically in a stable state; with the progress of the experiment, the total number of discharges increased rapidly in a linear trend within 12 to 14 hours. When the insulating paperboard was aged for 16 days, the sample exhibited a rapid fluctuating upward trend within 1 to 5 hours; the total number of discharges basically remained unchanged from the 6th to 11th hours, and the number of discharges in the 6th hour decreased greatly; when the discharge time was 12-13h, the number of discharges increased rapidly. When the insulating paperboard was aged for 32 days, the total discharge times change curve of the test product could basically be divided into three sections. The total discharge times fluctuate at a relatively rapid rate from 1 to 6 hours; the increase rate of the total discharge times starts to slow down in the 7th to 10th hours; from the 11th to 12th hours, the increase rate of the total number of discharges further increased. When the insulating paperboard was aged for 48 days, the rising rate of total discharge times in the whole discharge stage remained at a high level, but there is a certain downward trend within 5 to 8 hours.

In order to more intuitively see the relationship between the total discharge times of the oil-paper insulation tip and the ageing degree of the insulating paperboard in the initial stage of discharge, Figure 10-35 shows the change rule of the total discharge times of pressurized 1-7h. It can be clearly seen from the results in Figure 10-32 that the total number of discharges at the tip of the oil-paper insulation has no obvious relationship with the ageing degree of the insulation board at the initial stage of discharge.

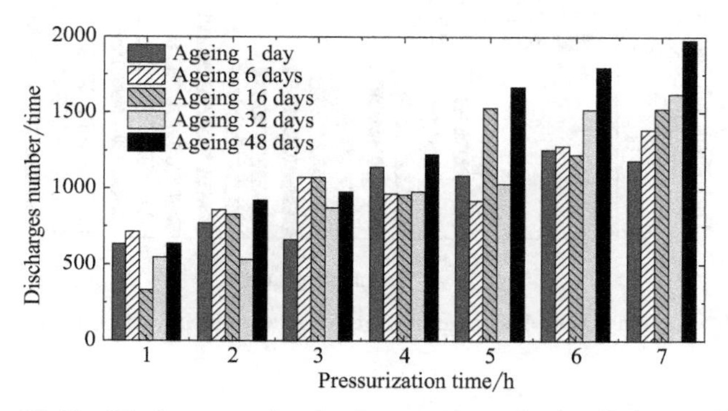

Figure 10-35 Discharge number development law of point discharge (1-8h)

10.4.3.3 The Maximum Discharge Variation of Tip Discharge

Taking 1h as the statistical period, calculate the maximum discharge volume of the oil-paper insulated tip discharge in each statistical period, and obtain the curve of the maximum discharge volume of the tip discharge with the pressured time, as shown in Figure10-36. Figure 10-37 shows the change rule of the maximum discharge volume of the tip discharge of each sample under pressure for 1-7h.

It can be seen from Figure 10-36 that the maximum discharge capacity of the tip discharge of each sample has a similar variation, and it basically fluctuates and rises in a straight line. Comparing the results of Figure 10-36 and Figure 10-37, it can be seen that in the initial stage of discharge, the ageing degree of the insulating paperboard has no obvious influence on the change of the maximum discharge capacity of the oil-paper insulation tip; the ageing degree of the insulating paperboard during the discharge development and discharge dangerous stage has a greater impact on the discharge of the oil-paper insulation tip. The more severe the ageing of the insulation board, the faster the maximum discharge of the oil-paper insulation tip will increase; the same pressured time, the maximum discharge of the oil-paper insulation tip will increase significantly with the increase in the ageing degree of the insulation board.

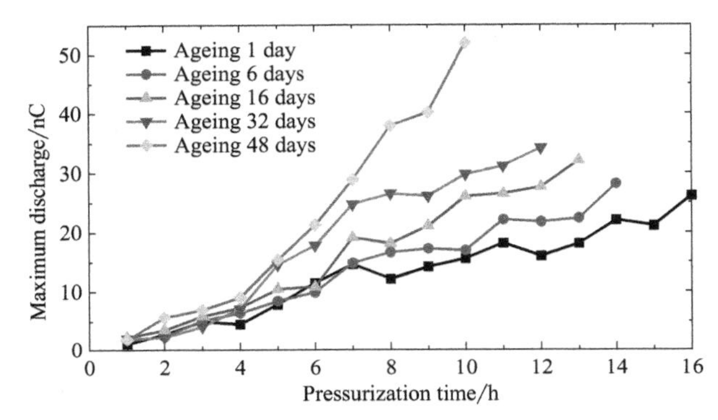

Figure 10-36 The maximum discharge quantity development law of point discharge

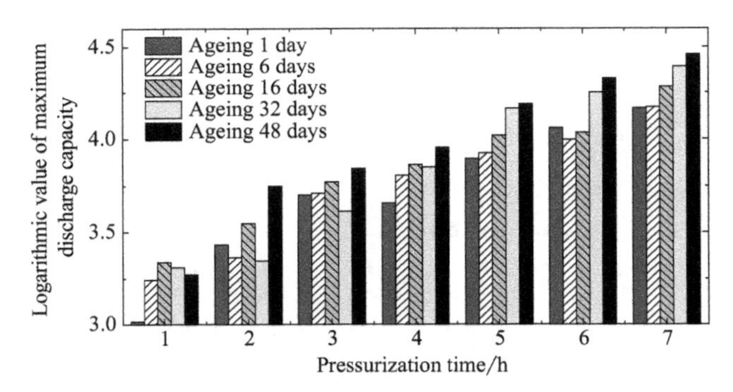

Figure 10-37 The maximum discharge quantity development law of point discharge (1-8h)

10.4.3.4 Phase Variation of Tip Discharge

This paper summarizes the discharge phase change rule of the surface discharge of the transformer oil-paper insulation under different ageing degrees of the insulating paperboard. Figure 10-38 to Figure 10-42 are the scatter diagrams of the tip discharge q-φ of the insulating cardboard for each ageing degree. It can be seen from the figure that in the initial stage of discharge, the discharge phase at the tip of each test sample is mainly concentrated in the vicinity of 60°-130° and 230°-

310°, namely, it is distributed near the positive and negative half-cycle peak. As the discharge time increases, the discharge phase begins to expand to both sides: for insulation paperboard aged 1 day, when it was pressurized to 16h near breakdown, its phase distribution basically extended to the entire phase period; for insulation paperboard aged 6 days, when it was pressurized to 14h near breakdown, its phase distribution mainly concentrated between 15°-175°, 185°-355°; for the 16-day ageing insulation paperboard, when it was pressurized to 13h near the breakdown, its phase distribution was mainly concentrated between 25°-170°, 200°-350°; for the 32-day ageing insulation paperboard, when it was pressurized to 12h near the breakdown, its phase distribution was mainly concentrated between 25°-170°, 205°-350°; for the 48-day ageing insulation paperboard, when it was pressurized to 10h near the breakdown, its phase distribution was mainly concentrated between 25°-170°, 200°-300°. It can be seen that as the degree of ageing of the insulating paperboard increases, the discharge of the oil-paper insulation tip develops until the entire process of impending breakdown, and the discharge does not cover the entire phase cycle.

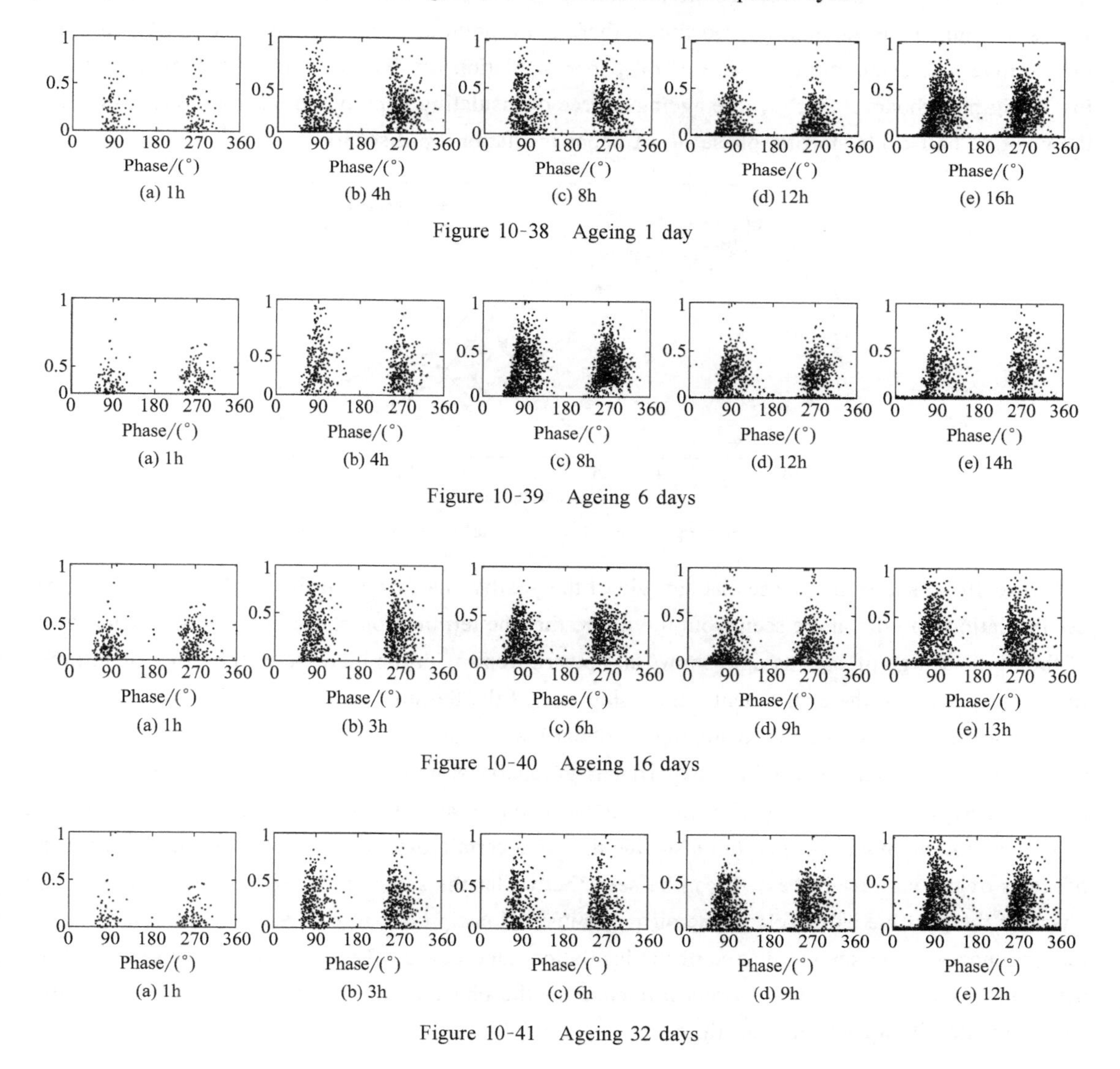

(a) 1h (b) 4h (c) 8h (d) 12h (e) 16h

Figure 10-38 Ageing 1 day

(a) 1h (b) 4h (c) 8h (d) 12h (e) 14h

Figure 10-39 Ageing 6 days

(a) 1h (b) 3h (c) 6h (d) 9h (e) 13h

Figure 10-40 Ageing 16 days

(a) 1h (b) 3h (c) 6h (d) 9h (e) 12h

Figure 10-41 Ageing 32 days

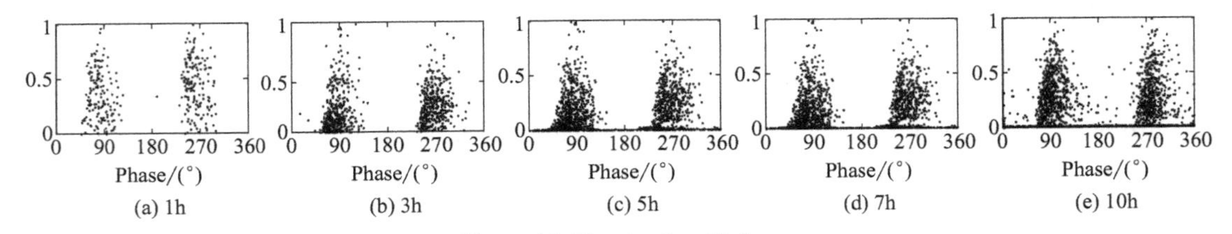

Figure 10-42 Ageing 48 days

Figure 10-43 shows the change rule of the positive half-cycle initial discharge phase of oil-paper insulation tip. It can be seen from the figure that the initial phase of the positive half-cycle of the tip discharge of each sample shows a downward trend as the experimental time increases. However, the relationship between the initial phase of the positive half cycle and the severity of the discharge is not strictly monotonous. At certain moments, the initial phase of the positive half cycle fluctuates greatly as the discharge time increases. At the same time, the comparison of the curves shows that under the same pressured time, there is no obvious corresponding relationship between the positive half-cycle starting phase of oil-paper insulation tip discharge and the ageing degree of insulating paperboard, that is, the ageing degree of insulating paperboard has no obvious effect on the positive half-cycle starting phase of oil-paper insulation tip discharge.

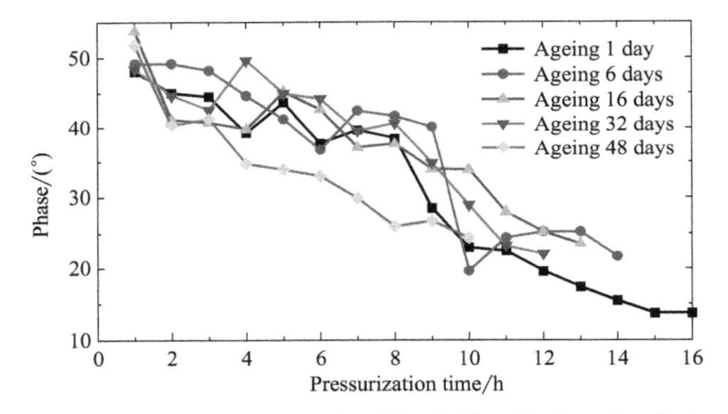

Figure 10-43 Starting phases of positive half cycle for point discharge

Figure 10-44 shows the phase change rule of the positive half cycle of the discharge of the oil-paper insulation tip. It can be seen from the figure that the termination phase of the positive half cycle of the tip discharge of each sample shows an upward trend with the increase of the experiment time. In the early stage of the experiment, the rising rate of the termination phase of the positive half cycle of the tip discharge of the insulating cardboard with various ageing degrees is faster, and the rising rate in the later stage of the experiment gradually slow down. Similarly, as the discharge time increases, the positive half-cycle termination phase and the discharge duration are not in a simple monotonic relationship, but also fluctuate to a certain extent. At the same time, comparing the laws of the various curves, it can be seen that under the same pressured time, the termination phase of the positive half cycle of the oil-paper insulation tip discharge does not show a simple correspondence with the ageing degree of the insulation paperboard. That is, the ageing degree of the insulating paperboard has no obvious influence on the phase of the positive half cycle of the discharge of the oil-paper insulating tip.

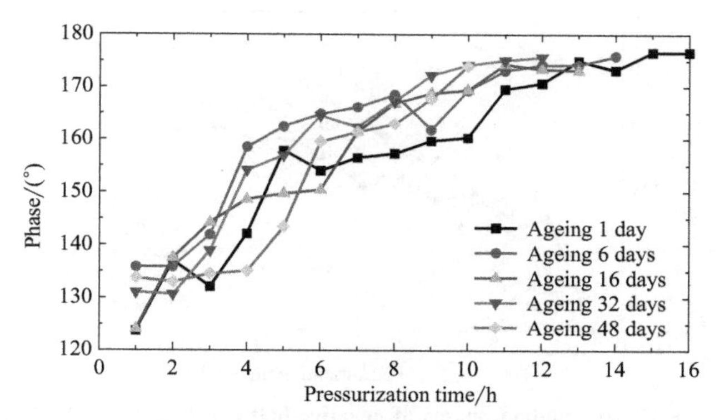

Figure 10-44 Ending phases of positive half cycle for point discharge

Figure 10-45 shows the initial phase change rule of the negative half period of the oil-paper insulation tip discharge. It can be seen from the figure that the negative half period starting phase of the tip discharge of each test sample fluctuates and decreases with the increase of experimental time, but at some moments, the negative half period starting phase increases with the increase of experimental time. At the same time, comparing the laws of the curves, it can be seen that under the same pressured time, the initial phase of the negative half cycle of the oil-paper insulation tip discharge has no obvious corresponding relationship with the ageing degree of the insulating paperboard, that is, the ageing degree of the insulating paperboard has no obvious relationship with the negative half cycle initial phase of the oil-paper insulation tip discharge.

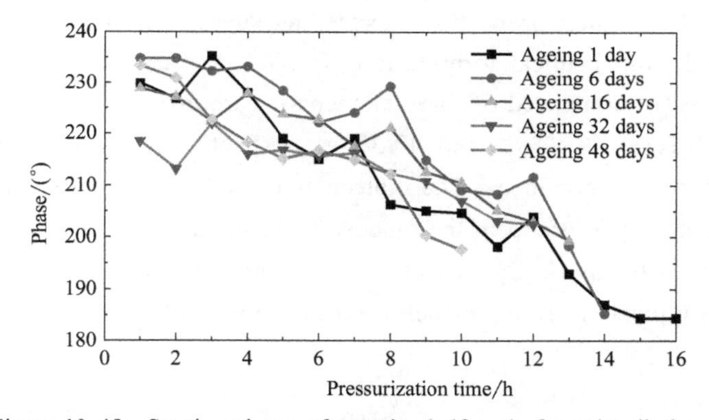

Figure 10-45 Starting phases of negative half cycle for point discharge

Figure 10-46 shows the change rule of the negative half-cycle termination phase of the oil-paper insulated tip discharge. It can be seen from Figure 10-46 that the end phase position of the negative half period of the tip discharge of each test sample fluctuates and rises with the increase of the experimental time. At the beginning of discharge, the ending phase position of the negative half period is between 310° and 320°. As the discharge progresses, approaching the breakdown phase, the ending phase position of the negative half period increases to between 350° and 360°. Similarly, in certain periods, the negative half-cycle termination phase decreases to a certain extent as the discharge time increases. Comparing the laws of the curves, it can be seen that the ageing degree of the insulating paperboard has no obvious influence on the negative half cycle termination phase of the oil-paper insulating tip discharge.

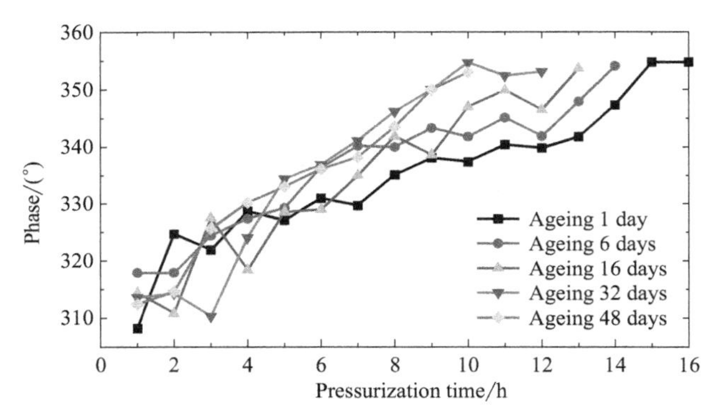

Figure 10-46 Ending phases of negative half cycle for point discharge

10.5 Reasons for the Influence of Ageing of Insulating Paperboard on the Development of Partial Discharge of Oil-Paper Insulation

10.5.1 Basic Structure of Insulation Paperboard

Insulating paperboard is mainly composed of cellulose, hemicellulose and lignin, among which cellulose accounts for the highest proportion, exceeding 90%. Cellulose is a polysaccharide natural polymer compound, its chemical formula is $C_6H_{10}O_5$, and its molecular structural formula is $(C_6H_{10}O_5)_n$. Cellulose is composed of three elements: carbon, hydrogen, and oxygen, and its mass fractions are 44.44%, 6.17%, and 49.39%, respectively. When cellulose is completely hydrolyzed, 99% glucose is obtained. The molecular formula of glucose is $C_6H_{10}O_5$. As shown in Figure 10-47, the aldehyde group in the glucose molecule and the alcohol hydroxyl group located at the C_5 position act to form a C_1—C_5 glycosidic bond, which in turn forms a hemiacetal ring structure, which is a typical six-ring, namely α-D-glucopyranose and β-D-glucopyranose.

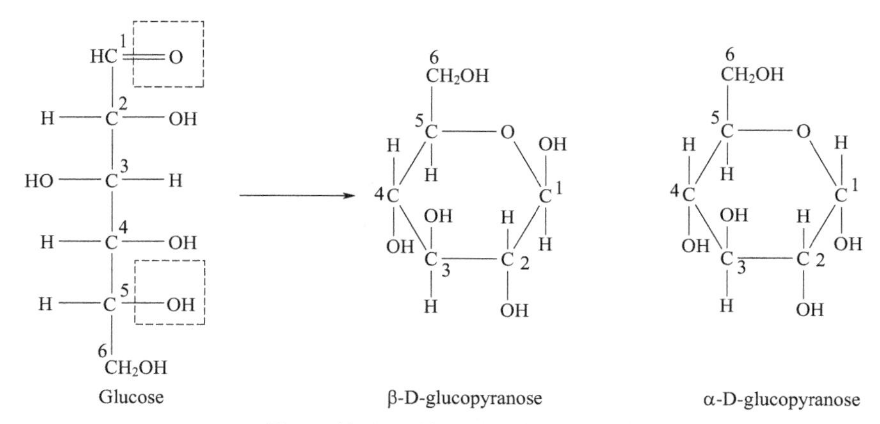

Figure 10-47 Glucose and glucopyranose

In order to maintain a stable structure, the sugar ring of glucose pyran cannot be in a plane. The hexasaccharide ring has 8 different conformations, among which the chair conformation has lower energy and is more stable. As shown in Figure10-48, the chair conformation includes $^4C_1$ and $^1C_4$ conformations. Both α-D-glucopyranose and β-D-glucopyranose are $^4C_1$ conformations [7].

(a) $^4C_1$ chair conformation (b) $^1C_4$ chair conformation

Figure 10-48 Chair conformation

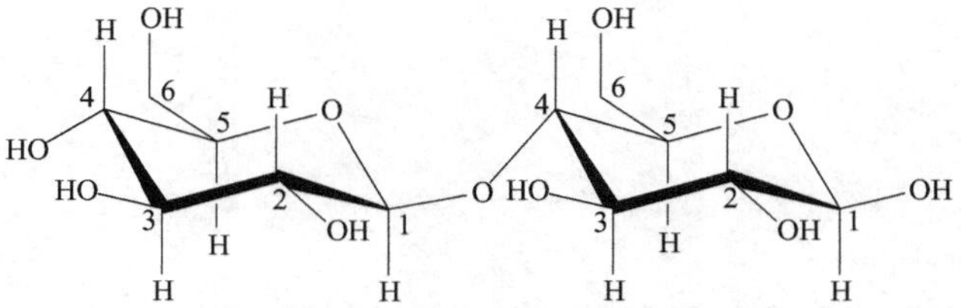

Figure10-49 1, 4-β glycosidic bond

Cellulose molecule is a high molecular polymer formed by multiple β-D-glucopyranosyl groups with 1, 4-β glycoside bonds (Figure 10-49) . Figure 10-50 shows the structure of cellulose molecular chain. In the figure, the cellulose molecule contains *n* β-D-glucopyranose groups, where *n* is called the degree of polymerization of cellulose. In the natural state, the length of the cellulose molecular chain in wood is about 5000nm, corresponding to about 10000 β-D-glucopyranose groups, and its degree of polymerization is about 10000. After cooking, bleaching and other processes, the degree of polymerization will decrease. According to the molecular chain structure of cellulose, each β-D-glucopyranose group in the cellulose molecule contains three free hydroxyl groups (—OH). When the distance between the hydrogen atom in these hydroxyls and the adjacent oxygen atom with strong electronegativity is less than 0.28nm, it is easy to form hydrogen bonds. In addition to van der Waal force and carbon-oxygen bonds, the bonds between cellulose macromolecules also include hydrogen bonds. The hydrogen bonds of cellulose molecular chain mainly include intramolecular hydrogen bonds and intermolecular hydrogen bonds. The intramolecular hydrogen bonds mainly include O_3—H⋯O_5 and O_2—H⋯O_6, and the intermolecular hydrogen bonds mainly exist in the form of O_6—H⋯O_3. The energy of van der Waals force is about 2-8kcal/mol, the bond energy of carbon-oxygen bond is about 80-90kcal/mol, and the hydrogen bond energy is about 5-8kcal/mol. Although the hydrogen bond energy is less than the carbon-oxygen bond energy, because cellulose is a polymer compound, its degree of polymeriza-

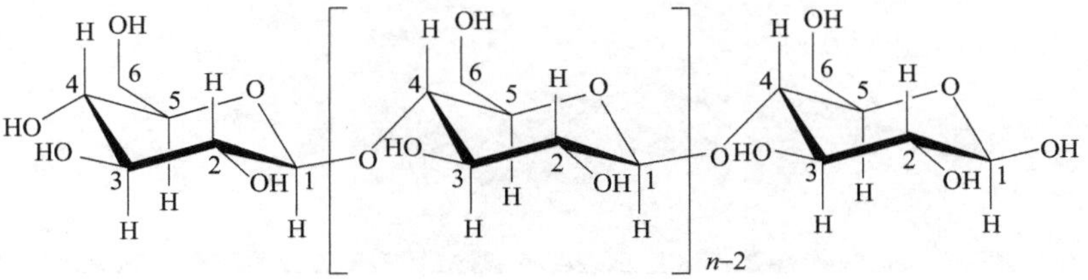

Figure 10-50 Cellulose molecular chain structure with degree of polymerization was *n*

tion is often larger and contains more hydrogen bonds, so the total hydrogen bond energy is much greater than Carbon-oxygen bond.

10. 5. 2 Analysis of the Microscopic Properties of Insulating Paperboard with Different Ageing Degrees

A scanning electron microscope (scanning electron microscope, SEM) was used to analyze the surface microscopic morphology of the insulating paperboard at various degrees of ageing. Figure 10-51 shows the SEM test results (amplified by 100 times).

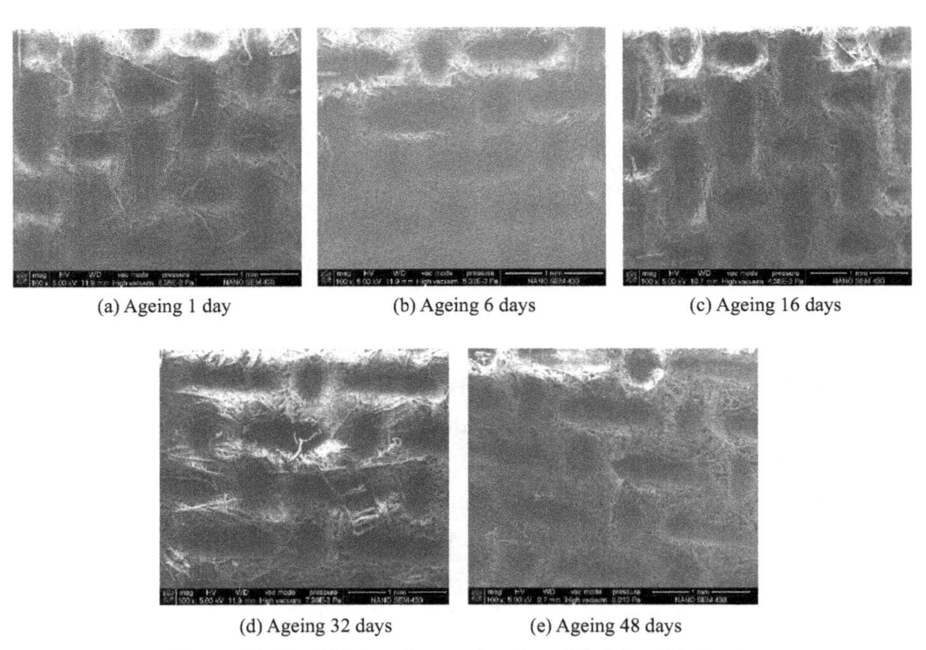

(a) Ageing 1 day (b) Ageing 6 days (c) Ageing 16 days

(d) Ageing 32 days (e) Ageing 48 days

Figure 10-51 SEM testing results (amplified by 100 times)

From the results in Figure 10-51, it can be seen intuitively that the insulating paperboard is pressed from plant fibers through a certain regular arrangement. Comparing the SEM results of insulating paperboard with different ageing degrees, it can be seen that as the ageing degree deepens, the plant fibers of the insulating paperboard still show a regular arrangement as a whole, and there is no obvious change. Set the SEM magnification to 1000 times to observe the surface morphology of a single fiber, as shown in Figure 10-52. It can be seen from the results in Figure 10-52 that the ageing did not cause obvious breakage of the fibers. However, as the degree of ageing deepens, the surface of the fiber begins to become rough, and some fibers under-

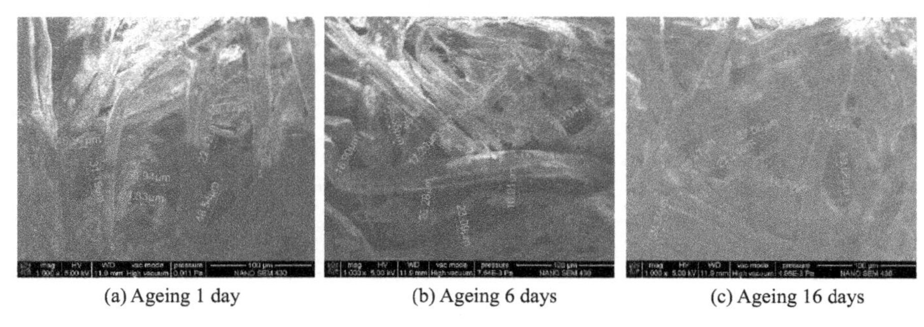

(a) Ageing 1 day (b) Ageing 6 days (c) Ageing 16 days

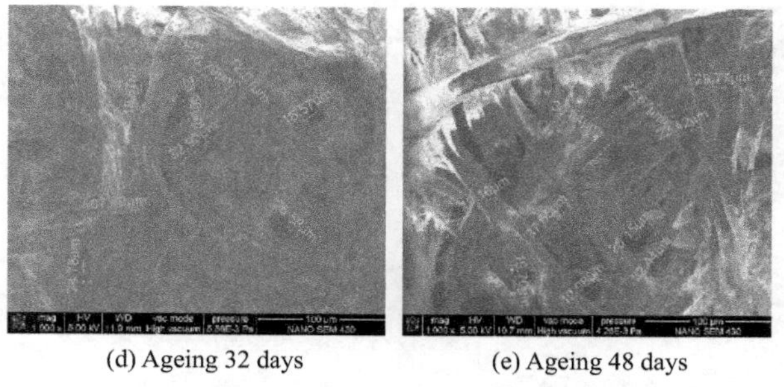

(d) Ageing 32 days (e) Ageing 48 days

Figure 10-52 SEM testing results (amplifiedby 1000 times)

go bending polymerization and other phenomena.

As shown in Figure 10-53, three sampling points are selected from the insulation paperboard of each ageing degree, and the fiber diameter at each sampling point is measured, and the average value of the fiber diameter of the insulation paperboard for each ageing degree is calculated, as shown in Table 10-10. From the results in Table 10-10, it can be seen that as the degree of ageing deepens, the diameter of the insulating cardboard fiber shows a decreasing trend.

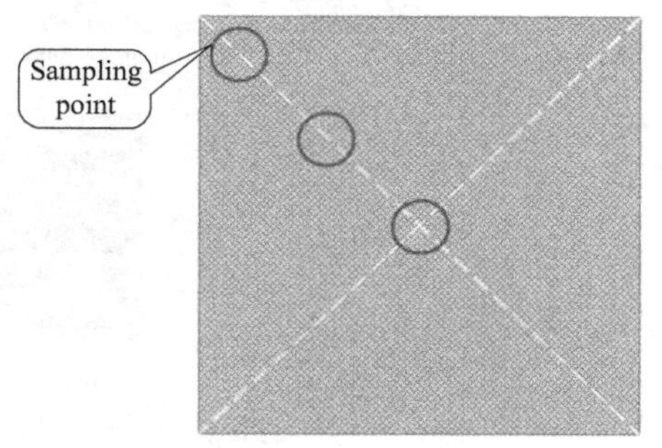

Figure 10-53 Sampling point for measuring the fibre diameter of the pressboard

Table 10-10 Fibre diameter of the pressboard of pressboard with various ageing degree

Ageing time / day	1	6	16	32	48
Fiber diameter/ μm	24.95	24.05	22.92	21.04	20.39

In summary, it can be seen from the SEM test results that as the degree of ageing deepens, the fiber diameter of the insulating paperboard shows a certain decreasing trend, and the fiber surface also becomes rough, but its surface structure does not change significantly, which explains why The reason why the initial discharge voltage of the oil-paper insulation partial discharge does not change much under different ageing degrees of the insulating paperboard.

The cross section of insulation paperboard with different ageing degrees was photographed by digital microscope with magnification of 1000 times, and the cross section microscopic images of insulation paperboard with different ageing degrees were obtained, as shown in Figure 10-54.

It can be seen from the results in Figure 10-54 that for the one-day-aged insulating paperboard, it can be seen from the cross-sectional micrograph that the structure is tight and no obvious pores are seen. With the deepening of the ageing degree of the insulating paperboard, there are more and more pores in its cross section.

In order to explain this phenomenon, this article uses Nicolet IS5 Fourier infrared spectrometer to perform infrared Fourier spectroscopy (fourier transform infrared spectrometer, FTIR Spectrometer) test on different degrees of ageing cardboard to analyze the changes of its main chemical functional groups. Before the test, the insulating paperboard samples were cleaned with acetone solution to filter out the remaining transformer oil samples. Figure 10-55 shows the infrared spectra of insulating paperboards of various ageing degrees.

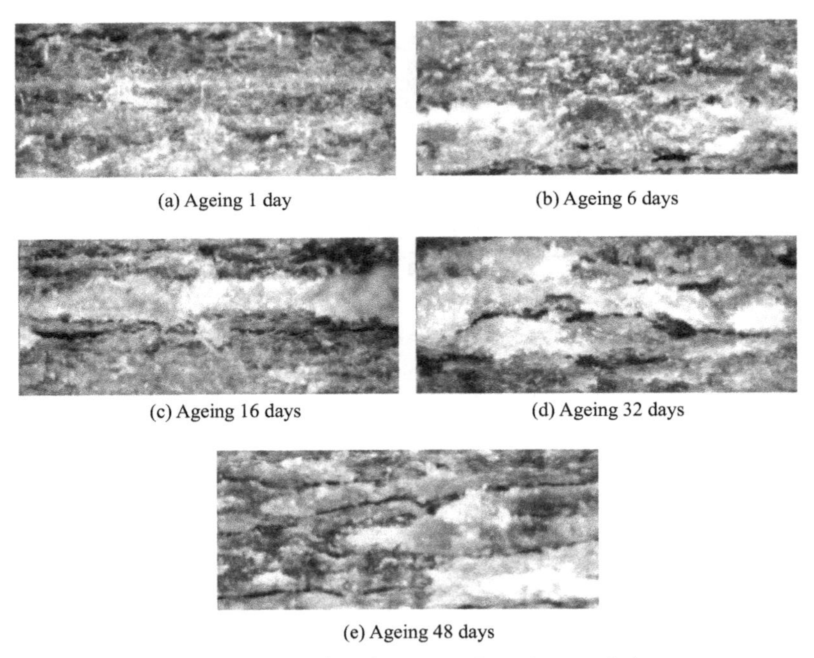

(a) Ageing 1 day (b) Ageing 6 days

(c) Ageing 16 days (d) Ageing 32 days

(e) Ageing 48 days

Figure 10-54 Pressboard cross section microscopic image

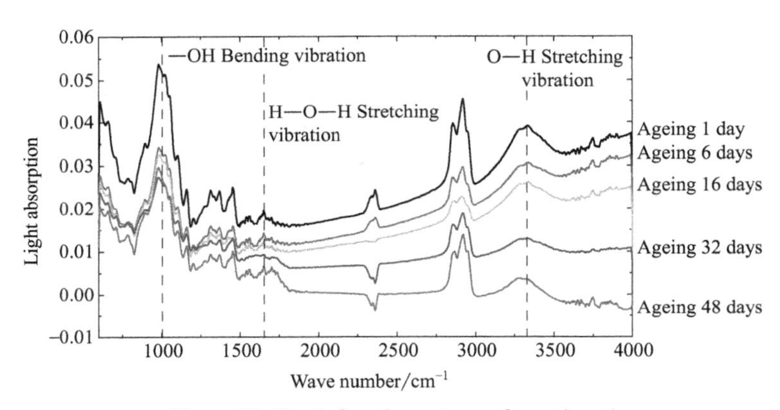

Figure 10-55 Infrared spectrum of pressboard

From the results in Figure 10-55, it can be seen that at the wave number of $1000cm^{-1}$, wave number of $1647cm^{-1}$ and wave number of $3346cm^{-1}$, the absorbance peak shows a downward trend. We can see that the wavenumber near $1000cm^{-1}$ represents O-H bond bending vibration, the wavenumber at $1647cm^{-1}$ represents H—O—H bond stretching vibration, and the wavenumber at $3346cm^{-1}$ represents O—H bond stretching vibration[8]. The drop of the three absorbance peaks indicates that the hydroxyl group in the insulating paperboard is broken, that is, the intra-molecular or intermolecular hydrogen bond is broken. From the results in Figure 10-55, it is not difficult to find that the greater the degree of ageing of the insulating paperboard, the lower the above three peaks, which means that the more serious the degree of ageing of the insulating paperboard, the more severe the rupture of hydrogen bonds within or between the cellulose molecules. As can be seen from the section 10.2 on the basic structure of insulating paperboard, the sum of hydrogen bond energy of insulating paperboard is much greater than that of carbon-oxygen bonds, and hydrogen bonds are the main factor that makes cellulose molecules bond. The breakage of the hydrogen bond will reduce the degree of cellulose bonding, resulting in a decline in the mechanical properties of the insulating paperboard, which is macroscopically expressed as a porous structure inside the insulating paperboard.

10.5.3 Explanation of the Influence of Ageing of Insulating Paperboard on the Development of Discharge

For the internal discharge process of the liquid dielectric, the initial charged particles are mainly caused by the following tworeasons: (1) the liquid dielectric contains impurities such as bubbles. The impurities are subject to a strong electric field, which is easy to produce initially charged particles, which will lead to discharge. (2) the unreasonable shape of the electrode causes the liquid dielectric to have a local high field strength area, which is likely to cause the liquid dielectric to produce initially charged particles. The oil-paper insulation test grade used in this paper is composed of ageing insulating paperboard combined with filtered new transformer oil, and the new transformer oil can be considered to be free of impurities such as moisture and bubbles. Therefore, the initial charged particles in this experiment should be mainly caused by the above point (2).

In order to analyze this phenomenon, the electric field is simulated and calculated based on the finite element simulationmethod. Firstly, analyze the surface discharge of oil-paper insulation. The electric field cloud diagram of surface discharge model is shown in Figure 10-56.

From the results in Figure 10-56, it can be seen that for the surface discharge model designed in this book, the interface between the column electrode and the insulating paperboard is a local high field strength area. This local high field strength will make the liquid dielectric (transformer oil) easy to ionize, and then generate initially charged particles.

Because the oil-paper insulation sample in this book has sufficient and good insulation paper immersion oil, the pores inside the insulation paperboard should be filled with filtered transformer oil before the discharge and the initial stage of the discharge. Since the relative dielectric constant of the insulating paperboard fully immersed in oil is generally about 3, and the relative dielectric constant of pure transformer oil is generally about 2, the difference between the two values is small. Therefore, the inside of the insulating paperboard is not discharged and the initial stage of the dischargethe increase in the electric field in the pores is not significant, as shown in the results in Figure 10-57. Therefore, the pores are not prone to discharge when the discharge is not started and

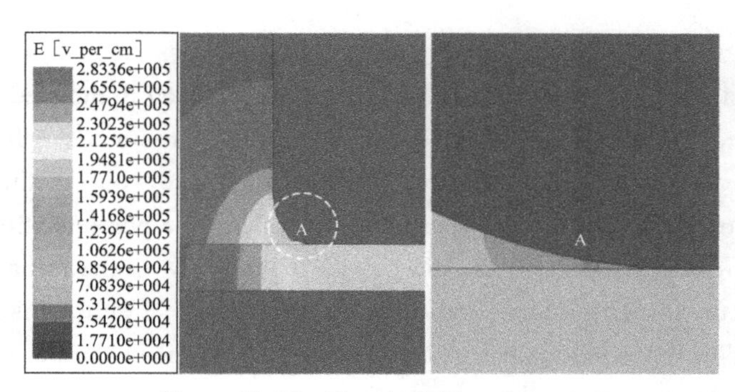

Figure 10-56 Electric field nephogram of surface discharge model

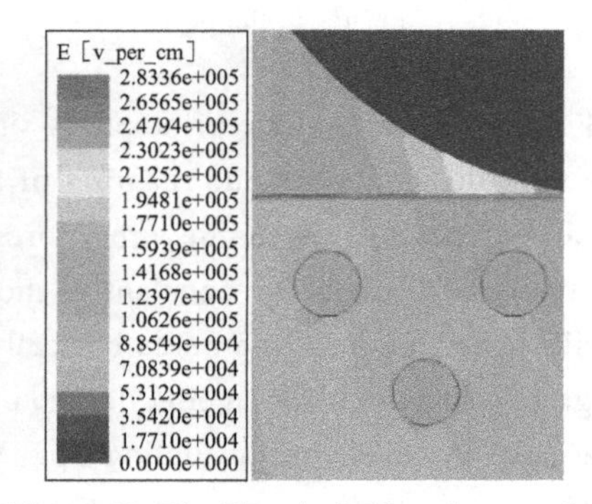

Figure 10-57 Electric field nephogram for oil in the pore

at the beginning of discharge. In addition, in the initial stage of the discharge, the cumulative effect of the discharge on the deterioration of the insulating paperboard is small, and the destruction of the cellulose of the paperboard is not obvious, so the gas content of the paperboard is also less.

In summary, in the initial stage of the discharge, the effect of the internal pores of the insulating paperboard on the discharge is not obvious, so at this stage, the ageing of the insulating paperboard has no obvious influence on the development of the discharge.

With the development of the discharge, the energy generated by the discharge continues to accumulate, causing the transformer oil in the pores of the insulating cardboard to decompose, and finally the pores are filled with residual gas. Since the relative dielectric constant of the gas is relatively low (it can be approximately regarded as 1), the electric field at the pores increases greatly. After simulation calculation, the pore electric field value at this time reaches about 21kV/cm, as shown in Figure 10-58. Under the action of this local high field strength, the residual gas in the pores is easier to discharge (the principle is similar to the bubble discharge principle), which intensifies the discharge process. At the same time, the bubbles generated by the discharge overflowed to the outside of the cardboard and gathered near the column electrode, which further aggravated the discharge process. Figure 10-59 shows the image of the bubble near the column electrode taken by the high-definition camera during the experiment. Because the higher the ageing degree, the more pore structure in the cardboard; at the same time, the higher the ageing degree of the cardboard, the more easily the internal cellulose will decompose gas under the action of discharge, therefore, in the discharge development and discharge dangerous stage, the ageing degree of the insulating paper board has a more obvious influence on the development of discharge.

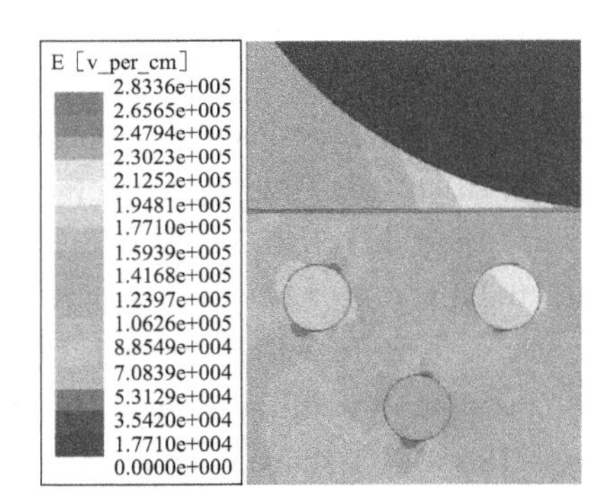

Figure 10-58 Electric field nephogram for gas in the pore

Figure 10-59 Bubble image in the discharge process

For tip discharge, the principle is similar to that of surface discharge. Figure 10-60 is the electric field distribution cloud diagram of the transformer oil-paper insulation tip discharge model. In the initial stage of discharge, the pores are filled with transformer oil. At this time, due to the relatively high dielectric constant of the transformer oil, the electric field value at the pores is still small. At this time, the pores basically have no effect on the discharge. Therefore, in the initial stage of discharge, the ageing degree of the insulating paperboard has no obvious effect on the discharge of the oil-paper insulating tip. With the development of the discharge, the oil in the pores decomposes, leaving gas in the pores. At this time, the electric field at the pores is stronger, which accelerates the discharge process. At the same time, the higher the ageing degree, the more

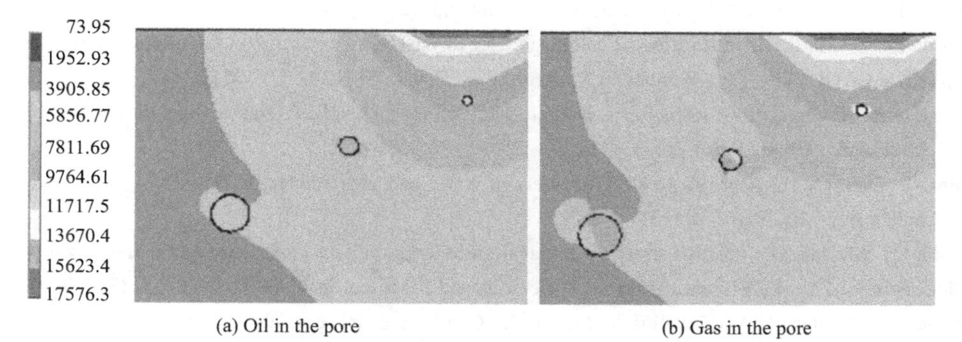

(a) Oil in the pore (b) Gas in the pore

Figure 10-60 Electric field nephogram of point discharge model

pores in the cardboard, which makes the development of discharge easier. Therefore, the ageing degree of the insulating paperboard has a greater influence on the discharge of the oil-paper insulating tip during the discharge development and discharge dangerous stage.

10.6 Summary of This Chapter

In this chapter, the insulation paperboard samples with different ageing degrees are obtained through the heating ageing method. Designed the oil-paper insulation surface discharge and tip discharge models, and conducted a pressure experiment using the constant voltage method to study the partial discharge degradation law of the transformer oil-paper insulation. The main conclusions of this chapter are as follows.

(1) According to the difference of discharge phenomena, the partial discharge of transformer oil-paper insulation can be divided into the initial stage of discharge, the development stage of discharge and the dangerous stage of discharge.

(2) With the increase of the ageing degree of the insulating paperboard, the creeping discharge and tip discharge withstand time of the oil-paper insulation of the transformer gradually become shorter, and the initial discharge voltage shows a smaller downward trend.

(3) In the initial stage of discharge, the ageing degree of the insulating paperboard has no obvious effect on the partial discharge of the oil-paper insulation; the ageing degree of the insulating paperboard has a greater influence on the partial discharge of the oil-paper insulation in the discharge development stage and the discharge dangerous stage. The higher the ageing degree, the more the discharge development. The faster the ageing degree, the larger the total discharge capacity, maximum discharge capacity, total discharge times and other parameters at the same pressurization time. Based on the method of combining microscopic testing and simulation, this phenomenon is explained. It is believed that this phenomenon is mainly caused by the pores generated in the insulating paperboard under the effect of ageing.

Notes and References

[1] Pei J C. Plant fiber chemistry [M]. Beijing: China Light Industry Press, 2012.

[2] Edwards H G M, Farwell D W, Webster D. FT Raman microscopy of untreated natural plant fibres [J].

Spectrochimica Acta Part A Molecular & Biomolecular Spectroscopy, 1997, 53 (13): 2383-2392.

[3] Garside P, Wyeth P. Identification of cellulosic fibres by FTIR spectroscopy: differentiation of flax and hemp by polarized ATR FTIR [J]. Studies in Conservation, 2006, 51 (3): 205-211.

[4] Tobazkon R. Prebreakdown phenomena in Dielectric Liquids [J]. IEEE Transactions on Dielectrics and Electrical insulation, 1994, 1 (6): 1132-1147.

[5] Kolb J E, Joshi R P, Xiao S, et al. Streamers in water and other dielectric liquids [J]. Journal of Physics D, Applied physics, 2010, 41 (23): 1-22.

[6] Lesaint O, Massala G. Positive streamer propagation in large oil gaps: experimental characterization of propagation modes [J]. IEEE Transactions on Dielectrics and Electrical insulation, 1998, 5 (3): 360-370.

[7] Butcher M, Ncuber A A, Ccvallos M D, et al. Conduction and breakdown mechanisms in transformer oil [J]. IEEE Transactions on Plasma Science, 2006, 31 (2): 467-475.

[8] Zhang Y B. Simulation study on partial discharge gas generation characteristics of oil-paper insulation transformer [D]. Harbin: Harbin University of Science and Technology, 2013.

Chapter 11
Intelligent Diagnosis Method of Advanced Electrical Equipment

11.1 Technology and Application of Intelligent Sensing and State Sensing for Transformation Equipment

11.1.1 Introduction

Power equipment is an important component of the power system. Different types and degrees of faults in operation of key equipment such as transformers, transmission lines, circuit breakers, etc. will directly affect the safe and stable operation of the entire power system. Comprehensively grasping, analyzing, and predicting the operating status and health level of power equipment, improving the level of fault diagnosis, and rationally arranging status maintenance or predictive maintenance are the necessary foundation for providing users with safe, high-quality, and economical power supply. Carrying out research on related technologies for power equipment maintenance, operation and maintenance has important basic theoretical significance and practical application value[1-3].

The power transmission and transformation equipment of our country's power grid presents the characteristics of various types of equipment, wide distribution range, and different structural parameters. In the process of long-term operation, the failure of power equipment is almost inevitable. The causes of the failure include the equipment left in the manufacturing process. Defects, problems in installation, overhaul and maintenance, insulation ageing and structural deterioration caused by long-term operation. However, the power transmission and transformation equipment itself is also an extremely complex system, with many characteristics that characterize its state, such as operating conditions, maintenance history, working environment, monitoring data, family quality history, etc., and these state information is uncertain and fuzziness, complex relationships among various parameters and mutual coupling influences, so it is very difficult to achieve effective and accurate evaluation of the operating status of power equipment[4-6]. At present, the operation and maintenance of power transmission and transformation equipment in my country is mainly carried out under the guidance of guidelines, regulations, expert experience or traditional ratios, waveform characteristics analysis methods and other methods[7-12]. This method is difficult to meet the demand for quantitative, differentiated, and differentiated power equipment. Refined operation and maintenance requirements may cause equipment to be "over-repaired" or "under-repaired", resulting in a huge waste of human and material resources[13]. With the continuous expansion of the scale of the power system, the wide deployment of various sensors and the rapid improvement of the level of power informatization, energy management systems, equipment management systems, and various information platforms have accumulated a large amount of multi-source heterogeneous power equipment data, including numbers, text, images, sounds, etc., and show significant "4V" characteristics, namely volume, variety, velocity, and value. On the one hand, with the explosive growth of power transmission and transformation equipment and its data volume, traditional methods have gradually shown some deficiencies in terms of evaluation accuracy, diagnosis efficiency, and knowledge update; on the other hand, massive data provides a data foundation for the development and application of big data, data mining, artificial intelligence and other technologies[14-16]. The theoretical breakthroughs of emerging artificial intelligence algorithms such as deep learning and knowledge graphs, as well as the development of high computing power technologies represented by GPUs and TPUs, provide technical support for the application of artificial intelligence in the operation and maintenance of power transmission and transformation equipment. As the most disruptive frontier technology at present, artificial intelligence technology is forming a new set of data science methodology, and profoundly affecting and changing the business model of the entire world including transportation, finance, medical care, industry, law, e-commerce and other fields and even the industry form. At present, various countries have formulated artificial intelligence development strategies to seize the commanding heights of the new round of technological revolution[17,18], and my country has also upgraded the new generation of artificial intelligence technology to a national strategy[19]. In this context, advancing the innovative application of artificial intelligence technology in the business of power transmission and transformation equipment status assessment, fault diagnosis, health management, etc., is to improve the intelligent and intensive level of power grid operation and maintenance, and promote the comprehensive operation and maintenance of equipment status. An important technical means to move forward in a precise and efficient direction. This article combines the specific business of artificial intelligence technology and the operation and maintenance of power transmission and transformation equipment, and expands the discussion layer by layer according to the data layer, algorithm layer and application layer. Among them, the data layer includes external data, ontology data, monitoring data, test data, inspection data, guidelines and regulations, etc. The algorithm layer includes traditional machine learning, heuristic intelligent algorithms, computer vision, natural language processing, knowledge reasoning and multi-modal learning technologies. The application layer includes the establishment of operation and maintenance knowledge base, comprehensive evaluation of the health status of power transmission and transformation equipment, equipment operation status prediction, equipment defect identification and fault diagnosis, equipment life evaluation and intelligent recommendation of operation inspection strategies. This chapter firstly introduces the data situation and current issues of power transmission and transformation equipment. Secondly, it explains the key business-related artificial intelligence technologies. Finally, it summarizes the current application scenarios and exploration practices of artificial intelligence technology in equipment operation and maintenance, and proposes artificial challenges and future development trends in intelligent research.

11.1.2 Data Situation

The continuous increase in the construction scale, voltage level, and complexity of our country's AC and DC power grids puts forward higher requirements for the omni-directional and full-cycle measurable and controllable and intelligent operation and inspection of core high-voltage electrical equipment. Sensors, computers, network communications and other technologies are widely used in the field of power transmission and transformation equipment, and a large number of new test methods and non-electricity measuring instruments have appeared, which provide a good foundation for the application of big data and deep artificial intelligence technology for power transmission and transformation equipment. This section will describe the current data situation of power transmission and transformation equipment and its existing problems.

11.1.2.1 Transmission and Transformation Equipment Data

The structured data of power transmission and transformation equipment mainly includes ontology data, online monitoring data, test data, meteorological and geographic environment data, etc., as shown in Figure 11-1. The ontology data refers to the ledger, design parameters, etc. of the equipment, which are entered into the system when the equipment is shipped or connected to the grid, and the data is usually relatively complete and accurate. Online monitoring data refers to continuous or periodic automatic monitoring and testing of the status of power equipment without power failure, with high monitoring frequency and large data volume. Test data comes from routine tests and diagnostic tests. Routine tests refer to various live detection and power outage tests performed regularly to obtain equipment status. Diagnostic test refers to the discovery of bad state of equipment through patrol inspection, online monitoring, routine test, etc., or experience of bad working conditions and warning of family defects. It is a test conducted to further evaluate the state of the equipment. The test data is generally relatively accurate, but affected by manual measurement and input, data integrity and recording frequency are difficult to guarantee. Among them, diagnostic tests are carried out for equipment that is suspected of being in poor condition, so the amount of data is extremely scarce.

The unstructured data of power transmission and transformation equipment mainly includes image/video data and text data, which come from inspection records, ultraviolet/infrared/visible light detection or monitoring, defect reports, situation descriptions, etc. The inspection record refers to the written or electronic document record obtained after inspecting various equipment according to the specified inspection content and inspection cycle during the operation of the equipment, which is often multi-modal data with a mixture of text and images. Ultraviolet/infrared/visible light detection or monitoring is often in the form of images or videos, and is marked with important detection information such as temperature according to the different detection/monitoring equipment.

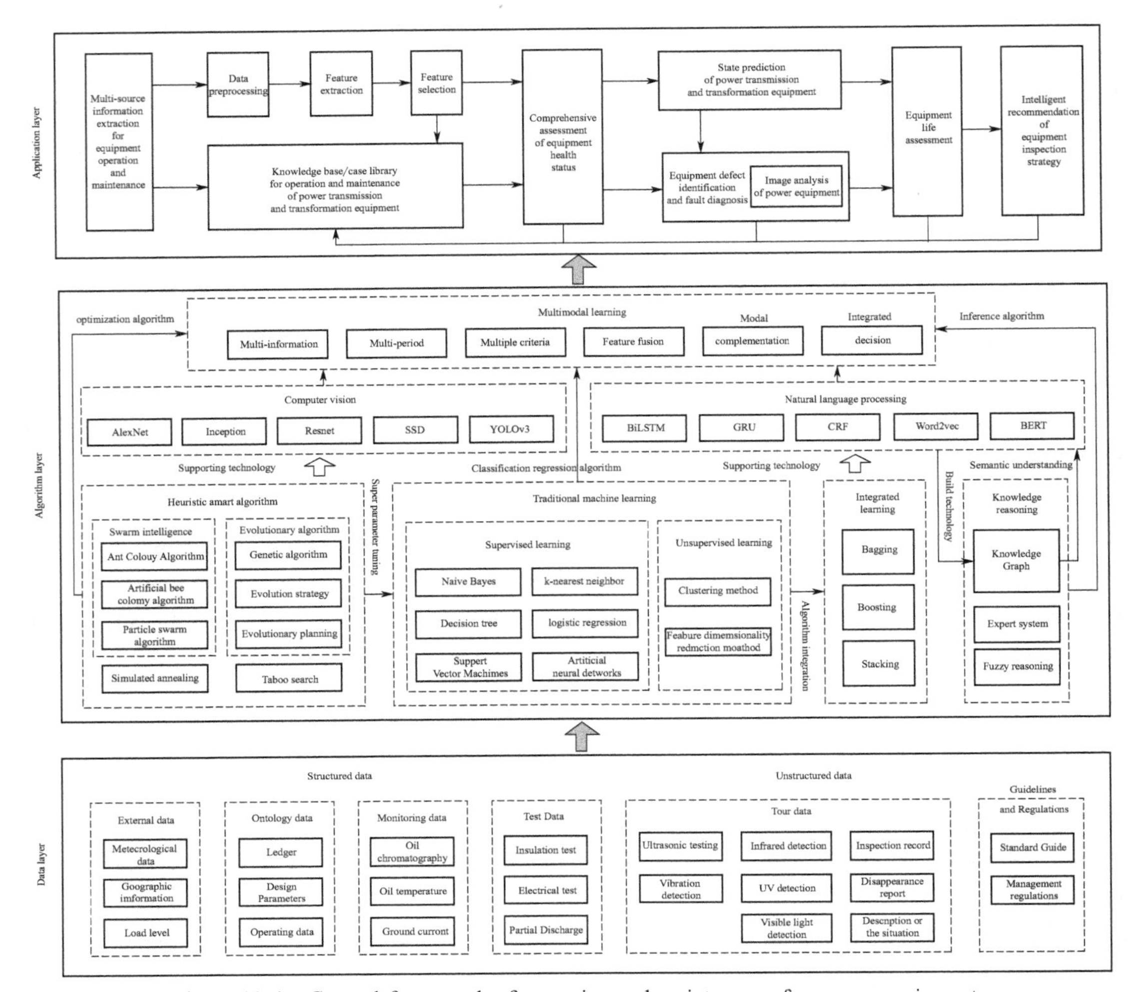

Figure 11-1 General framework of operation and maintenance for power equipment using artificial intelligence technology[20]

Deficiency report refers to the record report that the transportation inspection company of each network province handles equipment defects according to the inspection situation. It is also a multimodal data with a mixture of text and images, and lacks a unified standard format. The description refers to the defect description document prepared temporarily by the provincial transportation inspection company after the defect equipment is found. In addition, standard guidelines and management regulations (five-pass management regulations and rules) also provide a large amount of verified knowledge of general transmission and transformation equipment operation and inspection.

11.1.2.2 Existing Data Problems

With the continuous improvement of sensor components, monitoring devices, and detection methods, the normalized collection and transmission of multi-modal data of power transmission and transformation equipment is constantly developing in terms of volume, frequency, quality, and scope. However, the current data of power transmission and transformation equipment still has the following significant problems.

(1) Data quality issues. The quality of data related to power transmission and transformation

equipment will be affected to varying degrees in the collection, transmission, storage and other links, causing problems such as large measurement errors, abnormal data, data loss or duplication, and non-standard storage formats, which makes the effective data sample volume too small.

(2) Data imbalance problem. The number of fault data and cases accumulated by power transmission and transformation equipment is scarce. Compared with the ever-increasing normal operation monitoring data, it shows extreme data imbalance. It is difficult to portray global samples directly by applying artificial intelligence algorithms.

(3) Data imbalance problem. The difficulty of data fusion is mainly reflected in two aspects. One is the independent development and construction of different business systems in the early stage, which makes it difficult to effectively match the structured data of their equipment; Secondly, data such as documents and images often come from different people, manufacturers, and equipment, with extremely different features, making it difficult to achieve effective identification and fusion learning.

Therefore, when applying artificial intelligence technology for equipment operation and maintenance, on the one hand, it is necessary to use data completion, repair, augmentation, etc. , to try to maintain the completeness and balance of the data set; on the other hand, it is necessary to comprehensively consider various aspects such as the type of specific problem, data scale and structure, computing power, calculation accuracy and efficiency requirements, and select and design a suitable algorithm model. For extremely rare data, the use of rules to judge is often more effective than machine learning models, while support vector machines, *k*-nearest neighbors and other methods perform well in training sets with moderate amounts of data, and deep learning is more advantageous in the processing of images, text or large amounts of measurement data.

11.1.3 Key Technology

The core of intelligent operation and maintenance of power transmission and transformation equipment is to conduct deep fusion analysis of massive online and offline multi-source heterogeneous data, extract mapping relationships, key information and judgment rules, and establish data-driven prediction, evaluation and diagnosis models. In this way, the real-time and accuracy of equipment status evaluation, fault diagnosis and prediction can be improved.

Machine learning (ML) is an algorithm that generates a model from data on a computer, that is, "learning algorithm". For example, to classify the fault types of power equipment, ML trains the model based on empirical data, finds a suitable function f, and makes it the optimal mapping $Y=f(X)$ between the input variable X(such as weather, equipment body data, operating data, fault information, etc.) and the output variable Y(fault category) according to the principle of minimum error.

Scholars and research institutions at home and abroad have carried out a lot of research on machine learning methods for power equipment state evaluation and fault diagnosis. In recent years, as big data technology, computer computing power, and the theoretical level of learning algorithms have greatly improved, artificial intelligence technology based on deep learning has also been rapidly developed and widely used. This section will mainly explain the key artificial intelligence technologies involved in each link of the operation and maintenance of power transmission and transformation equipment, including traditional methods such as supervised and unsupervised machine learning, heuristic intelligent algorithms, and knowledge reasoning, with machine learning and deep learning as the bottom layer two types of advanced applications supporting technology, computer vision and natural language processing, and multi-modal learning that integrates various algorithms and data. Big data technologies such as data cleaning, outlier detection, and data completion of power transmission and transformation equipment have been discussed in many documents, so this article will not discuss too much.

11. 1. 3. 1 Traditional Machine Learning

(1) Supervised Learning

Supervised learning is to train machine learning with a certain inference function model from labeled data, which mainly includes classification and regression tasks. Among them, the most widely used classification method is the fault diagnosis of power equipment. Common classification methods include Naive Bayes, *k*-nearest neighbor, decision tree, Logistic (Softmax) regression, support vector machine, artificial neural network and other methods.

The most common model of regression (prediction) method is linear regression, which solves the over-fitting problem by introducing a regularization method of penalties, such as ridge regression, minimum absolute contraction and operator selection, and elastic nets, etc. The commonly used method for nonlinear regression problems is polynomial regression. In addition, many of the above classification methods can also be used for regression analysis, such as *K*-nearest neighbor method, classification and regression trees, support vector regression, and artificial neural networks (including fully connected layers model, long and short-term memory neural network), etc. Machine learning regression methods are widely used in the state prediction and fault prediction problems of power equipment.

With the continuous development of machine learning, a method of combining multiple weak learners has gradually emerged, which is called ensemble learning. This method makes the model have stronger generalization ability. Common integrated thought frameworks are mainly divided into 3 categories: Bagging, Boosting and Stacking. The basic principle of the Stacking method is shown in Figure 11-2. By adding a layer of learner, the learning result of the weak learner of the training set is used as input, and a learner is retrained to obtain the final result. Random forest is a typical representative of ensemble learning. It generates multiple training sets through self-sampling, and finally integrates the results, thereby effectively solving the problem of single decision tree prone to overfitting. At present, the random forest method has been applied to the fault diagnosis of power equipment, and has achieved good results.

(2) Unsupervised Learning

Unsupervised learning is to train the model in the absence of class information (or expected output value). Currently, unsupervised learning mainly refers to clustering methods. Typical clustering methods include K-means clustering, density-based clustering method, Gaussian mixture model, hierarchical clustering and so on. The clustering method can perform tasks such as status assessment, abnormality detection, and fault diagnosis without device sample annotation.

Generalized unsupervised learning also includes dimensionality reduction methods, which reduce the number of features in the data set through feature extraction, thereby reducing the "dimension disaster" and improving computational efficiency. Principal component analysis (PCA) is the most common dimensionality reduction method. It is generally used in the data preprocessing stage

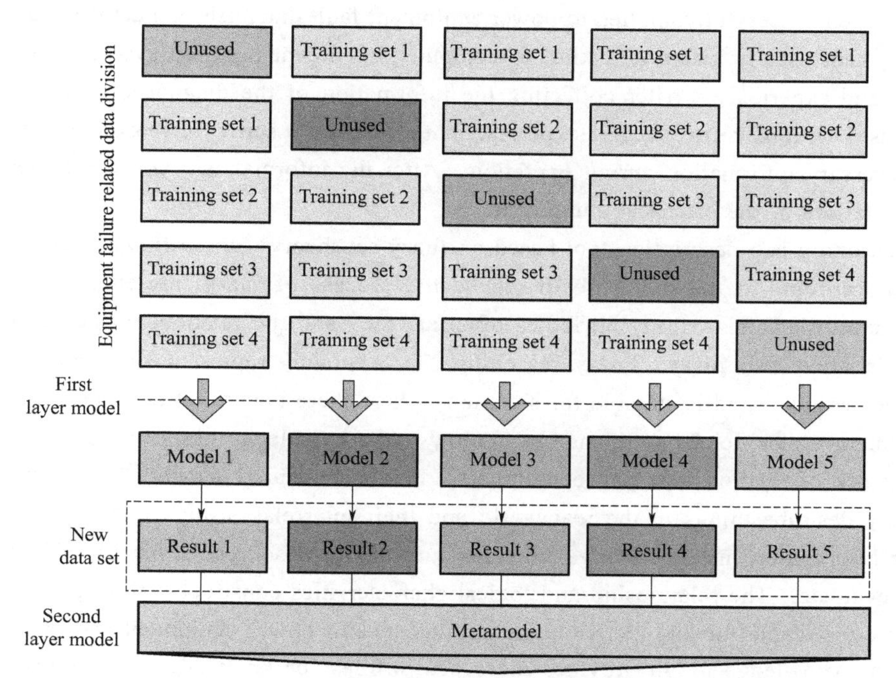

Figure 11-2 Principle of Stacking algorithm

of the machine learning model to analyze the main factors that affect the state of the equipment or cause failures.

Heuristic intelligent algorithms mainly refer to algorithms designed and inspired by certain laws of nature, including swarm intelligence (SI) and evolutionary algorithms (EA). The theory of SI refers to the characteristics of simple intelligence that individuals with simple intelligence exhibit group intelligence behavior that surpasses individual abilities through mutual cooperation and organization. Representative algorithms include particle swarm algorithm, artificial bee colony algorithm, and ant colony algorithm. EA is developed from the theory of evolution. Typical algorithms include genetic algorithms, evolution strategies, evolutionary planning, and so on. In addition, generalized heuristic algorithms also include tabu search algorithms and simulated annealing methods. Heuristic intelligent algorithm is a relatively mature global optimization method with high robustness and wide applicability, which can effectively deal with non-convex and nonlinear problems that are difficult to solve by traditional optimization algorithms.

In the fault diagnosis of power equipment, heuristic intelligent algorithms are mainly applied to two types of problems. One is to optimize the hyperparameters of other classifiers to find the classification model with the best performance. The other is to solve optimization problems, especially the optimization of intelligent maintenance strategies for power equipment.

(3) Knowledge Reasoning

The goal of artificial intelligence reasoning technology is to simulate human thinking mode, and to carry out logical reasoning in combination with relevant experience and professional theories. Reasoning technology can find the causal relationship between equipment's bad working conditions, abnormal symptoms, failure modes, etc., such as expert system, fuzzy theory and other technologies have been successfully applied to power equipment fault diagnosis and achieved good results.

Expert system usually refers to a kind of computer intelligent program system with specialized knowledge and experience. After collecting the information of the diagnosed object, it comprehensively uses various rules (expert experience) to conduct a series of reasoning, and asks the user for necessary information during operation. After the information, you can quickly find the most likely failure of the electrical equipment.

Fuzzy reasoning is a description tool based on fuzzy set theory, and a fuzzy inference engine is established according to the rule base to realize the process of fuzzy inference. In the field of equipment fault diagnosis, fuzzy attributes often appear, and the relationship between faults and symptoms is often also fuzzy. Fuzzy reasoning can accurately represent signs and failures with fuzzy characteristics, and is suitable for situations where there is conflict between evidence.

As an emerging artificial intelligence technology, the knowledge graph (KG) reveals the semantic network of relationships between entities. It is a structured semantic knowledge base that can formally describe things in the real world and their interrelationships. The knowledge graph can gather information into knowledge, making information resources easier to calculate, understand and evaluate. The relationship and logical characteristics constructed through the knowledge graph are very suitable for the operation and maintenance of power equipment and fault reasoning. Although the knowledge graph provides more possibilities for knowledge reasoning, related research is still in its infancy. Retrieval of defect records of power equipment through knowledge graph is a useful exploration in related fields.

11.1.3.2 Computer Vision

Early computer vision (CV) required a lot of experience to manually design features, and the framework was not very versatile. With the breakthrough of deep learning theory, convolutional neural networks (CNN) has greatly improved the accuracy of image recognition, and even surpassed the human level in the recognition of certain scenes. It has been widely used in image classification, Target recognition, target detection, semantic segmentation and other tasks. Convolutional neural network is a deep neural network composed of input layer, convolutional layer, pooling layer, fully connected layer and softmax layer. This method uses a convolution kernel to automatically extract image features, greatly reduces network parameters through receptive fields and weight sharing, and avoids over-fitting problems caused by too many parameters. Typical image processing models based on convolutional networks include LeNet-5, AlexNet, Inception, VGG19, Resnet, SSD, YOLOv3, etc.

With the continuous improvement of the standardization, informatization, and intelligence of power grid operation and maintenance work, the condition monitoring and operation and maintenance of power equipment based on computer vision have become an emerging industry that has developed in recent years. Its specific application in power inspection mainly for target recognition and defect detection of surveillance images. For the video and images collected by drones, helicopters, robots and other inspection equipment, computer vision technology can be used for power transmission equipment (such as lines, towers, fittings, etc.) and substation equipment (such as transformers, isolation switches, insulators, meters, etc.) appearance defect identification, thereby greatly reducing the risk of personnel working in high-risk environments such as high-voltage lines and substations, and improving the quality and efficiency of inspections in dangerous environments.

However, various objects have different appearances, shapes, postures, and data imbalance problems caused by the scarcity of defective samples, coupled with the interference of factors such as illumination and occlusion during imaging, the accuracy and efficiency of the equipment's target detection and defect recognition tasks are not yet completely meet the requirements of production requirements.

11.1.3.3 Natural Language Processing

Natural language processing (NLP) is a technology for systematic analysis, understanding and information extraction of text data in an intelligent and efficient way. By using NLP and its components, you can manage very large blocks of electrical text data, or perform a large number of automated tasks. Natural language processing related technologies include word segmentation, named entity recognition, part-of-speech tagging, dependency syntax analysis, word vector representation technology, semantic similarity calculation, text analysis, etc. From a technical point of view, there are still many difficulties in text information mining: on the one hand, the text is unstructured data, and there is no clear expression form, and it needs to be transformed into structured data that is computable and easy for machine understanding; on the other hand, natural language itself has ambiguity and ambiguity, and there is no unified judgment standard (i. e. loss function) for the accuracy of the result. In addition, text knowledge itself contains complex human thinking content such as logic, emotion, and speculation, which is difficult to understand and use accurately.

During the long-term operation, overhaul and maintenance of power equipment, a large number of unstructured text data such as defect and fault reports, test inspection records, and overhaul and elimination documents have been accumulated. These data contain a wealth of key features such as failure information, failure causes, and maintenance methods. Mining the semantic information and causal links in them is of great significance for guiding equipment status evaluation, operation, maintenance, and maintenance.

11.1.3.4 Multimodal Machine Learning

The data sources related to the operation and maintenance of power transmission and transformation equipment are wide and varied, including physical signals, images, videos, texts, audios and other heterogeneous data from multiple sources. Each source and form of information can be called a modality. Comprehensive analysis of various modalities constitutes multimodal machine learning (MMML). Multi-modal learning aims to improve the ability to process and understand multi-source modal information through machine learning methods. Multi-modal machine learning integrates multi-modal information for better feature representation, extraction and recognition. It should be noted that the multi-modal learning model does not simply splice different models and turn on their respective "switches" in different scenes, but truly integrates and learns multi-source features from the model mechanism.

Compared with single-modal learning, multi-modal learning eliminates the redundancy between modalities through the complementarity between multiple modalities, so as to achieve a better learning effect. Transfer learning (TL) is a typical multi-modal learning method. This method uses a resource-rich modal information to assist the learning of another modal with relatively poor resources, and has a good development prospect in small-sample learning. Figure 11-3 is a schematic diagram of a typical deep adaptation network migration learning. In the equipment operation, maintenance and repair business, due to various factors such as equipment, technology, resources, and procedures, some of the collected state variable data resources are relatively abundant, while other information is relatively scarce. If modal complementation can be carried out, comprehensive analysis of equipment status or faults from different sides will further improve the accuracy of judgment. At present, some scholars have explored equipment operation and maintenance information fusion technology at the levels of multi-period, multi-information, and multi-criteria, and have achieved good results. However, multi-modal machine learning in the true sense is still in its infancy.

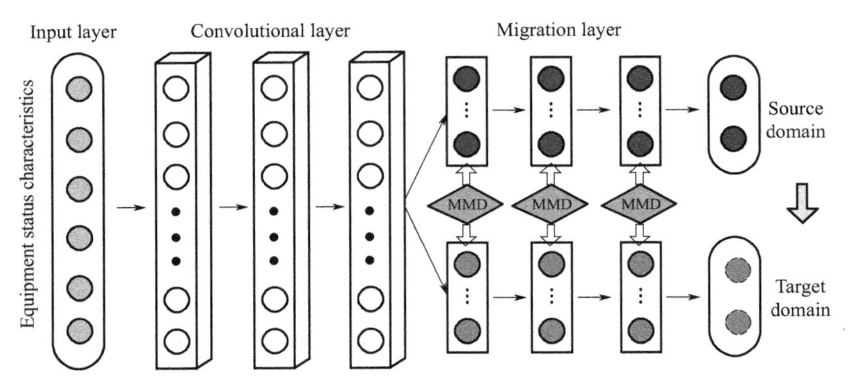

Figure 11-3 Principle of deep adaptation network for transfer learning

11.1.4 Application Scenarios

Based on the data of power transmission and transformation equipment and the key technologies of artificial intelligence, this section sorts out the intelligent operation and maintenance system of power transmission and transformation equipment composed of 7 main application scenarios. Among them, operation and maintenance information extraction and knowledge base establishment provide knowledge graph relationships and equipment failure cases for subsequent applications. Application modules such as equipment health evaluation, equipment status prediction, defect identification and fault diagnosis, equipment inspection image analysis, equipment life evaluation, and intelligent recommendation of operational inspection strategies are used to locate, analyze and process equipment in poor condition through a hierarchical progression.

11.1.4.1 Operation and Maintenance Information Extraction and Knowledge Base Establishment

At present, the power company has deployed a large number of sensing devices in the station, along the line, poles and towers for the transmission and transformation equipment, and arranged manual inspections, inspections and tests on a regular or irregular basis, forming a large number of information systems, document images class records, and through long-term rule summary and experience induction, a large number of international, industry, and company-level standards and regulations have been formulated. It is necessary to collect historically accumulated failure cases from these multi-modal data, and extract the industry knowledge and expert experience contained therein.

On the one hand, according to the standard templates of failure cases of different types of equipment, a structured case library is formed by sorting out basic equipment information, related abnormal information when defects or failures occur, and test data. In this way, it provides positive samples of model learning for application modules such as follow-up comprehensive assessment of health status, defect identification and fault warning, and life prediction.

On the other hand, natural language processing technology is used to extract equipment operation and maintenance information from unstructured, multi-modal data such as standard guidelines, management regulations, and elimination reports, to establish relationships between different concepts or entities, and then use the knowledge graph technology builds a knowledge base and provides a basis for intelligent search and recommendation of operational inspection strategies for equipment in different states. For example, based on some of the corpus in the relevant guidelines for oil-immersed transformers, such as "the result of degradation of paper insulation, first is the rupture of cellulose macromolecules, which is manifested as a decrease in the degree of polymerization and a decrease in mechanical strength; secondly, along with the degradation process, a variety of ageing products dissolved in oil can be obtained, such as CO, CO_2, and furfural". Paper insulation, marking products and other entities and the relationship between them are extracted. The design is shown in Figure 11-4.

In addition, with the popularization of infrared/ultraviolet/visible light and other power equipment monitoring devices in the system, the research on the analysis of images and videos based on artificial intelligence technologies such as image recognition and deep learning, and the extraction of power equipment information is also deepening. A large number of images and related identification information should also be included in the knowledge base.

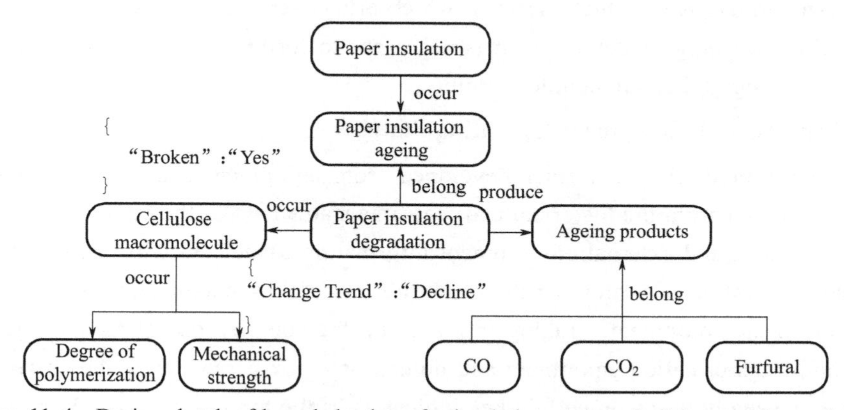

Figure 11-4 Design sketch of knowledge base for insulation ageing of oil immersed transformers

11.1.4.2 Equipment health assessment

In order to effectively ensure the power supply reliability of the power grid and reduce the waste of manpower and material resources caused by regular maintenance, the domestic and foreign power industry has extensively carried out the health assessment and maintenance of AC and DC transmission and transformation equipment such as power transformers, reactors, transformers, and circuit breakers. Work and formulated relevant work guidelines, forming a lot of expert experience. However, the equipment represented by power transformers has complex structure, high cost and key role, and its distribution area and working characteristics are different, and there are more quantities to characterize its state, which has a high degree of uncertainty and ambiguity. In the development process, there is a lack of universal, objective and comprehensive evaluation standards.

In order to solve the drawbacks of traditional business evaluation work based on guidelines and expert experience, academia has carried out research work on equipment state evaluation models based on multi-source equipment state data, applying mathematical analysis methods and machine learning algorithms, hoping to fully and accurately reflect the true state of the transformer. These models are mainly divided into 2 categories. One is to use mathematical models to objectively calculate evaluation weights. By analyzing the relationship between various status indicators and transformer status, determine the closely related key feature indicators, relative importance and evaluation weights, and then evaluate the transformer status. Currently commonly used mathematical analysis methods include analytic hierarchy process, entropy method, etc. The second is based on training samples, using machine learning algorithms to directly construct a predictive model between state quantities and transformer state evaluation. Currently commonly used machine learning algorithms include artificial neural networks, Bayesian probability, and cluster analysis.

It should be noted that equipment health assessment is a problem of imbalanced samples and small samples caused by the scarcity of the absolute number of typical positive samples (abnormal state data). The more expensive and critical equipment, such as large power transformers, the more attention the power company pays to its health in the daily operation and maintenance process. In order to ensure reliable power supply, it will even adopt the strategy of early retirement and high-frequency equipment update. Therefore, the number of samples of historical cases of abnormal transformers is extremely scarce, which brings serious overfitting risks to the training process of machine learning models, and it is necessary to further develop methods such as unbalanced sample learning and small sample learning.

11. 1. 4. 3 Prediction of Equipment Operating Status

Equipment status prediction is further developed from equipment status monitoring and status evaluation. It can start from the historical and real-time status data of the equipment, and combine the grid operation data and external environment information associated with the equipment to discover equipment operation indicators or the changing law of key parameters, so as to predict the future operation of the equipment. Taking into account the complex operating conditions of power transmission and transformation equipment and numerous index parameters, the current equipment state prediction is usually based on certain key indicators as the forecast target. With the help of artificial intelligence in dealing with highly nonlinear and multiple correlation problems, it can be established time series or association prediction models, common methods include support vector machines, deep belief networks, recurrent neural networks, long short-term memory (LSTM), etc. According to different prediction targets, the current research objects of the state prediction of power transmission and transformation equipment based on artificial intelligence include winding state, insulating oil chromatogram, oil temperature, load level, etc.

Load current prediction can predict the equipment load level, which directly affects the stable and safe transmission of electric energy, and has important reference value for reducing the cost of transmission. LSTM has a strong ability to learn mid-term or long-term data in terms of model principles, and is suitable for realizing time series data prediction.

At present, time series data prediction for a single indicator is the most widely used in equipment state prediction, and there is little in-depth exploration of the relationship and mutual influence between the internal parameters of the equipment, resulting in the performance of the prediction model being overly dependent on the parameter tuning of the model and the design of input variables. In the future, cluster analysis and association relationship mining can be considered in the prediction model link to improve the optimization efficiency and accuracy of the prediction model.

11.1.4.4 Defect Identification and Fault Diagnosis

At present, the failure cases and precursor cases of power transmission and transformation equipment mainly record basic information of the equipment, abnormal information related to defects/faults, and test data, and can be classified according to the type, location, and inducement of the failure. A large number of researches use artificial intelligence technology to diagnose faults in power equipment, including power transformers, high-voltage cables, high-voltage circuit breakers, and relay protection.

However, the fault diagnosis of power equipment still has problems such as single data source, unbalanced sample data, rough classification rules, and insufficient knowledge utilization. To this end, firstly, based on the current case data of power transmission and transformation equipment, combined with equipment family defects, operating data, meteorological data and other multi-source information, methods such as generative adversarial networks, sample synthesis, etc. need to be used for data proliferation to balance the ratio of positive and negative samples. Then, according to whether the data is labeled, whether the time series record is complete, etc., try classification, clustering, prediction and other algorithms, and introduce relevant empirical rules in the operational inspection knowledge base for learning guidance. Finally, the hierarchical identification and diagnosis of defects are given. The first level mainly judges the type of equipment defects, the second level mainly judges the defect parts of the equipment, and the fault warning is carried out according to the probabilistic ranking of the suspected defects.

11.1.4.5 Image Analysis of Power Equipment Inspection

With the leap-forward development of computer vision technology, intelligent analysis technology of power equipment images has gradually achieved new breakthroughs. As an important part of equipment operation and maintenance, this section will focus on the inspection image analysis based on deep learning.

The image data analysis of the inspection of power transmission and transformation equipment mainly combines inspection images and videos of drones, inspection robots and other equipment to identify defects in the appearance of transmission lines, towers, and substations. The focus includes wires, insulators, fittings, towers, ground wires, grounding devices, foundations, etc. So as to solve the contradiction between the increasing scale of the power grid and the insufficient deployment of operation and inspection personnel. Inspection drones are generally equipped with stable visible light detectors and imagers, which can take close-up shots of transmission lines, especially important equipment, through functions such as hovering and fixed-point photography, so as to realize the inspection of the damage, deformation, theft of the tower, the damage and contamination of the insulator, the loosening of the clamp, the falling off of the pin, the hanging of foreign objects, the broken strand of the wire, the poor contact of the joint, the local hot spot and other faults; or fly along the upper side of the line and take pictures of the passage from top to bottom. Generally, faults such as illegal buildings in the corridors of the line, icing and falling towers can be found.

At present, many research institutions have begun to use computer vision technology to analyze the inspection images, so as to realize the detection of defects and failures, and realize the automatic release of reports. The application scope includes identification of power transmission and transformation equipment, image segmentation, condition monitoring, fault diagnosis, insulator covering Fields such as ice, gray density recognition, natural language description of images, etc.

In recent years, the image and video recognition technology of substation inspection robots has also been developed rapidly. Inspection robots generally carry visible light cameras, ultrasonic sensors, infrared thermal imaging cameras, audio collection and other equipment, and have the advantages of high inspection efficiency and the ability to conduct inspections in harsh environments such as rain and snow. Among them, image detection objects include equipment defects, status of opening and closing actuators, appearance abnormalities, foreign objects, meter readings and oil level gauge position and road scene recognition, etc.

During the inspection process of power transmission and transformation equipment, due to weather conditions, environmental factors, detection methods, etc., the collected images may have low resolution. Moreover, the small sample size of images with faults and defects, coupled with the complex and changeable background, brings great challenges to the task of target recognition and fault detection and judgment. On the one hand, it is necessary to gradually accumulate inspection images of power equipment, especially to strengthen the collection and sorting of defective samples, and to improve the quality of samples through image enhancement and image defogging/raining technology; on the other hand, it is necessary to further develop image recognition technologies such as transfer learning and small sample learning to improve the recognition accuracy under unbalanced samples.

11.1.4.6 Equipment Life Assessment

Taking into account the overall economy of the power grid, on the basis of ensuring the reliable operation of the power system, extending the operating time of power transformers can maximize the use value of power equipment. The life of power equipment depends on the comprehensive effect of its own ageing degree, load level and operating environment. The life types are mainly divided into three categories: physical (or natural) life, technical life, and economic life.

How to formulate maintenance and renewal plans economically and reasonably has become the focus of attention of power companies and related research institutions around the world. The American Electric Power Research Institute adopts the "three-level evaluation method" and has developed a relatively complete "comprehensive life management program" as a general guideline for the life evaluation of enterprise electrical equipment. Japan evaluates the remaining life of electrical equipment according to the guidelines customized by the Natural Resources and Energy Agency and the Ministry of International Trade and Industry. Related researches on the assessment of the remaining life of power equipment by relevant domestic institutions are also gradually advancing.

At present, the machine learning methods that have been used for the life assessment of power equipment include Bayesian models, artificial neural network methods, fuzzy inference, support vector machines, ensemble learning, and so on. These algorithms train models on the basis of a sample set composed of various loss factor data such as equipment delivery information, online monitoring information during operation, periodic maintenance record information, preventive test statistics information, etc., to analyze the loss and use of power equipment structural components life.

Research on life assessment based on artificial intelligence technology is still in its infancy. Due to the many factors that affect the life of equipment, future machine learning models need to integrate multi-source heterogeneous data from the entire life cycle of the equipment in operation for health assessment and life prediction, so as to make differentiated replacement plans for power transmission and transformation equipment.

11.1.4.7 Intelligent Recommendation of Maintenance Strategy

With the rapid development of the power grid, its scale is getting bigger and bigger, and its complexity is getting higher and higher. There are more and more restrictive conditions when the maintenance plan is arranged, and it is getting more and more complicated, which brings new challenge to the production and operation departments of the power system. At present, the decision-making mode of the maintenance plan of the production and operation department of the power grid is still relatively extensive. The traditional maintenance decision-making methods and related technologies can no longer fully meet the many requirements of the new situation of the power grid development. It is necessary to develop effective automation auxiliary software and systems.

Intelligent maintenance of power equipment is based on equipment status assessment and prediction results, and according to certain optimization goals, the best maintenance strategy including maintenance time, maintenance order and maintenance methods is recommended through intelligent algorithms. Maintenance decision optimization is essentially a multi-objective multi-constraint optimization problem. The optimization targets mainly include indicators such as reliability, economy, and practicability. The solution method of this optimization problem includes mathematical programming method and heuristic intelligent algorithm. Mathematical programming methods include integer programming and linear programming methods. However, for problems with high dimensionality, strong nonlinearity and many uncertain factors, traditional mathematical optimization algorithms have greater limitations, and heuristic intelligent algorithms are widely used in the decision-making process of equipment maintenance due to their strong versatility.

At present, the heuristic algorithms used for the optimization of power equipment maintenance strategy include genetic algorithm, tabu search algorithm, particle swarm algorithm, evolutionary algorithm and so on.

Constructing a knowledge graph in the field of power equipment operation and maintenance is another way of recommending maintenance strategies, which has broad application prospects. A typical construction idea of the equipment operation and maintenance knowledge graph is shown in Figure 11-5. The construction of knowledge graph is mainly based on natural language processing technology, mining the entities, attributes and relationships related to equipment operation and maintenance, so as to construct a knowledge network. At present, some scholars have explored text knowledge mining in the field of equipment operation and maintenance, including: new word discovery technology and word meaning network mining technology based on power equipment related texts, extract power text features and build power thesaurus, accurately extract equipment defect information, automatically mine case information in fault texts, evaluate equipment status, defect text quality improvement, equipment defect classification, equipment life state evaluation, etc.

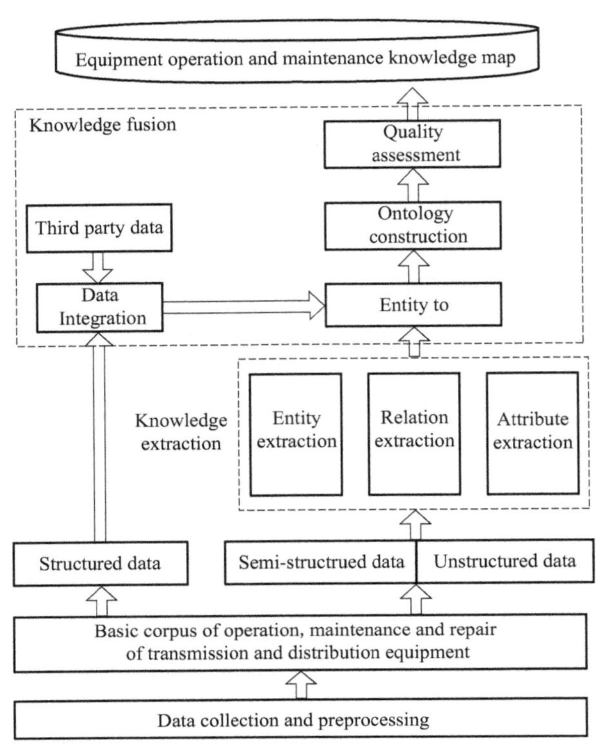

Figure 11-5 Basic procedure of knowledge graph construction for power equipment maintenance

11.1.5 Facing Challenges and Future Trends

Early machine learning technology has been carried out for many years in the field of operation, maintenance and repair of power transmission and transformation equipment. Emerging artificial intelligence technologies represented by deep learning provide new ideas and methods for solving power business problems under the opportunity of significant improvements in data, algorithms, and computing power. However, the application of artificial intelligence technology still faces many difficulties. Many scholars have conducted in-depth discussions from the perspectives of data quality, sample scarcity, data barriers, data acquisition sources, and sensing technology. These discussions mainly focused on external objective conditions such as data, environment and systems. This article attempts to start from the internal factors of artificial intelligence technology itself, and summarize the current technical problems and research directions in the application of artificial intelligence technology in the operation and maintenance of power equipment, mainly including the following aspects:

(1) At present, most machine learning models (especially deep neural networks) are typical

"black box" models. The algorithm cannot give a convincing explanation for its results. On the one hand, business personnel cannot fully trust artificial intelligence technology. On the other hand, it may lead to machine decision-making risks due to insufficient interpretability and transparency. In the future, the research and development of explainable machine learning will become one of the important technical powers to promote the application of artificial intelligence technology.

(2) The equipment information under the fault condition has a natural imbalance between positive and negative samples. Even if the amount of data collection is increased, this problem cannot be solved fundamentally. In the future, it is necessary to develop technologies such as multi-modal machine learning and transfer learning to solve the problem of lack of samples from the algorithmic mechanism. For example, research on machine learning methods such as Zero-Shot Learning and Few-Shot Learning.

(3) At present, services such as equipment condition assessment, fault diagnosis and maintenance are based on expert knowledge and empirical formulas, or based on pure data-driven machine learning models, and the method of fusing the two has not been discussed and studied in depth. Research on the dual-drive model of knowledge and data to reduce excessive dependence on data and enhance the robustness of machine learning is a future development direction.

11.2 Typical Application and Prospect of Digital Twin Technology in Power Grid Operation

11.2.1 Introduction of Digital Twin Technology

At present, a new round of scientific and technological revolution and industrial change has swept the world. With the emergence of new technologies such as big data, cloud computing, Internet of Things, mobile Internet, artificial intelligence, blockchain (big cloud intelligence chain), digital economy is constantly changing the way of human production and life.

As a pillar industry of the country, the power industry plays an important role in the sustainable development of the nationaleconomy[21]. In order to meet the needs of national economic take-off and social progress, the scale of power grid construction is becoming larger and larger, so the requirements of power grid safe and stable operation and enterprise quality and efficiency are more urgent. In the context of digital informatization, the promotion of comprehensive perception of power grid status, full online business of enterprises, and full connection of operational data has become an effective way to achieve stable operation of the power grid and improvement of corporate mechanisms, and to build a highly digital and highly intelligent energy Internet enterprise.

As an emerging and rapidly developing digital information technology, digital twinning provides new ideas for advancing the omni-directional perception, network connection and stable operation of power grid construction. It takes digitization as the carrier, and realizes the real-time perception of the state of the equipment or system in the real space by establishing the mapping from the real space to the virtual space, and by feeding back the data carrying instructions to the equipment or system to guide its decision-making. Through the construction of the digital twin power grid system, the operation, management and service of the power grid can be changed from reality to virtual, and through the modeling, simulation, deduction and manipulation in virtual space, the virtual control of reality has strengthened the grid's self-awareness, self-decision-making, and self-reliance. Evolve capabilities to support the digital operation of various grid businesses, revolutionary changes have been made to the traditional operation mode and operation mode, the construction and management mode of a new digital smart grid has been opened up, and the digitalization and intelligent transformation of the power grid has been promoted. It is an inevitable stage and necessary way to build an energy Internet enterprise.

The digital twin conceptual model was first proposed by Professor Grieves M. in 2003 in the product life cycle management course of the University of Michigan in the United States, this conceptual model was called the "mirror space model" at the time[22]. Later, it was formally defined as "digital twin" in his work. It also points out that the digital twin system includes three aspects: physical space entities, virtual products in virtual space, and data and information interaction interfaces between physical space and virtual space[23]. In 2011, NASA applied digital twin technology to aircraft maintenance and quality monitoring for the first time. In the same year, the US Air Force Research Laboratory proposed a digital twin conceptual model for predicting the life of aircraft structures. Since then, with the development of technologies such as artificial intelligence, virtual reality, and the Internet of Things, digital twin technology has been widely used in the entire life cycle process of equipment design and manufacturing, factory testing, operation and maintenance, scrap disposal and asset management in the industrial field.

Compared with the industrial manufacturing field, the power system is larger in scale, more complex, and has higher requirements for real-time data collection, analysis and feedback. Therefore, the current digital twin technology has not yet been fully developed in the power industry. However, with the continuous development of the digitalization and intelligentization of power grids in recent years, research and construction of digital models and systems have been carried out. The D5000 smart grid dispatching control system[24], which is widely used in dispatching departments, can realize the functions of power system dispatching planning, dispatching management, real-time monitoring and early warning, and is an important manifestation of the digital application on the grid side. The equipment side carries out equipment status assessment based on artificial intelligence, big data analysis and other technologies, the construction of fault warning models, and the construction of intelligent operation inspection management and control systems, which can realize the comprehensive perception and status analysis of power grid equipment, operation inspection equipment and operation inspection personnel. In addition, building information (BIM) models used in the engineering field have also been gradually explored and applied in power grid construction in recent years[25]. By establishing three-dimensional substation models and introducing infrastructure project management, project design management, construction management and schedule management are realized. Although the application of these models and systems does not fully realize the functions of inter-professional intercommunication and data-guided decision-making, they can be used as the basic model of the digital twin power grid system to build digital twin power grids, promote the virtual and real integration of digital power grids and physical power grids, and realize data drive the operation of the power grid to lay a theoretical and model basis.

Based on the above research foundation, this paper introduces the meaning, architecture, characteristics and key technologies of digital twins, and outlines the connotation of digital twin power grids based on the characteristics of power grids, forms the framework of digital twin power grids, and explains the operation mode of digital twin power grids. Finally, the typical applications and prospects of digital twin power grids are proposed from the four levels of equipment layer, power grid layer, business layer and operation management layer, which provide theoretical support and guiding ideas for the construction of digital twin power grids.

11.2.2 Digital Twin

11.2.2.1 The Meaning of Digital Twins

Since the digital twin technology was proposed, various scholars have defined digital twins differently according to the different physical objects it describes. Summarizing its various definitions and contents, digital twins can be described as: using data such as physical models, sensor updates, and operating history to integrate multi-disciplinary, multi-physical, multi-scale, and multi-probability simulation processes to complete the mapping in virtual space. The mapping creates a virtual model for a physical entity object in a digital way to simulate its behavior in a real environment. Real-time perception, diagnosis, and future prediction of the state of physical entities are realized through actual measurement, simulation and data analysis. The model itself is evolved through mutual learning between digital twin models, and the behavior of physical entities is regulated by optimizing instructions.

11.2.2.2 The Architecture of the Digital Twin System

The digital twin system includes five layers of content: physical realm, digital twin, measurement and control entity, user domain, and cross-domain functional entity, as shown in Figure 11-6.

The first layer is a realistic physical domain. In the process of industrial production and enterprise operation, various physical entities including personnel, equipment, test, products, materials, processes, environment and services are included in the design stage, production operation stage and maintenance service stage.

The second layer is the digital twin corresponding to the physical realm of reality. Every object in the real physical domain will be mapped to the virtual side, and there will be a corresponding digital twin, called a digital twin component. It is a digital model that reflects a certain feature of a physical entity, covering three types of functions: modeling management, simulation analysis and twin intelligence. Modeling management is the digital modeling and model display of physical entity objects, synchronization and operation management with physical entity object models. Simulation services include model simulation, analysis services, report generation and platform support. Twin co-intelligence involves the interface, interoperability, online plug and play, and secure access of resources such as digital twins.

The third layer is the measurement and control entity, which is responsible for the interactive function between the real physical domain and the digital twin, and can realize the state perception and control function of the physical entity object. Measurement perception is that the digital twin collects various parameters such as the design and operation of the device from the physical entity, performs data preprocessing and data identification; object control is to transfer the control strategy issued by the digital twin to the physical entity.

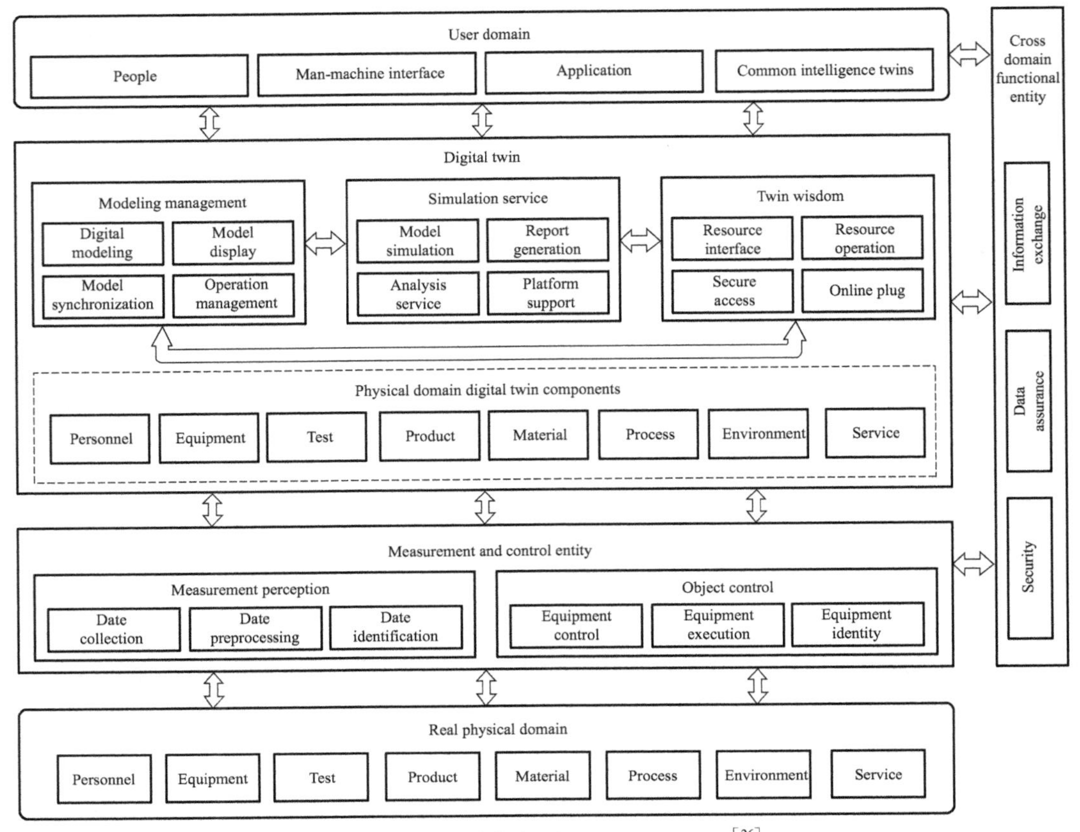

Figure 11-6 Digital twin system structure[26]

The fourth layer is the user domain using digital twins, including humans, human-machine interfaces, application software, and shared intelligence twins.

The fifth layer is a cross-domain functional entity, which provides support for information exchange, data assurance, and security assurance for the data flow and information flow transfer between the measurement and control entity, the digital twin, and the user domain.

11.2.2.3 Characteristics of the Digital Twin System

The digital twin system not only forms a mirror image of physical entities in the real physical domain, but more importantly, realizes two-way data transmission with physical entities. On the one hand, it must accept real-time information from the physical domain, and on the other hand, it must drive the real physics in reverse. Domain, realizing the foresight of physical entity objects. It can be seen that the digital twin system has the following characteristics:

(1) Interactivity. Interactivity refers to the real-time dynamic interaction between digital virtual objects and between physical entities. The goal of the digital twin is to predict and optimize, and feedback the optimization results to the physical entity object to guide its decision-making; the new state information of the physical entity object also needs to be transmitted to the digital twin in real time for iterative update and correction.

(2) Deductive. Deductivity refers to the inversion, forward deduction and dynamic prediction of the state of physical entity objects in virtual space through modeling and simulation in digital

twins. Modeling and simulation in digital twinning are not only the digital model of physical entity objects, but also the calculation, analysis and prediction of the future state of physical entity objects by integrating physical mathematical laws and mechanisms into the model according to past and current states.

(3) Sharing. Sharing refers to data sharing in digital twin system. The digital twin system requires the use of data structures and forms of a single source, and the formation of data sharing and exchange between single or multiple digital twins through unified standardized data, so as to achieve seamless collaboration between simulation and analysis.

(4) Sociality. Sociality means that digital twins can automatically extract knowledge from massive physical connection data, independently evolve and upgrade, and in turn guide the operation and operation of physical entity objects. Relying on artificial intelligence, internet of things, blockchain, big data analysis, cloud computing and other technologies, digital twins have "social" thinking ability and can achieve the goal of "data-driven".

11.2.3 Digital Twin Power Grid System

11.2.3.1 The Connotation of the Digital Twin Power Grid

The digital twin power grid is the future development form of the power grid in which the physical power grid in the physical dimension and the virtual power grid in the information dimension coexist and blend together. It creates a digital power grid corresponding to the physical entity power grid in the digital space, and reflects the state of the physical entity power grid in the real environment through holographic simulation, dynamic monitoring, real-time diagnosis and accurate prediction, thereby promoting the digitalization and virtualization of the total elements of the power grid, the real-time and visualization of the whole state, and the collaborative and intelligent operation and management of the power grid, so as to realize the collaborative interaction and parallel operation of the physical power grid and the digital power grid.

The essence of the digital twin power grid is a grid-level data closed-loop empowerment system. Through data global identification, accurate status perception, real-time data analysis, scientific model decision-making, and intelligent and precise execution. Realize the simulation, monitoring, diagnosis, prediction and control of the power grid, improve the configuration efficiency and operation status of the material resources, intellectual resources, and information resources of the power grid, and develop a new digital smart grid construction and operation management mode.

In terms of power grid construction, the digital power grid is planned and constructed synchronously with the physical power grid. The planning stage begins to model, the construction stage continuously imports data, and the operation stage relies on the digital power grid model and the full data management of the physical power grid. For a power grid that has been built and operated for many years, a digital twin power grid can be constructed and managed through the comprehensive deployment of Internet of Things facilities and digital modeling of the power grid.

In terms of power grid operation and management, the digital power grid and the physical power grid interact with virtual reality, twinning in parallel, and virtual controlling reality. Through the perception and information transmission of the Internet of Things, the realization of the transition from reality to the virtual, and then through scientific decision-making and intelligent control, from the virtual to the real, so as to realize the optimization of the operation and management of the physical power grid. The optimized physical power grid and digital power grid continue to iterate on virtual and real, continuous optimization, and gradually form an endogenous development model of deep learning and self-optimization, and realize the independent management of power grid operation.

11. 2. 3. 2 Framework of Digital Twin Power Grid

The overall architecture layout of the digital twin power grid is composed of smart infrastructure, smart information hub, and smart application scenarios, as shown in Figure 11-7.

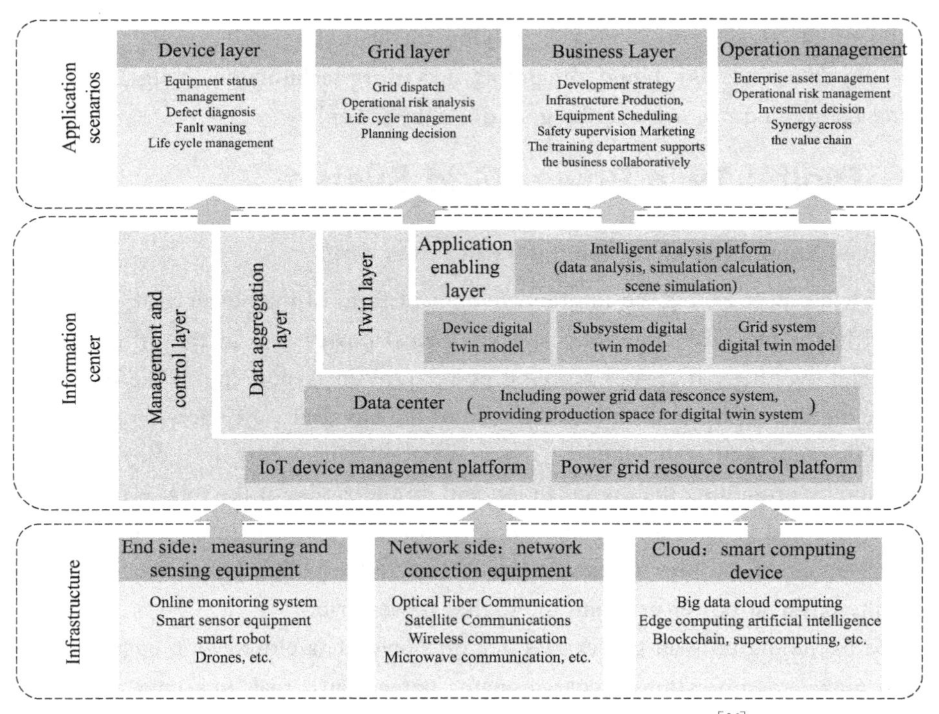

Figure 11-7 The framework of digital twin power grid[26]

Smart infrastructure is based on cloud, network, and terminal, forming a global perception of the power grid on the terminal side, and in-depth description of the operation status of the power grid; a high-speed network is formed on the network side to complete efficient two-way transmission of data and lay the foundation for intelligent interaction; collaborative intelligent computing is formed in the cloud server to realize intelligent decision-making and control of the power grid.

The intelligent information hub is also the brain of the digital twin power grid, which consists of four layers: management and control layer, data aggregation layer, twin model layer, and application empowerment layer.

(1) Management and control layer. Including the IoT device management platform, whose role is to perform unified access, management and reverse control of intelligent IoT devices; the grid resource control platform, whose role is to perform unified access, management and reverse control.

(2) Data aggregation layer. The data center, its role is to aggregate the entire amount of data in

the entire domain.

(3) Digital twin model layer. Including equipment digital twin model, subsystem digital twin model and power grid digital twin model. Its function is to merge with the data center to form a digital twin model corresponding to the physical entities in the power grid. It is the core of precise mapping of virtual and real interaction.

(4) Application enabling layer. The intelligent analysis platform, whose role is to use big data, artificial intelligence, blockchain, virtual reality and other technologies to provide data analysis, simulation calculation, scene simulation and other support for the application of the digital twin power grid.

The application of digital twins in power grid operation is divided into equipment layer, grid layer, business layer and operation management layer according to scenarios. Its foundation is the equipment layer that revolves around equipment independent management and independent evaluation, and the core is around power grid automation dispatch management and independent planning. The grid layer of decision-making, the key support is the business layer that centers on cross-departmental collaborative support business development, such as development strategy, infrastructure, production, equipment, scheduling, safety supervision, marketing, and training, and the top-level application is the operation management team that focuses on corporate operation optimization and value chain collaboration.

11.2.3.3 The Operation Mode of the Digital Twin Power Grid

The digital twin power grid integrates diverse and heterogeneous data such as global perception, historical accumulation, and operation monitoring in various scenarios, integrates multi-disciplinary and multi-scale simulation processes, and integrates applications such as equipment management, grid dispatching and operation, and power services. The complex system coexisting with the real power grid and blending virtual reality and reality reflects the whole process of real power grid operation. Its operation mode is shown in Figure 11-8.

The formation of global digital identification in physical entity grid provides preconditions for accurate matching, connection and control of digital twin grid and physical entity grid information. The intelligent sensing monitoring system and data acquisition device are used to accurately sense and collect real-time equipment operation data, grid operation data, and personnel behavior and management data and other physical entity grid status information, and transmit data to the digital twin through the digital twin link, providing data sources for intelligent analysis and decision-making of digital twin power grid.

In the digital space, a power grid brain composed of a data resource system, a digital twin model and a machine intelligence platform is formed. The whole-domain data resource system is the foundation for building a digital twin power grid. Through the establishment of a data center, data collection, collection, unified management and use are realized, and data is provided for the construction of digital twin models and machine intelligence platforms for intelligent operation and decision-making. The real-time mapping digital twin model is the core of the digital twin power grid. The multi-dimensional data space of the power grid is constructed by loading the full data resources of the whole domain, and the digital portrait of the power grid is constructed using the building information model (BIM model) and the engineering information model (EIM model). Space simulation builds virtual and real mapping equipment, subsystems and power grid digital twin models.

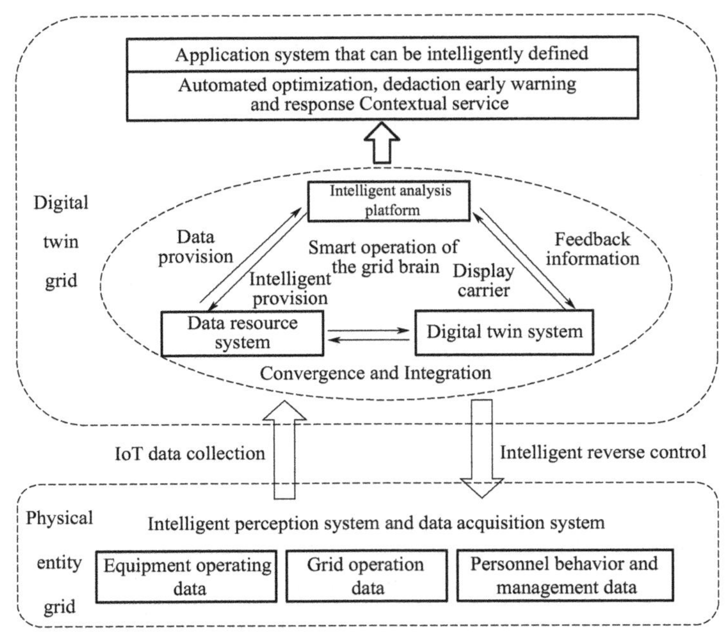

Figure 11-8 The operation mode of the digital twin power grid[26]

The intelligent analysis platform provides intelligent operation decision-making for the digital twin power grid. Through the construction of a deep learning intelligent analysis platform that integrates advanced technologies such as the "big cloud mobile smart chain", machine intelligent algorithms are used to perform data analysis and simulation calculations on the digital twin model, and real-time feedback to the digital twin model, optimization and evolution of the model, forming a self-optimizing intelligent operation mode.

Through the digital twin grid, the "self-learning and self-optimization" function is integrated into the grid operation management, which can realize the functions of equipment and grid autonomous diagnosis and early warning, intelligent identification, state evaluation, planning and decision-making, and then reverse control of the physical entity grid equipment and related The main body (such as people) enables the grid resources to be deployed in a timely manner and problems can be dealt with quickly, so as to achieve the effect of the overall optimization of the grid.

11.2.3.4 Key Technologies of Digital Twin Grid Construction

The digital twin power grid is a complex technology and application system for the new digital smart grid. The integration of multiple types of technologies, the integration of multiple sources of data, and the integration of various platform functions are key elements for the construction of digital twin power grids. Digital twin power grids are not about overthrowing existing power grids and building new power grids, but need to combine existing platforms and technical means to support the digitization of power grids. The key technologies and technical difficulties for the construction of a digital twin power grid include:

(1) Digital identification technology. Fully digital identification of equipment standardized coding design and establishment of a standardized digital identification system are the basis for distinguishing entity identities in the digital space, as well as the basis for realizing the one-to-one map-

ping between digital power grids and physical power grids. At present, State Grid Corporation has vigorously promoted the physical "ID" coding and labeling of incremental inventory equipment, but it has not yet realized the digital identification of all equipment in the entire domain. In order to achieve the accurate matching of the digital twin power grid model and the physical entity power grid equipment, it is necessary to continue to carry out the incremental inventory equipment coding and labeling and data traceability based on the physical "ID" of the existing equipment, and establish a unified standard digital identification system.

(2) Intelligent sensing technology. The intelligent perception system is the entrance and channel of the digital twin power grid perception interconnection. There are a large number of power equipment, their own parameters are complex, static parameters and dynamic parameters coexist, but the equipment types are clear, highly correlated, and easy to classify and manage. Deploying sensor devices in the entire territory requires a large amount of capital and manpower investment for the purchase, installation, management and maintenance of sensor devices. Therefore, for the existing equipment, select representative equipment to deploy miniaturized, integration, and high-precision sensing devices, integrate low-power communication technology to achieve the integration of sensing and transmission, and then build an integrated model of the same type of equipment based on modeling technology, Through simulation verification and continuous iteration of the model, the effect of deploying sensor devices across the entire domain can be achieved. For newly-added equipment, the necessary sensing and communication devices can be integrated in accordance with the corresponding standards and specifications during the manufacturing stage to realize the intelligent perception of the equipment.

(3) Equipment management technology. The IoT device management platform is a platform that provides capabilities such as device connection and follow-up services. There are problems such as language barriers, different operating systems, and different communication protocols when connecting various intelligent sensor devices to the power grid at the equipment layer, grid layer, business layer, and operation management layer. Therefore, the device IoT management technology under the digital twin power grid needs to adapt to the access and data communication needs of different terminal devices with multiple languages and multiple operating systems, and to ensure communication security, real-time and stability, and a variety of developments. The tool can accept data sent by any device that has a protocol driver installed.

(4) Data resource aggregation technology. The construction and iterative optimization of digital twin models of equipment, subsystems and power grids need to rely on the collection of data collected by intelligent sensing equipment across the entire domain. Based on the data center currently being constructed, the collection, management and sharing of data resources from various sources and structures can be realized. Through metadata specification and unified conversion format, as well as a unified data service interface, it can break the barriers that heterogeneous data and cross-professional data cannot be used collaboratively, and realize the data sharing, analysis and mining and integration needs of horizontal, cross-professional and vertical different levels.

(5) Construction technology of digital twin model. The core of the digital twin power grid is a high-precision, multi-coupled digital twin power grid model. The traditional power system modeling is based on physical models, based on physical field modeling and simulation, taking assumption-modeling-setting parameters-simulation-verification as the steps, reflecting the equipment and

power grid information through the physical mechanism and operation process, but it is difficult to integrate the equipment, the simulation process of integrating historical data of power grid operation into the physical model. Modern power systems are faced with massive amounts of data. Equipment and systems can be regarded as black boxes. They are analyzed through external input-output data, and data-driven equipment and system models are established based on algorithms such as machine learning, deep learning, and data mining, which can integrate long-term historical data. And experience input model, and continuously iteratively improve the model to reflect the operating status of equipment and power grid. Digital twin power grid model construction technology adopts a combination of model-driven modeling and data-driven modeling. Initialization modeling comprehensively uses actual perception measurement, equipment three-dimensional modeling, digital modeling and other means to obtain raw data, through data fusion processing, superimposed system modeling and dynamic modeling, to generate a digital twin model. The modeling strategy is based on equipment modeling, and then connects the digital twin model of the subsystem and the digital twin model of the entire power grid.

11.2.4 Typical Applications and Prospects of Digital Twin Power Grids

The digital twin power grid requires the integration of the physical entity power grid and its digital mirror image to realize its various functions and applications, as shown in Figure 11-9. By uploading the sensor data, historical data and related derivative data in the physical entity power grid to the data center, and the collected input digital mirror model, the model is updated iteratively through data analysis and simulation calculation, and the updated power grid information is displayed in real time. The digital twin can not only reflect real-time information, but also has the function of active prediction. It can predict the future development trend of power grids or equipment through a series of repeatable, variable parameter, and accelerated simulation experiments. Based on the prediction results, relying on IoT data sharing and full access to operation data, the grid operation and maintenance strategy is optimized to complete the coordinated operation and maintenance function, and at the same time provide guidance and decision-making in the enterprise operation and management, so as to realize the coordination of the entire value chain of the grid.

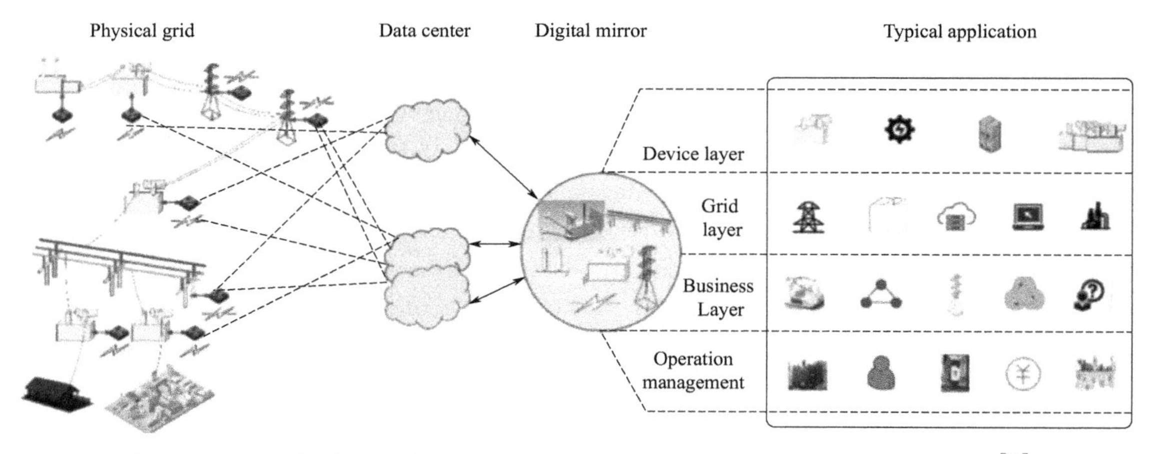

Figure 11-9 The implementation process and typical applications of digital twin power grid[26]

According to the potential application scenarios at the device layer, grid layer, business layer, and operation management layer described in the overall framework of the digital twin power grid, the following typical applications can be realized.

11.2.4.1 Device Layer

The typical application in the equipment layer is mainly based on the fine digital modeling and integrated digital modeling technology to construct the digital twin model of the equipment, and the collected operation monitoring data, equipment life cycle data and environmental data are input into the model. The real-time state information of the physical entity equipment can be accurately mapped to the digital twin model to realize the visualization of the equipment state. The operator can access and operate the digital twin model to achieve friendly interaction between the equipment on-site and remotely; through the application of artificial intelligence algorithms such as machine learning, the perception data is in the model continuously iterating and verifying, and then driving the equipment to achieve autonomous state management, defect diagnosis, fault warning and full life cycle management; based on the refined digital twin model of existing typical equipment, the equipment of the same type, region, and scenario is integrated. The digital twin model is based on the established machine learning algorithm to guide the cluster management and autonomous decision-making of the global equipment, as shown in Figure 11-10. The specific application scenarios are as follows.

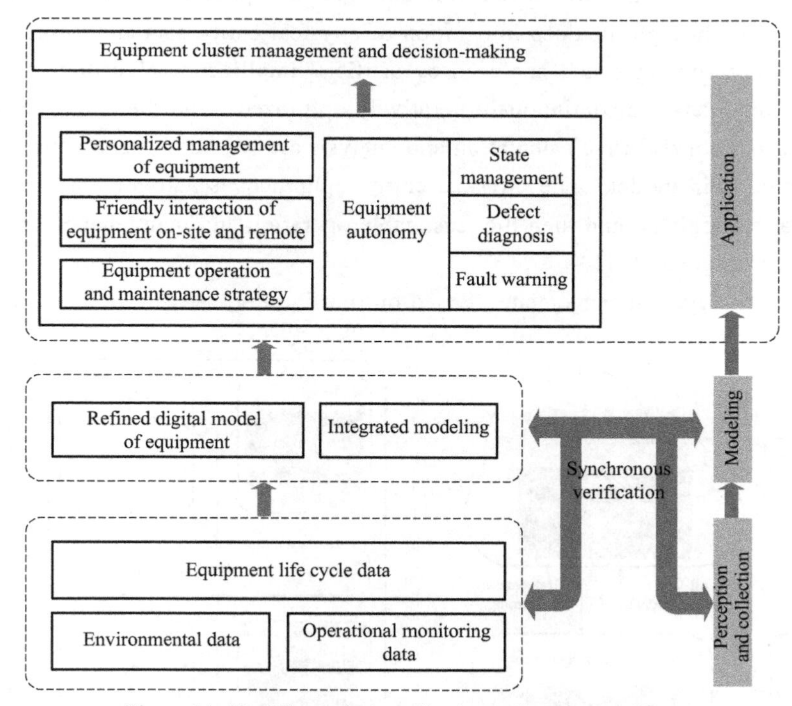

Figure 11-10 The typical applications of equipment layer

(1) Friendly interaction on site and remote of equipment. According to different on-site interaction scenarios, intelligent IoT terminals based on RFID, QR code and other physical "ID" coding and labeling technology are adopted to scan the code to build a database and identify the equipment to ensure the physical entity equipment and the digital twin model and data One-to-one correspon-

dence. Based on the equipment-based three-dimensional model and data, the equipment's all-element perception monitoring data collected by the platform (including data reflecting the physical and spatial characteristics of the equipment, as well as real-time measurement data such as electricity, sound, light, chemical, and heat and historical data), to build a refined digital twin model of equipment that combines model-driven and data-driven equipment. By accessing the digital twin model, it is possible to obtain equipment operating data, evaluate equipment operating status, and query equipment life cycle data and information in real time on remote terminals; in addition, VR technology and three-dimensional visualization display components are used to display equipment three-dimensional models in real time, and operators The digital twin model can be operated to simulate the operation and control of field equipment, and realize the rapid interactive management of on-site and remote key equipment.

(2) Equipment independent defect diagnosis and fault early warning. Based on the integrated sensing monitoring system of digital twin power grid, the physical entity equipment perceives its operation status and environmental data in real time, realizes the real-time monitoring of power equipment in the operation and maintenance stage, and collects the monitoring data of equipment full factor perception in the data center. Based on the three-dimensional model of equipment and the sensing monitoring data, a refined digital twin model of equipment is constructed by combining model-driven and data-driven. The device digital twin model runs synchronously with the physical equipment driven by twin data, which can form data such as equipment evaluation and fault prediction. On this basis, through the deep integration of physical entity data and virtual twin data, the digital twin model is simulated and analyzed by artificial intelligence algorithms such as machine learning. The model data are continuously iteratively optimized, and the data information and instructions obtained from real-time calculation and analysis are fed back to the physical entity equipment by the digital twin model. The physical entity equipment is autonomously driven for defect diagnosis and fault warning, and then the reasonable operation and maintenance strategy is formulated, as shown in Figure 11-11.

(3) Equipment life cycle management. Based on the fine digital twin model of equipment, the

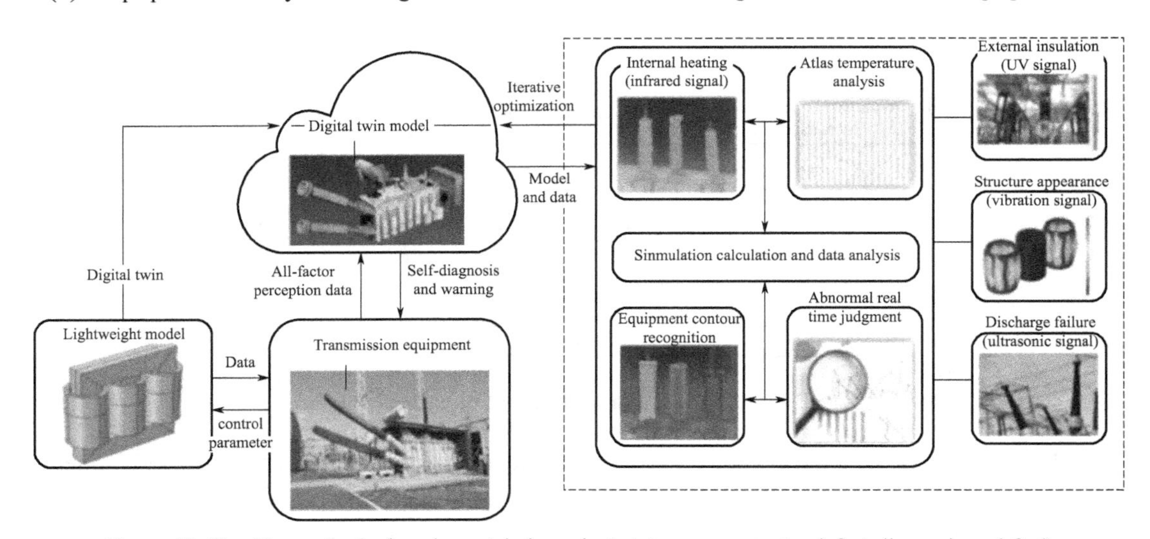

Figure 11-11 The method of equipment independent state management, defect diagnosis and fault early warning based on digital twin

historical data of equipment and real-time monitoring data collected by the data center are input into the digital twin model, and the input massive data are sliced, layered and combed to realize the real-time recording of multi-dimensional data of equipment data. Using big data analysis technology and machine learning algorithm, the historical data and real-time data of the equipment are analyzed, and the future state of the equipment can be predicted rollingly. The digitaltwin model of the equipment feedbacks the rolling prediction information to the physical entity equipment, which can drive the physical entity equipment to carry out independent evaluation and decision-making in the whole life cycle. Based on the RFID, two-dimensional code and other physical "ID" labeling technology, the physical entity equipment and the digital twin model are accurately matched. The digital twin model of the equipment and the life cycle data can be accessed by scanning the code, and the life cycle intelligent management of the equipment and spare parts is carried out.

(4) Equipment cluster management and decision-making. In actual grid operation, it is impossible to carry out refined modeling for all equipment. Based on the established refined digital twin model of typical equipment, through the horizontal comparison of the refined digital twin model of individual equipment in the same type, same region, and the same scenario, combined with the operation data of the equipment in the area (subsystem), cluster Analysis and simulation calculation classify equipment, deduce atypical equipment information, and build an integrated digital twin model of equipment of the same type, same region, and same scene. Based on the established methods of autonomous defect diagnosis, failure early warning and autonomous state assessment of typical equipment, cluster analysis of the operating status, potential failures and countermeasures of the equipment in the area (subsystem), and guide the corresponding decision-making of various types of equipment, realize the evolution of equipment diagnosis decision from individualization to clustering.

At present, the application of digital twin technology at the device layer is mainly focused on the autonomous state management and fault diagnosis of the device. The Shanghai 35kV Cailun Substation is equipped with a digital mirroring evaluation system, which establishes a digital twin model of the equipment in the substation, and collects real-time data from front-end sensors (such as real-time load, current, oil chromatogram, etc.) and historical data (historical sensor acquisition, defect data, maintenance times and other data) input model to carry out equipment operation status analysis, through real-time update of equipment self-status and life cycle data, online assessment of the health status of each equipment in the station and rolling update of targeted maintenance strategies. Tianjin 110kV Youlegang Smart Substation has completed the establishment of the holographic model of the substation. Through the multi-data fusion algorithm, the cloud data and the edge IoT agent device state perception data are extracted and comprehensively processed to form a virtual space digital twin model; through comprehensive collection and real-time update of objects networking perception information and cloud life information, using machine learning and big data analysis methods to achieve active early warning of equipment defects, intelligent fault decision-making and life cycle monitoring.

The application of digital twins at the equipment level can solve the current problems of wasting manpower and financial resources in operation and maintenance based on planned maintenance, as well as the problems of poor equipment status evaluation indicators and inaccurate evaluation results. The preventive operation and maintenance is shifted from predictive operation and maintenance. Effectively reduce unnecessary on-site operations, improve the accuracy of equipment status assessment, fault diagnosis and the level of safety management and control, and realize intelligent and lean equipment management.

11. 2. 4. 2 Grid Layer

The typical application at the power grid layer is mainly based on the digital twin model established by equipment and subsystems to form the digital twin model of the entire power grid. Transmit system online monitoring data, operating status data, equipment life cycle data, and environmental data to the power grid digital twin model in real time, and apply artificial intelligence algorithms such as machine learning to iteratively optimize the model and make rolling predictions to enable the power grid to independently predict future operations real-time feedback to the physical entity grid to realize the autonomous assessment of grid operation risks and the monitoring and management of the grid's life cycle; the real-time operation status prediction results of the grid are combined with power flow control and other operating data for simulation analysis in the grid digital twin model to drive the realization of the grid considering the automatic dispatching and predictive dispatching of equipment operation status, it provides guidance for the online analysis and decision-making of the power grid, and at the same time guides the power grid to carry out independent planning, as shown in Figure 11-12. The specific application scenarios are as follows:

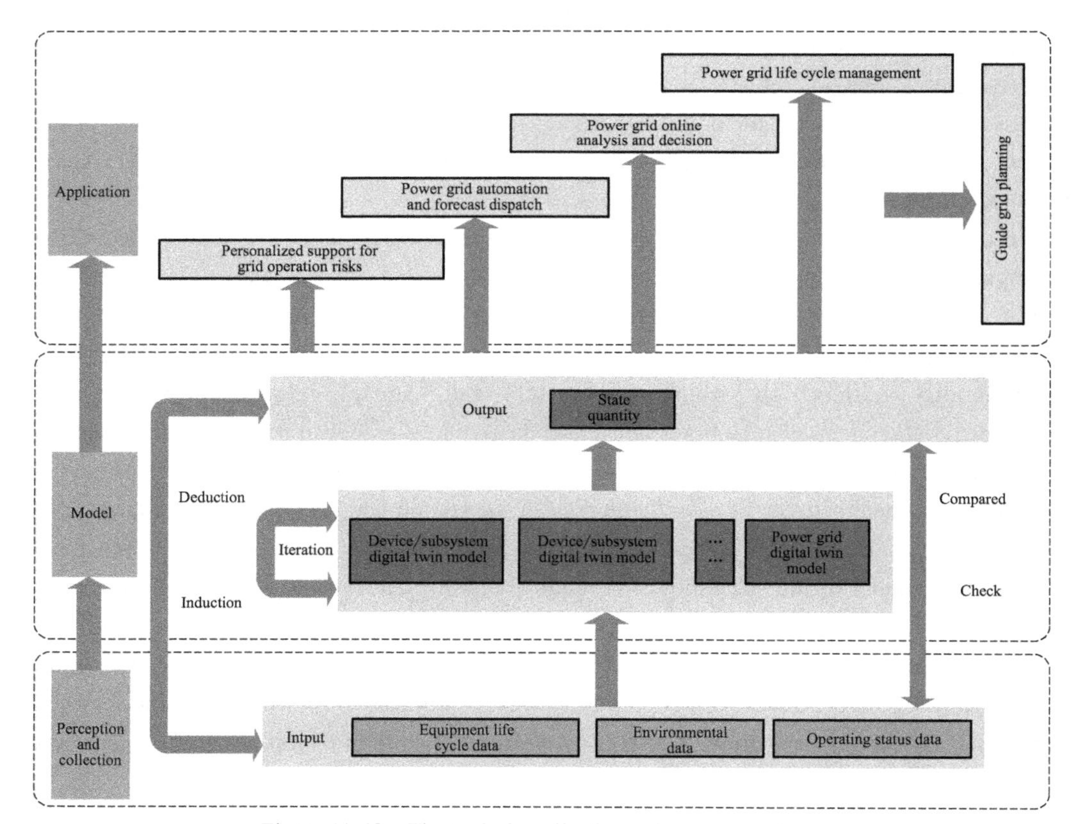

Figure 11-12 The typical applications of power grid layer

(1) Personalized support for grid operation risk assessment. The reliability of each equipment on the line is not the same. Based on the line equipment operating state data, environmental data,

and life cycle data collected by the data center, the line equipment digital twin model is constructed, and the real-time collected sensing and monitoring data are input into the digital twin model. Apply machine learning and other artificial intelligence algorithms to iteratively optimize and predict the model to obtain the operating status and operating trend of the line equipment. Apply cluster analysis to analyze historical data and future forecast data in the digital twin model, and summarize the operating characteristics of different types or different manufacturers. Based on the classification results, the real-time sensing and monitoring data are input into the corresponding digital twin model for iterative optimization and rolling prediction, and the simulation results and response strategies are fed back to the physical line equipment in real time to realize the independent risk assessment of each line equipment, and then independently determine the maintenance time, maintenance frequency and maintenance task allocation, etc., to provide support for power grid planning management to dispatching departments and planning departments.

(2) Automated power grid dispatching and predictive dispatching considering equipment operation status. Traditional power grid automation dispatching, especially the main grid dispatching, is mainly based on the operation data of grid nodes and main line power flow, without considering the factors of the operation status of the power grid equipment. The power grid automation dispatching and predictive dispatching based on the digital twin technology are the basis of the traditional power grid automation dispatching above, integrate real-time data such as equipment operating status parameters and environmental parameters. Based on the current D5000 power grid control system, the power grid digital twin model built by the digital twin model of each subsystem of the power grid is nested, and the real-time operating status data of the equipment collected in the data center, environmental data, and the real-time power flow in the D5000 system are shared input the power grid digital twin model, apply machine learning and other artificial intelligence algorithms, continuously iterate and optimize the dispatching model, autonomously predict the future operating state of the power grid, and guide the dispatching department to more accurately grasp the true operating state of the power grid to avoid decision-making errors.

(3) Power grid life cycle management. The power grid digital twin model constructed by inputting the historical data of power grid equipment and real-time monitoring data collected by the data center, applying machine learning and other artificial intelligence algorithms, continuously iteratively optimizes the power grid digital twin model, and makes the power grid digital twin model and the physical power grid one-to-one mapping, simultaneous operation; applying big data analysis technology and machine learning algorithms to analyze historical and real-time data of power grid equipment, and perform rolling predictions to realize the prediction from equipment to grid operation trend; the power grid digital twin model feeds back the prediction information to the physics The physical power grid can drive the physical physical power grid to conduct independent assessment and decision-making throughout its life cycle. Linking the entire life cycle of the digital power grid and the physical power grid, the maintenance strategy, the management strategy of spare parts and the grid operation management strategy can be optimized as a whole to guide the power grid planning and decision-making.

(4) Online analysis and decision-making of power grid. At present, the online analysis response speed of the D5000 power grid control system is minute level. Through the power grid digital twin model built in the D5000 system, the implementation of second level online analysis and

decision-making can be realized. In the D5000 system, the power grid measurement information is formed through data acquisition and monitoring control system (SCADA system) and state estimation processing to form a power flow profile. The power flow profile is data integration and calculation control (the process delay is minutes), and then the power grid online analysis is realized. Using the power grid digital twin model, the power grid measurement information collected by the SCADA system no longer forms a power flow profile through state estimation, but is input into the power grid digital twin model (second-level delay), and artificial intelligence algorithms such as machine learning are applied to continuously iteratively optimize through the model. Update the model itself in real time, realize real-time synchronization between the physical entity grid and the grid digital twin model, and drive the online analysis and decision-making of the grid.

At present, the application of digital twin technology at the power grid layer is mainly concentrated in the online analysis and decision-making of the power grid. The D5000 platform of Hunan Provincial Electric Power Commission adds a power grid analysis model consisting of a physical model and a calculation model, and inputs real-time power grid measurement information collected by SCADA into the physical model, and relies on the calculation model to perform state estimation calculations, based on real-time analysis models (physical models and mathematical models) conduct power grid security and stability assessment. The entire process is less than 300ms, which is a hundred times shorter than the 60s on-line analysis cycle of the original D5000 system grid data.

The application of the digital twin at the grid layer can predict the operation risk of the grid equipment in advance, arrange the maintenance plan, and improve the monitoring efficiency and maintenance of the dispatching operation based on the life cycle data of the digital twin model of the grid equipment without performing on-site inspections. Grid automation and predictive scheduling considering the operating status of grid equipment can enable the dispatching department to more accurately grasp the true status of the grid and make scheduling decisions more accurate; in addition, the grid digital twin model can accelerate the speed of simulation and evaluation, and more quickly realize online analysis and decision-making of power grid.

11.2.4.3 Business Layer

At the business level, the construction process of the digital twin system is also the process of realizing the comprehensive management of the device data of the entire network. Relying on the synchronization construction of the data center, the data format is unified and editable, the data can be shared, and it has the ability of visual interaction. The digital model in the entire region can be effectively applied from the policy department, the infrastructure department, the equipment management department, the safety supervision department, the dispatching department to the marketing department, and realize the data exchange and business collaboration of various business departments. The specific application scenarios are as follows:

(1) Batch business processing in the distribution station area. In the same way as the refined modeling and integrated modeling of the equipment layer, based on the physical "ID" coding and labeling technology such as RFID and QR codes, a digital twin model of the equipment in the distribution station area that maps one-to-one with physical equipment is constructed. Based on artificial intelligence algorithms such as machine learning, the all-element perception data collected by the data center is simulated and iteratively analyzed in the digital twin model, and the digital twin model of the device is updated in real time to realize the synchronous operation of the digital twin

model and the physical physical device. Code and other methods can access the digital twin model of the equipment and perform simulation operations on the model to realize the batch management of visual monitoring, active control and active early warning functions of various equipment in the station area.

Due to the uniformity and editability of the data format, the digital twin model of the entire station area can be effectively applied from the development and policy department to the dispatching department to provide support for the construction of the distribution station area. The development and policy department can carry out the evaluation and planning of distribution stations according to the existing digital twin model. The infrastructure sector can add elements on the basis of the existing digital twin model of the station area, so that physical equipment and digital twin model can be constructed simultaneously; the infrastructure sector can add elements on the basis of the existing digital twin model of the station area, so that physical equipment and digital twin model can be constructed simultaneously; the equipment management department realizes real-time monitoring and independent management of equipment by accessing the digital twin model of the equipment in the station area; the safety supervision department realizes the whole life cycle safety management and control and fault handling of the equipment by visiting the digital twin model in the station area; the dispatching department can independently regulate the distributed energy in the station area by visiting the digital twin model in the station area.

(2) Operational inspection business support. Based on the refined digital twin model of equipment, combined with its two applications at the equipment level, equipment autonomous defect diagnosis and fault early warning, equipment on-site and remote friendly interaction, it can realize active early warning of equipment failure and remote operation and maintenance; in addition, the application of VR technology and The three-dimensional visualization display component simulates the installation of safety operation fences in the digital twin model, displays them to operation and maintenance personnel in a visual way, and matches the position of the on-site safety operation fences. If the safety operation requirements are not met, remote warning is given in time to protect the site Work safety.

(3) Load classification forecast. Input the historical data and real-time data of regional load data, power grid information data, distributed power data, environmental data (temperature, light, wind speed, altitude, geographic location, etc.) collected by the data center into the built power grid digital twin model, apply the method of big data analysis and cluster analysis to analyze historical data and real-time data, extract effective feature quantities that affect different types of loads, and accurately identify different types of loads; and apply artificial intelligence algorithms such as machine learning, based on the extracted characteristic historical data and real-time data, iterative optimization and rolling forecast of the load in the digital twin model, guide the distributed power output and grid dispatch in the area, and realize the effective energy allocation and energy use identification.

(4) Holographic training. Based on the fine digital twin model of equipment synchronized with physical entity equipment, the real-time operation state of equipment is simulated. VR technology and three-dimensional visualization display components are used to form immersive learning experience for trainers. The various working conditions and daily operation and maintenance operations of equipment, as well as substation fault location and accident treatment process are displayed to

trainers. In addition, the trainers can also operate the digital twin model of equipment through interactive equipment to simulate the operation of physical entity equipment on the scene. The digital twin model feedbacks the simulation results to the trainers through visual display components, so as to realize the interactive function between the trainers and 3D virtual equipment, effectively improve the business skills of the operation and maintenance personnel, and prevent accidents caused by on-site misoperation.

The application of digital twins in the business layer, relying on the synchronization construction of the data center, can solve the current grid cross-professional and cross-departmental data format inconsistency, unable to integrate applications, and business that cannot be coordinated. It can realize cross-professional and cross-departmental business data and digital twin models shared utilization; in addition, derivative applications (such as operation inspection support, holographic training, etc.) developed by applying equipment and power grid digital twin models at the business layer can assist on-site operation and maintenance personnel to simulate the operating environment, improve operating efficiency, and ensure operating safety.

11.2.4.4 Operation Management

Based on digital twin technology, the functions of digitization and intelligent construction can be extended to the operation and management level of the entire company, while reducing management and control manpower, while improving management and control efficiency, realizing asset digitization, knowledge digitization, and business digitization. The specific application scenarios are as follows.

(1) Asset management optimization. By constructing a refined digital twin model of typical equipment, using this type of typical equipment to represent all equipment of the same type, based on the equipment life cycle management method of applying digital twin technology at the equipment layer, the equipment digital twin model is iteratively optimized, and the future of the equipment is improved. Carry out rolling forecast of the state to realize the autonomous state assessment and prediction of the equipment during the whole life cycle, guide the operation and maintenance personnel to carry out predictive maintenance and asset performance evaluation according to the equipment state, and guide them to give asset management suggestions at each stage of the equipment life cycle (operation and maintenance, return to factory maintenance, scrapping, etc.), assist the company to optimize asset management and achieve lean management of the company.

(2) Operational risk management. In terms of reducing power grid security risks, on the equipment side based on the built integrated digital twin model for equipment clustering autonomous fault diagnosis and prediction, on the power grid side based on the grid digital twin model for autonomous operation risk assessment, and real-time transmission of equipment and grid risks. Forming risk self-diagnosis and early warning modes, guiding enterprises to propose personalized management suggestions for different equipment and systems; in terms of reducing business risks, online power grid analysis and predictive dispatch based on the constructed power grid digital twin model can be used to evaluate and predict power grids in real time operational benefits, forming an autonomous early warning model for business risks, and guiding enterprises to make business decisions.

(3) Service enterprise investment decision. Based on the refined digital twin model of equipment built at the equipment level, and the realization of the equipment status autonomous evaluation and full life cycle autonomous management applications, according to the equipment in the life

cycle status, targeted investment recommendations are proposed to allocate equipment for transmission and transformation. Provide reference for precise investment planning in procurement, transformation, construction, operation and maintenance management, etc. Based on the grid digital twin model built on the grid side and the implementation of autonomous evaluation applications of the grid status under different operating conditions, it guides the formulation of investment planning recommendations for grid construction based on the history, current operating status and future operating status predictions of the power grid.

(4) Convergence of service three networks. On the basis of constructing the digital twin model of the power grid, the digital twin model of the transportation network and the digital twin model of the charging service network are constructed. The real-time road flow data, environmental data and historical data are input into the digital twin model of the transportation network, and the real-time charging data, environmental data and historical data are input into the digital twin model of the charging service network. The artificial intelligence algorithms such as machine learning are applied to continuously iteratively optimize the model to form the digital twin model that is real-time mapped to the physical entity transportation network and the charging service network. The simulation calculation and big data analysis method are used to obtain the correlation between power grid, charging service network and traffic network. Through the synchronous prediction and correlation of three network models, the independent evaluation of three network matching and the rolling planning of charging facilities are realized, and the new ecology of intelligent new energy vehicle service is created.

(5) Synergy across the entire value chain of the power grid. Based on Internet of Things technologies such as "Big Cloud IoT Smart Chain", it enables full access to operational data, cross-professional data intercommunication, and breaks professional data barriers in enterprise planning, material procurement management, engineering management, power grid operation management, and power grid operation and maintenance. Digital twin technology is applied in various business links such as maintenance, power marketing management, and various business data are collected in real time, and analyzed and predicted in the model, so as to realize self-awareness of power grid status, coordinated operation of power grid and equipment, new customer service experience, and open sharing of energy ecology synergy model across the value chain.

The application of digital twins in the operation management layer relies on the comprehensive application of the device-side and grid-side digital twin models, connects the business applications of the enterprise's back-end, and guides the formulation of enterprise operation-level strategies; in addition, through the sharing and integration of internal and external information and data analysis can improve the economic efficiency and market competitiveness of the enterprise, and promote the synergy of the entire value chain of the enterprise.

11.2.5 Conclusion

This section introduces the meaning, architecture and characteristics of digital twins. Based on the digital twin technology, it puts forward the concept, connotation, construction framework, operation mode and key technologies of digital twin power grids, and proposes typical applications that can be realized by digital twin power grids based on the characteristics of the power industry scenes.

(1) At the equipment level, based on the digital twin, it can realize the friendly interaction of equipment on-site and remote, equipment independent state management and defect diagnosis, equipment life cycle management, equipment cluster management and decision-making, and intelligent equipment monitoring and autonomous operation inspection and decision-making and other related businesses to provide support, which can effectively improve work efficiency and accuracy of decision-making.

(2) At the grid level, based on digital twins, personalized support for grid operation risk assessment, automated grid dispatching and predictive dispatching considering equipment status, grid life cycle management, grid online analysis and decision-making can be realized, thereby guiding the grid to carry out independent planning decision making.

(3) At the business level, based on the digital twin, relying on the synchronization construction of the data center, it can realize the batch business processing in the distribution station area, the operation and inspection business support, the load classification forecasting, and the holographic training. Breaking down professional barriers and departmental barriers, realizing the collaboration of various professional data and the business of various departments, and providing business support for planning, construction, operation inspection, dispatching, and safety supervision.

(4) At the operation management level, based on digital twins, asset management optimization, operational risk management services, enterprise investment decision-making services, three-network integration services, and coordination of the entire value chain of the power grid can be realized. Significance.

Through the typical applications of digital twin technology in the equipment layer, grid layer, business layer and operation management layer, it provides theoretical support and construction ideas for the construction of the digital twin power grid, and provides a new way for the digital and intelligent development of the power grid.

11.3 Technologies and Solutions of Blockchain Application in Power Equipment Ubiquitous Internet of Things

11.3.1 Introduction

Blockchain is a distributed database. It is a series of data blocks that are related using cryptographic methods. Each data block contains information about a network transaction to verify the validity of itsinformation[27]. The characteristics of the blockchain are decentralization (distribution), trustlessness, and non-tampering of data. The core technology is database, encryption algorithm, fast search, access, etc[28]. The specific connotations of its characteristics are as follows.

(1) Decentralization. Different from a centralized network, in the process of building a blockchain, each node has the same authority and obligation. Even if a node is destroyed, it will not affect the operation of the system.

(2) De-trust. The nodes of the blockchain do not need to trust each other, and are only based

on the common recognition of a certain rule. They cannot deceive and tamper with each other. Even if there is human intervention, it is difficult to affect the entire system.

(3) Cannot be tampered with. The transactions in the blockchain will be permanently recorded. Unless more than half of the nodes are intervened, it is difficult to tamper with and easy to trace.

Therefore, the core of the blockchain is a distributed database, which aims to achieve scene focus, consensus building, information security, cost saving and efficiency improvement.

Blockchain technology originated from the foundational paper "Bitcoin: A Peer-to-Peer Electronic Cash System"[29] published in 2008 by a scholar with the pseudonym "Satoshi Nakamoto". After that, it gradually developed and grown in the open source community, and formed an open source project based on the public chain represented by Bitcoin and Ethereum. In recent years, developed countries in Europe and the United States have successively released a series of reports on blockchain, exploring blockchain technology and its applications. The European Union established the European Blockchain Observation Forum in February 2018, and carried out blockchain policy determination, industry-university-research linkage, and standard formulation. In the United States, due to policy differences among states, the promotion of industrial policies has been slow, but blockchain is still a boom among American start-ups. Japan uses NTT as the mainstay and the government provides support, while South Korea uses finance as an entry point to explore blockchain applications.

In recent years, blockchain technology has also received great attention and development in China. The State Council of China has issued the "Thirteenth Five-Year" National Informatization Plan, blockchain and big data, artificial intelligence, machine deep learning and other new technologies become the focus of national layout[30]. The People's Bank of China has issued the "Thirteenth Five-Year Development Plan for Information Technology in China's Financial Industry", which clearly proposes to actively promote research on the application of new technologies such as blockchain and artificial intelligence, and organizes pilot projects for national digital currencies[31]. The Ministry of Industry and Information Technology released the "White Paper on China's Blockchain Technology and Application Development", which is the first official guidance document for blockchain[32]. In particular, on October 24, 2019, the Political Bureau of the CPC Central Committee conducted a collective study on the current status and trends of blockchain technology. Xi Jinping, General Secretary of the CPC Central Committee, proposed that the integrated application of blockchain technology in new technological innovation and industrial transformation It plays an important job in. With the development of blockchain technology, its application has surpassed the financial field, and has been gradually implemented in the Internet of Things, supply chain, credit investigation, identity authentication, energy and other fields. This article will focus on the application of blockchain technology in the ubiquitous Internet of Things for power equipment[33].

The development of the Internet of Things technology has guided the evolution path of global information infrastructure, and has promoted the development of traditional products, equipment, processes, and services towards digitalization, networking, and intelligence. At present, the development of the Internet of Things presents new trends such as "edge intelligence, connection ubiquity, service platformization, and data extension". The proposal of the ubiquitous power Internet of Things is the inevitable development of the deepening of the Internet of Things technology[34-35].

However, the large-scale development of the Internet of Things faces some key challenges. On the one hand, it is an inevitable demand to support the access and service request response of a large number of terminal sensors and information in the future, which requires wider network coverage, larger network carrying capacity and faster response speed. If the power equipment ubiquitous Internet of Things follows the traditional thinking, it will soon face the same problem. On the other hand, the traditional Internet of Things is threatened in terms of data security, privacy protection, authentication and access control. For example, information is vulnerable to attacks and leaks during transmission; the cloud storage method causes the server to be overloaded and the data lacks backup; traditional channels cannot meet the needs of privacy protection; devices such as sensors and positioning systems cannot fully protect user privacy data; The decryption of the secret key causes the identity authentication to become invalid, and the data flow monitoring causes the message authentication to become invalid. The blockchain technology with the characteristics of decentralization, trustlessness and data encryption provides an opportunity for the Internet of Things to solve the problems of massive terminal access, data security and privacy protection in the traditional Internet of Things. This section will explain how to use blockchain thinking to combine blockchain technology with the Internet of Things to improve the performance of power equipment ubiquitous Internet of Things in terms of network structure, cost reduction, establishment of trust, multi-agent collaboration, and edge computing.

11.3.2 Ubiquitous IoT Architecture Design for Power Equipment based on Blockchain

11.3.2.1 Problems That Exist When a Large Number of Terminals Are Connected to the Ubiquitous Power Internet of Things

According to the Internet of Things platform construction plan for power transmission and transformation equipment of the National Grid Corporation of America as shown in Figure 11-13, the sensing data obtained by massive sensors will be uploaded to the access node through the aggregation node, and finally through the access controller and control of the network layer the gateway enters the server of the cloud platform. The existing solution is still a centralized distributed network structure in essence. When faced with the access and data transmission of hundreds of millions of massive IoT terminal devices, the ubiquitous power IoT will encounter the following problems.

(1) Rigid network structure. The data flow of power equipment ubiquitous Internet of Things is finally aggregated to the server of the central control system. With the continuous evolution of the construction of the ubiquitous Internet of Things, it is foreseeable that the future Internet of Things devices will grow geometrically. Centralized services will cause the load of their servers and databases to increase sharply and become unsustainable.

(2) Equipment safety. As the value of data grows, electric power IoT terminal devices will become potential targets for malicious attacks by hackers. For example, the famous botnets of things (botnets of things), according to publicly reported data, the botnet has cumulatively infected more than 2 million Internet of Things devices such as cameras, and launched DDoS attacks, which led to the paralysis of the American domain name resolution service provider Dyn.

(3) Information security. Streaming data in the power Internet of Things, especially when

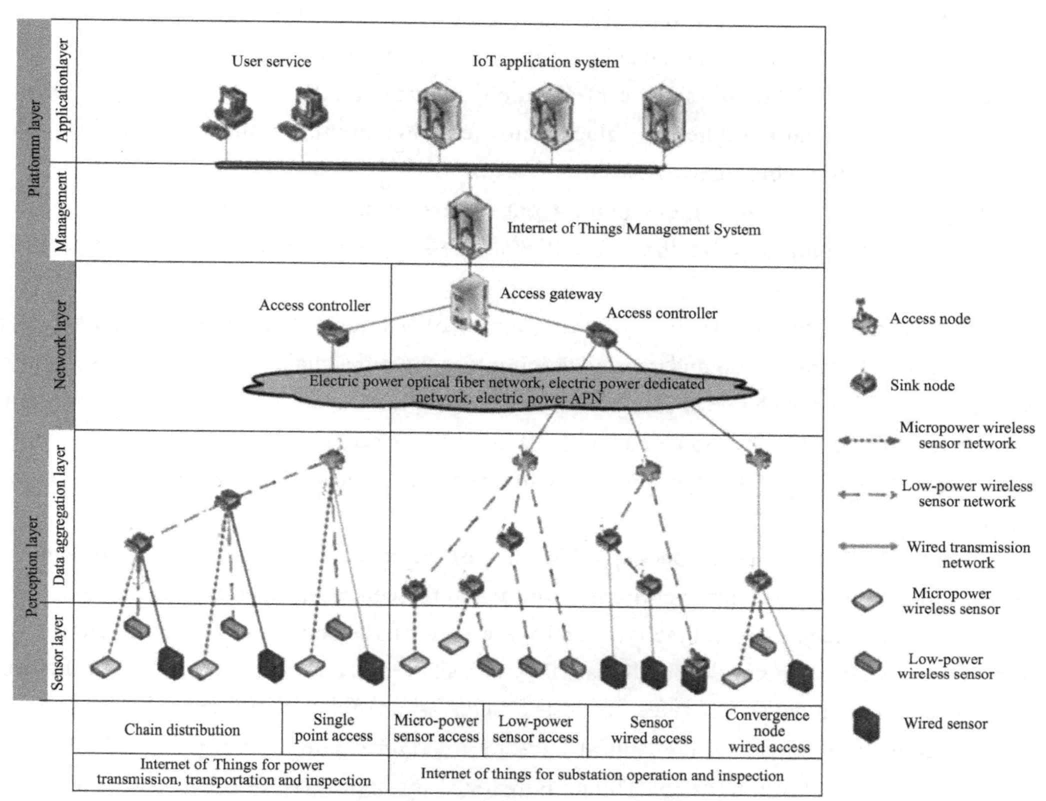

Figure 11-13 Framework of ubiquitous power internet of things for power transmission and distribution equipment[26]

transmitted via wireless networks, can be easily stolen. At the same time, with the combination of big data and the Internet of Things, massive data is stored in a small number of central nodes, data lacks backup, and data security cannot be guaranteed.

(4) Personal privacy protection. The traditional Internet of Things does not have application-based privacy protection capabilities such as resistance to key sharing attacks. Whether the GPS positioning system of various sensors in the Internet of Things of power equipment can completely keep the user's private data confidential, and whether the information is monitored by manufacturers, are important issues that need to be faced in the security of the Internet of Things.

(5) Communication problems. In a centralized network, even if the communication between two sensors is close at hand, the central node of the application layer is required as an intermediary to complete identity establishment, data transmission, etc., which additionally increases the central node load and information transmission delay.

(6) Multi-agent collaboration. The ubiquitous Internet of power equipment is a self-organizing network within an enterprise. When it involves collaboration across multiple peers, the cost of establishing credit is high.

11.3.2.2 Ubiquitous Power Internet of Things Solutions under Blockchain Thinking

To use blockchain technology and platforms to improve the construction of the ubiquitous Internet of Things for power equipment, it is necessary to establisha blockchain thinking.

(1) Use the core idea of "decentralization" in the blockchain to solve the inherent problems of "centralization" such as insufficient capacity, high cost, and low trust.

(2) Balance "decentralization" and "centralization" to realize ABCD, that is, the intersection and integration of artificial intelligence, blockchain, cloud computing, and big data (AI, blockchain, cloud computing, big data, ABCD).

(3) Using smart contracts to achieve multi-agent collaboration in a decentralized environment.

Use blockchain thinking to solve the difficulties of the above-mentioned power equipment in the Internet of Things.

(1) Optimize the network structure of the Internet of Things. Balancing the advantages and disadvantages of decentralization and centralization under the blockchain thinking, this section proposes to adopt the alliance chain scheme as the blockchain platform for the ubiquitous power equipment in the Internet of Things. Some convergence nodes (or access nodes) in the perception layer and access gateways (or access controllers) in the network layer will also store and maintain the same block chain.

On the one hand, it realizes the backup of the data of the central node of the platform layer, and avoids the significant impact on the entire network when the central node fails. On the other hand, the nodes in the blockchain are subjective, and these nodes have complete data (big data). Combined with their edge computing capabilities, they can share some of the original sent to the central node of the application layer. Service requests, such as artificial intelligence (AI) and cloud computing, alleviate the load on the central node to a certain extent. Simply put, the horizontal expansion of ABCD on the Internet of Things is realized through the blockchain, which improves the flexibility of the entire system and effectively avoids the high operation and maintenance costs brought about by the previous centralized ACD vertical expansion.

(2) Improve the security of the Internet of Things. After power equipment ubiquitous Internet of Things combined with blockchain technology, the storage of data needs to be encrypted by asymmetric cryptography. If there is no private key, the data cannot be deciphered even if it is obtained. Each block uses a hash value generated by a certain algorithm to identify its uniqueness. If you want to tamper with the data, you must modify the data in all blocks, which fundamentally improves the data security of the Internet of Things. In addition, the dissemination of data in the power Internet of Things blockchain needs to be digitally signed, that is, any operation on the data must be authenticated, authorized to decrypt, and audit trails, which helps to improve the security of data sharing.

Based on the blockchain, a device identity management system can be constructed to establish a mapping relationship from the identity of the individual entity to the identity of the access device, so as to achieve two-way traceability verification between the device and the user, and prevent the access of malicious nodes.

(3) Break through the data and collaboration barriers between nodes. Smart contracts are computer programs that are stored and run on the blockchain. When the conditions for the establishment of the contract are met, the contract code will be automatically executed. In other words, the blockchain is just a distributed ledger, and the smart contract further combines different application scenarios to stipulate who and who, under what circumstances, and what kind of ledger will be generated. Therefore, the application of smart contracts can help break data barriers, promote

the horizontal flow of information, and enhance the ability of multiple parties to collaborate. For example, smart contracts can be used to mine idle nodes of power equipment ubiquitous in the Internet of Things, to achieve intelligent matching with computing resource demanders, so as to achieve the purpose of distributed computing.

11.3.2.3 Ubiquitous IoT Architecture of Power Equipment based on Blockchain

Following the blockchain thinking, this section proposes a blockchain-based power equipment ubiquitous IoT system architecture, as shown in Figure 11-14, which is a partially centralized partitioned alliance chain. As Chen Chun, an academician of the Chinese Academy of Engineering, believes, the alliance blockchain is one of the research hotspots of China's blockchain technology.

(1) Massive terminal sensors. Massive sensors at the end of the ubiquitous Internet of Things for power equipment are responsible for the initiation of transactions. The transaction initiated by the terminal is mainly defined as the upload of various equipment status monitoring data. Terminal sensors generally have problems such as low computing power, small storage capacity, and short battery life. Therefore, the terminal only records transaction accounts and does not participate in maintaining the blockchain. Of course, with the development of technology, some comprehensive smart sensors can not only undertake the sensing function but also provide enough computing power and storage space, which can also participate in the maintenance of the blockchain in the Internet of Things terminal.

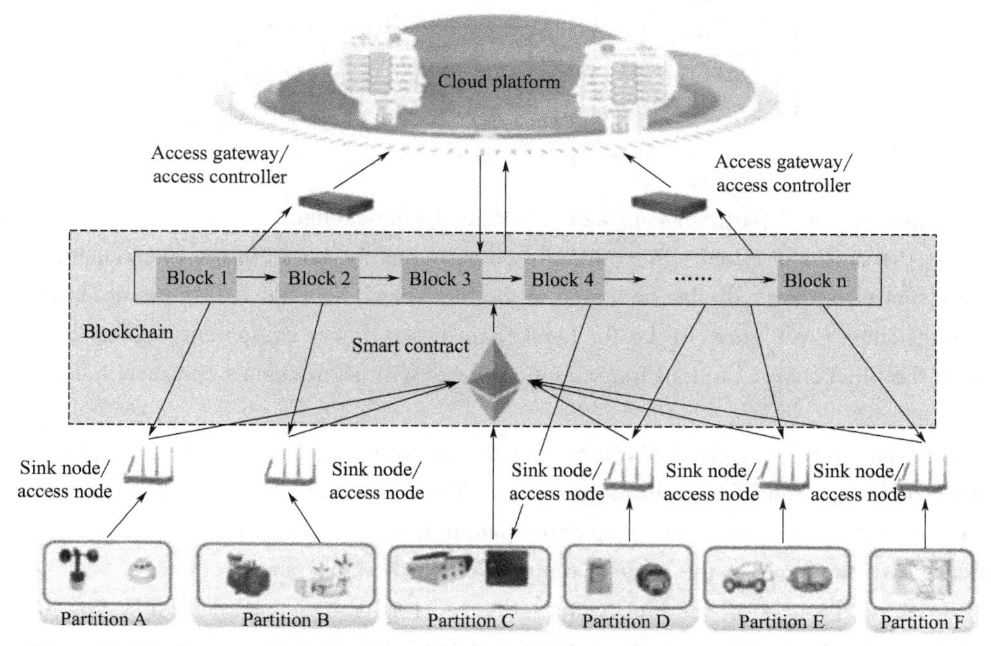

Figure 11-14 Framework of blockchain-based ubiquitous Internet of Things for power equipment

(2) Blockchain gateway. The blockchain gateway can be the convergence node and access node of the ubiquitous IoT perception layer of the original power equipment, or the access controller and access gateway of the network layer. In addition to the terminal access management, the blockchain gateway also needs to have transaction management functions, which are responsible for

forwarding transactions initiated by the terminal within the network, verifying the legality of new transactions in the ledger, and maintaining a unified ledger. Therefore, objectively, the blockchain gateway is required to have a certain amount of computing power and larger storage space.

In particular, blockchain gateways can also initiate transactions, where transactions are more service-related, such as data exchange, computing power sharing, etc.

(3) IoT blockchain. The basic structure of the Internet of Things blockchain is shown in Figure 11-15. Each block includes a block header and a block body, and the hash value of the block header is used as the unique identifier of the block. Each block references the hash value of the block header of the previous block to form a chain-like data block structure. All data generated during the creation time are recorded in the block body.

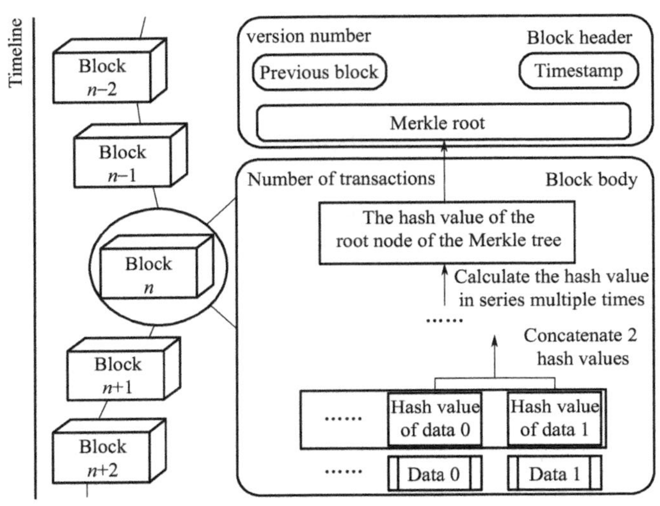

Figure 11-15 Structure of block

The IoT gateway will participate in and maintain the blockchain, and the various status data collected by the sensor terminal will eventually be permanently stored in the blockchain. In particular, the smart contract will also be stored in the blockchain as a block. In the ubiquitous Internet of Things shown in Figure 11-14 for blockchain-based power equipment, smart contracts are the core of the blockchain, and all transaction requests will trigger smart contracts before transactions.

(4) Cloud platform. The cloud platform corresponds to the original ubiquitous platform layer in the Internet of Things and inherits all its functions. The difference is that the cloud platform is also a node in the blockchain, which needs to participate in the maintenance of the blockchain, and can complete its own functions in the form of transactions in the blockchain.

(5) Transaction process of IoT blockchain. Figure 11-16 shows the transaction flow of data in the blockchain-based power equipment ubiquitous in the Internet of Things. For each new transaction, each node in the Internet of Things will use the public key of the transaction information to verify the transaction information, and only the transactions that pass verification by most nodes will be recognized and recorded in the ledger. The blockchain is jointly maintained by all nodes in the network, and each node keeps a copy of the unified ledger. Eventually a distributed ledger was formed.

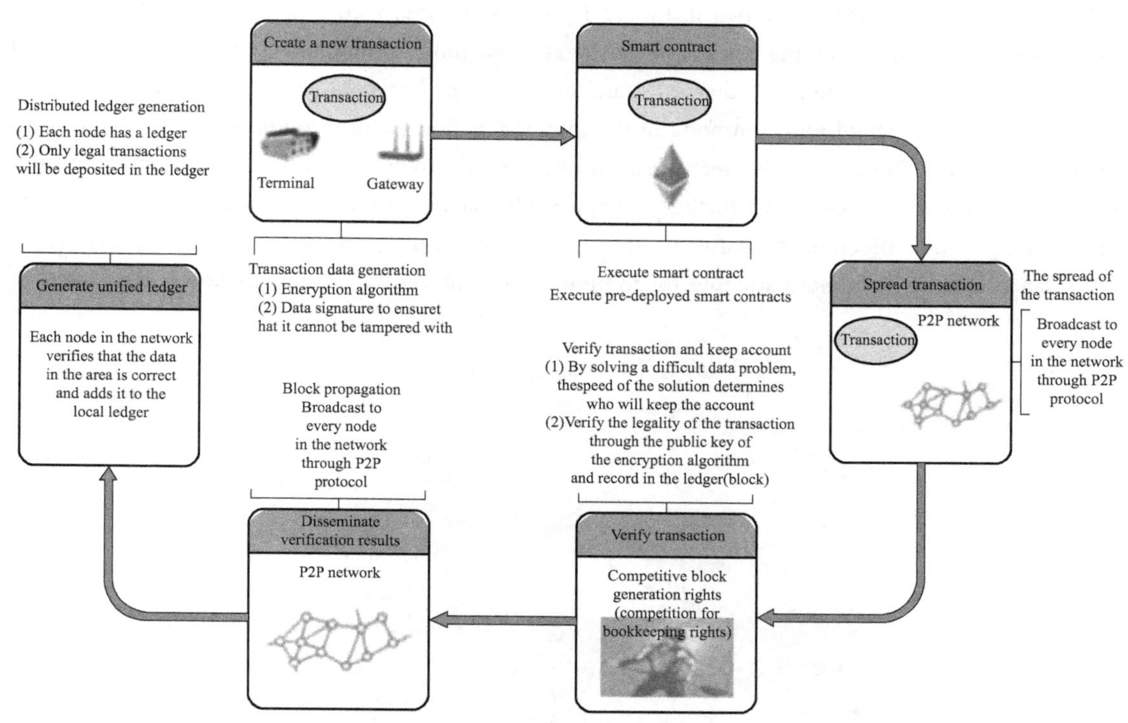

Figure 11-16 Data transaction process in blockchain-based ubiquitous Internet of Things for power equipment

11.3.3 Key Technology

In order to improve the performance and practicability of blockchain-based power equipment, the following key technologies need to be broken through.

11.3.3.1 Design of Partitioned Parallel High-Throughput Alliance Chain

After the ubiquitous power equipment integrates blockchain technology into the Internet of Things, it will inevitably face the performance problems of the blockchain itself. There are already widely used public chains (such as Bitcoin, Ethereum) and alliance chains (such as Hyperledger), and there is an order of magnitude gap between the throughput and the actual demand of the large-scale Internet of Things with massive deviceaccess[36,37]. The development of high-throughput blockchain to optimize the performance of blockchain transactions has become one of the research hotspots of blockchain technology. Existing research has been carried out at the level of related protocols[38], algorithms[39] and underlying architecture[40].

The Ubiquitous Internet of Power Equipment is essentially a centralized distributed network. Therefore, a partially centralized alliance blockchain was selected in its blockchain architecture design. Another feature of the ubiquitous power Internet of Things is to facilitate the division of areas. As small as a substation, all the sensors installed in the station, such as partial discharge, gas composition, temperature and humidity, video, infrared, etc., rely on the convergence node or access node to build a small blockchain. Several substations are then connected to a controller or gateway to build a larger-scale blockchain. In this way, the original blockchain is divided into many partitions, and transaction records will be processed in parallel in different partitions.

Therefore, the architecture shown in Figure 11-17 can be improved to the partitioned parallel alliance chain. According to the region or business, the power equipment ubiquitous Internet of Things is reasonably partitioned, that is, parallelized, and then data and service transactions are carried out with the cloud platform node in the interconnection chain through the cross-chain interconnection node. The consensus mechanism during on-chip transactions uses streamlining technology to optimize efficiency, and further improves efficiency through the random rotation of the bookkeeping node collection mechanism, so as to finally build a partitioned parallel high-throughput alliance chain that is more suitable for the ubiquitous Internet of Things of power equipment.

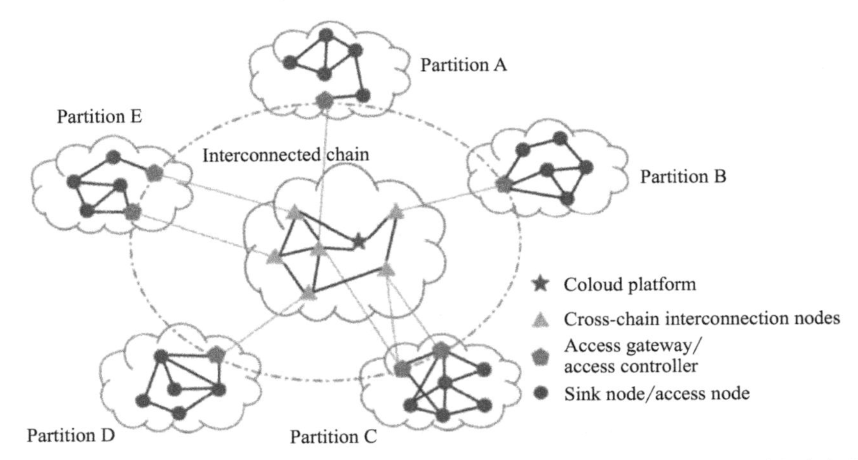

Figure 11-17 Structure of partitioning and parallel high-throughput consortium blockchain

11.3.3.2 Cross-Chain Communication

After the blockchain is partitioned, it will inevitably face the problem of cross-regional transactions. A reasonable solution is to achieve cross-regional communication of transactions through the collaboration of partition gateways and interconnected chain nodes. This requires a secure, credible, and real-time cross-chain communication technology. Issues such as validity, scalability, and atomicity in cross-chain transactions are the current research focus of blockchain cross-chain communication technology. The "Decentralized Internet of Things Communication Architecture Based on ICN and Blockchain Technology" proposed by Beijing University of Posts and Telecommunications, Huawei Technologies Co., Ltd. and other units has become an international standard. The standard combines ICN and blockchain technology to provide decentralized, secure, credible, and real-time data communication technology for the Internet of Things, and can be applied to cross-slice blockchain data transactions after the parallelization of power equipment ubiquitous Internet of Things partitions.

11.3.3.3 Improved Consensus Algorithm

Large resource consumption is one of the important factors restricting the application of blockchain. As shown in Figure 11-16 for the ubiquitous IoT blockchain data transaction process, the traditional method of mining-based accounting rights competition is a typical consensus mechanism that consumes huge amounts of freedom and requires targeted optimization. Common consensus algorithms include Proof of Work (PoW), Proof of Stake (PoS), Proof of Share Authorization (DPoS), and Byzantine Fault Tolerance (PBFT).

From the perspective of the improved partitioned parallel alliance chain structure, voting-based consensus mechanisms, such as the PBFT algorithm, can reduce resource consumption and effectively increase transaction speed and throughput, and are more suitable for ubiquitous power IoT environments. However, the PBFT algorithm also has shortcomings: on the one hand, as the number of consensus nodes in the system increases, the communication overhead of the system will greatly increase; on the other hand, all nodes have the same probability of acting as the master node. If a malicious node acts as the master node, it will destroy the consensus of the nodes in the entire network of the system, thus reducing security.

The improvement method is to divide all nodes into several groups of equal number, and the nodes in each group take turns as the main node of the group. The effect is as follows:

(1) After grouping, the influence of malicious nodes can be limited within the group, thereby reducing the harm caused by malicious master nodes.

(2) The algorithm transmits messages in the form of broadcast, and the communication complexity of a complete consensus is $O(n^2)$. Assuming that the nodes are divided into m groups, the nodes of the entire system only need to perform a PBFT consensus within the group, that is, the communication complexity of a complete consensus is reduced to $O(n^2/m)$, which significantly reduces the communication overhead.

As mentioned earlier, node grouping is easy to implement in the ubiquitous Internet of Things for power equipment. The improved PBFT algorithm is used as the consensus algorithm in each partition of the partitioned parallel alliance chain, which plays an important role in building a high-throughput blockchain.

11.3.3.4 Ubiquitous Smart Contracts under the Internet of Things

Smart contracts play a key role in opening up data and collaboration barriers between nodes in the ubiquitous Internet of Things proposed in this section. Its execution logic is shown in Figure 11-18. The created smart contract can be deployed as a block to the blockchain. The contract cannot be edited again after deployment. To update or replace the contract, the original contract must be "dead" to deploy a new contract. The block structure determines that smart contracts can be easily replaced without affecting the operation of other contracts.

A simple application is used to illustrate the operating mechanism of smart contracts in the IoT blockchain. Assuming that the smart contract deploys an alarm threshold for bus temperature data, when a sensor initiates a data transaction, the data is firstly sent by the IoT gateway to the smart contract in the blockchain, and the smart contract determines whether the data is temperature data and whether it exceeds Threshold and send the result to the agreed IoT gateway or cloud platform. At the same time, the smart contract broadcasts the data to the entire network to complete the subsequent data transaction process.

11.3.3.5 Encryption Algorithm

At present, the encryption algorithms used by major large blockchain platforms mostly adopt international standards, such as SHA256. From the perspective of national security, data security should be considered at the beginning of the construction of the ubiquitous power Internet of Things. The National Cryptographic Algorithm is a series of algorithms formulated by the National Cryptographic Bureau. The development of a blockchain platform based on the national secret al-

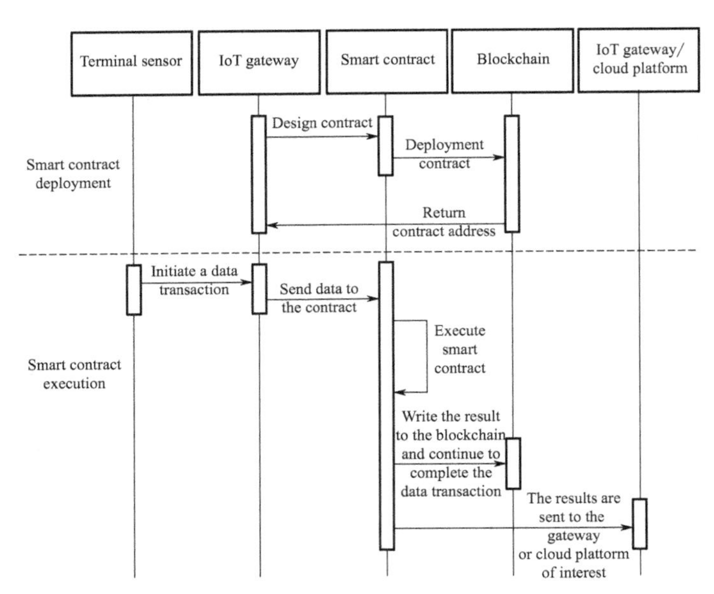

Figure 11-18 Logical process of smart contract

gorithm in the ubiquitous Internet of power equipment can effectively guarantee the security, confidentiality, credibility and reliability of the ubiquitous power Internet of Things data.

11.3.3.6 Data Compression

Each node participating in the maintenance of the blockchain needs to store a copy of the ledger locally. With the progress of data transactions, data expansion is inevitable. This is not a problem unique to the blockchain. Data has exploded in the Internet era, but at the same time storage capacity has also grown simultaneously. Specifically, when the blockchain is used in the ubiquitous Internet of Things application of power equipment:

(1) The access gateway or access controller with stronger hardware performance can be divided into heavy nodes, and the aggregation node or access node with weaker performance can be divided into light nodes. The heavy node is responsible for storing the full amount of data of the blockchain, while the light node only stores the 256 hash value of the root node of the Merkle Tree, that is, it only participates in transaction verification and does not store data.

(2) For stored data, the Merkel tree algorithm design itself guarantees that each transaction can be deleted (or transferred) individually, and only the hash value of this transaction is retained, which can be flexibly processed according to the actual storage situation.

11.3.4 Technology Outlook

At present, the application research of blockchain technology in the power system has just started, and preliminary research results have been obtained in the fields of power market transactions, which provide new ideas and means for the development of smart grids and ubiquitous power Internet of Things. There are still many problems to be solved in management and technology. The main challenges include:

(1) Compared with the traditional information system, the blockchain system's centralization

and regional data sharing mechanism involves multiple business departments of the power system, and the requirements for security are more difficult. The establishment, security and safety of the blockchain system anti-attackrequires a lot of collaborative work and technical research.

(2) Any good technology or tool needs to be used correctly in order to exert greater value. The blockchain system consumes a lot of resources for distributed storage and processing, and the centralized management costs are lower. How to weigh the comprehensive efficiency of the system, analyze the application effects of different scenarios, and play the value of the blockchain also requires in-depth and detailed the study.

(3) Electric power professionals currently have insufficient understanding of the basic concepts, implementation methods, and application value of blockchain technology. The application of blockchain technology in the Internet of Things of power equipment requires the deep integration of high voltage, power systems, computers, and information processing. It can only play a role through innovative thinking and groundbreaking research work. The lack of interdisciplinary talents may affect the development and application of technology.

The development trend of blockchain technology in the application of Internet of Things in power equipment is mainly reflected in the following aspects:

(1) With the development of blockchain on-chain and off-chain data synergy technology, the Internet of power equipment and various business systems will be integrated, so as to achieve more stable, safer, and more efficient multi-source interaction of grid equipment status data comprehensive analysis.

(2) The use of blockchain technology to achieve quality control and accurate traceability of multi-source data will provide multi-party trust and more refined data support for grid fault analysis and business management.

(3) Blockchain technology should not be independent of advanced information processing technologies such as big data, artificial intelligence, and 5G communications. They are in a complementary development relationship. Realizing the seamless connection of high-speed data communication, efficient and safe storage and intelligent analysis will play a greater role in the construction and development of the ubiquitous power Internet of Things.

11.3.5 Conclusion

(1) The integration of blockchain and the ubiquitous Internet of power equipment will help the ubiquitous power Internet of Things realize a truly distributed system and form a new cross-border data sharing and collaboration model. On the contrary, it will also accelerate the deep integration of blockchain and the real economy, and help the digital transformation of the economy.

(2) The partitioned parallel high-throughput alliance blockchain can be better compatible with the relationship between decentralization and centralization, which is an important research direction of blockchain in the ubiquitous Internet of Things application of power equipment. However, in order to promote the real application of blockchain technology in the power Internet of Things, it is also necessary to strengthen targeted research on key technologies such as consensus algorithms, smart contracts, data compression, and cross-chain communication.

(3) What needs to be pointed out is that the blockchain is essentially a decentralized distributed data processing architecture. It is an incentive and restraint mechanism based on a combination of

multiple technologies, not a special core technology. In the process of research and application, we must pay attention to grasp the relationship between decentralization and centralization, and make all judgments based on actual results.

Notes and References

[1] Chinese Society for Electrical Engineering Information Commttee. Chinese electric power big data development white paper. Beijing: China Electric Power Press, 2013.

[2] Li G, Yu C H, et al. Challenges and prospects of fault prognostic and health management for power transformer. Automation of Electric Power Systems, 2017, 41 (23): 156-167.

[3] Xue Y S, Lai Y N. Integration of macro energy thinking and big data think: part one big data and power big data. Automation of Electric Power Systems, 2016, 40 (1): 1-8.

[4] Yan Y J, Sheng G H, et al. The key state assessment method of power transmission equipment using big data analyzing model based on large dimensional random matrix. Proceedings of the CSEE, 2016, 36 (2): 435-445.

[5] Liao R J, Wang Y Y, Liu H, et al. Research status of condition assessment method for power equipment. High Voltage Engineering, 2018, 44 (11): 3454-3464.

[6] Li G, Zhang B, et al. Data science issues in state evaluation of power equipment: challenges and prospects. Automation of Electric Power Systems, 2018, 42 (21): 10-21.

[7] Guide for condition evaluation of power equipment: Q/GDW 171-2008. Beijing: State Grid Corporation of China, 2008.

[8] 35 kV～500 kV guide for condition evaluation of oil-immersed power transformers (reactors): Q/CSG11001-2014. Guangzhou: China Southern Power Grid Corporation, 2014.

[9] Guide for the diagnosis of insulation ageing in oil-immersed power transformer: DL/T 984-2005. Beijing: National Development and Reform Commission, 2005.

[10] State Grid Corporation of Production Technology. Summary of technical standards for the maintenance of power transmission equipment. Beijing: China Electric Power Press, 2011.

[11] Production Equipment Management Department, China Southern Power Grid Corporation. Guidelines for risk assessment of power equipment. Nanning: China Southern Power Grid Corporation, 2014.

[12] Qiu Z B, et al. Mechanical fault diagnosis of high voltage disconnector based on motor current detection. Proceedings of the CSEE, 2015, 35 (13): 3459-3466.

[13] Jia J W, et al. Optimum maintenance policy for power delivery equipment based on Markov decision process. High Voltage Engineering, 2017, 43 (7): 2323-2330.

[14] Zhou Z H. Machine learning. Beijing: Tsinghua University Press, 2016.

[15] Dai Y, et al. A brief survey on applications of new generation artificial intelligence in smart grids. Electric Power Construction, 2018, 39 (10): 1-11.

[16] Liu Y P, et al. Review on applications of artificial intelligence driven data analysis technology in condition based maintenance of power transformers. High Voltage Engineering, 2019, 45 (2): 7-18.

[17] National Science and Technology Council. National artificial intelligence research and development strategic plan. America: National Science and Technology Council, 2016.

[18] Government Office for Science. Artificial intelligence: opportunities and implications for the future of decision making. UK: Government Office for Science, 2016.

[19] State Department. Plan for the development of the new generation of artificial intelligence. Beijing, China: State Department, 2017.

[20] Pu T J, et al. Research and application of artificial intelligence technology in the operation and maintenance of power equipment. High Voltage Technology, 2020, 46 (02): 369-383.

[21] Li X R, et al. Constructing digital grid and informatizedenterprise. Automation of Electric Power Systems,

2007, 31 (17): 1-5, 44.

[22] Grieves M. Product lifecycle management: The new paradigm forenterprises. International Journal of Product Development, 2005, 2 (1-2): 71-84.

[23] Grieves M, Vickers J. Digital twin: mitigating unpredictable, undesirable emergent behavior in complex systems. Transdisciplinary Perspectives on Complex Systems, Berlin, Germany: Springe-Verlag, 2017: 85-113.

[24] Li G X, et al. Intergrated dispatch control and anti-misoperation system based on D5000 platform. Lectric Power Automation Equipment, 2014, 34 (7): 168-173.

[25] Xiang C M, Zeng S M, Yan P, et al. Typical applications and prospects of digital twin technology in power grid operation. High voltage technology, 2020, 1838.

[26] Zhou M, Feng D H, Yan J F, et al. A Software Platform for Second-order Responsiveness Power Grid On-lineAnalysis. Power System Technology, 2020, 44 (9): 3474-3480.

[27] SW AN M M. Blockchain: blueprint for a neweconomy. USA: O'Reilly, 2015.

[28] Tan L, Chen G. Blockchain 2.0. Beijing: Publishing House of Electronics Industry, 2015.

[29] Nakamoto S. Bitcoin: a peer-to-peer electronic cashsystem. 2019.

[30] China State Council. The "13th five-year plan" national informatization plan. Beijing: China State Council, 2016.

[31] The People's Bank of China. The 13th five-year plan for the development of information technology in China's financial industry. Beijing: The People's Bank of China, 2017.

[32] Ministry of Industry and Information Technology of the PRC. White paper on the development and application of blockchain technology in China. Beijing: Ministry of Industry and Information Technology of the PRC, 2016.

[33] Jiang X C, et al. Construction ideas and development trends of transmission and distribution equipment of the ubiquitous power internet ofthings. High Voltage Engineering, 2019, 45 (5): 1345-1351.

[34] Wu S S, et al. Discussion on application of distribution internet of things in new industryform. High Voltage Engineering, 2019, 45 (6): 1723-1728.

[35] Cai Y M, et al. Novel edge-ware adaptive data processing method for the ubiquitous electric power internet of things. High Voltage Engineering, 2019, 45 (6): 1715-1722.

[36] Jiang X C, et al. Key technologies and solutions for the ubiquitous IoT application of blockchain in power e-quipment. High Voltage Technology, 2019, 45 (11): 3393-3400.

[37] Shao Q F, Jin C Q, et al. Blockchain: architecture and research progress. Chinese Journal of Computers, 2018, 41 (5): 969-988.

[38] Gilad Y, Hemo R, Micali S, et al. Algorand: scaling byzantine agreements for cryptocurrencies. Proceed-ings of the 26th Symposium on Operating Systems Principles, Shanghai: ACM, 2017: 51-68.

[39] Gencer A E, et al. Bitcoin-NG: A scalable blockchain protocol. Santa Clara: USENIX NSDI, 2016.

[40] Huang T C, et al. Communication network structure and topology analysis of internet of things for power transmission and transformationequipment. High Voltage Engineering, 2015, 41 (12): 3922-3928.